# Non-venomous or slightly venomous snakes

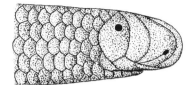

Worm Snake
*Xerotyphlops vermicularis*

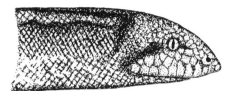

Western Sand Boa
*Eryx jaculus*

Dahl's Whip Snake
*Platyceps najadum*

Aesculapian Snake
*Zamenis longissimus*

European Cat Snake
*Telescopus fallax*

Western Montpellier Snake
*Malpolon monspessulanus*

# SNAKES
## OF EUROPE,
## NORTH AFRICA &
## THE MIDDLE EAST

### A PHOTOGRAPHIC GUIDE

# SNAKES
## OF EUROPE,
## NORTH AFRICA &
## THE MIDDLE EAST

A PHOTOGRAPHIC GUIDE

## PHILIPPE GENIEZ

**Translated by Tony Williams**

**Princeton University Press**
**Princeton and Oxford**

In the United Kingdom: Princeton University Press, 6 Oxford Street,
Woodstock, Oxfordshire OX20 1TR

press.princeton.edu

Jacket image courtesy of Christopher Mattison

First published as *Serpents d'Europe, d'Afrique du Nord et du Moyen-Orient* © Delachaux et Niestlé SA, Paris, 2015

ISBN 978-0-691-17239-2
Library of Congress Control Number: 2017935145
British Library Cataloging-in-Publication Data is available

This book has been composed in The Sans: Sans Plain, Semi Light,
Bold and Extra Bold.

Printed on acid-free paper. ∞

English language edition typeset by D & N Publishing, Wiltshire, UK
Printed in France
10 9 8 7 6 5 4 3 2 1

# Contents

# Foreword

This book is primarily intended for amateur naturalists, but we hope that professional zoologists will also find it useful, because of the quality of the information and the latest scientific research it incorporates, especially recent advances in molecular biology. At the same time, we hope it serves another purpose: increasing interest in and helping with efforts to conserve a very fascinating group of animals.

The geographic area covered by this book is very close to that covered by the well-known *Handbook of the Birds of Europe, the Middle East and North Africa: The Birds of the Western Palaearctic*, initiated by Cramp and Simmons in 1977, and that of the *Guide des Mammifères d'Europe, d'Afrique du Nord et du Moyen-Orient*, by Aulagnier et al., published in 2008. Indeed, we could have named this work *The Snakes of the Western Palaearctic*.

A general overview (page 7) considers the anatomy, biology, ecology, behaviour, venom and care of snakes in captivity, as well as their protection and systematic classification. The book presents all of the 123 snake species recorded in the Western Palaearctic. The species appear in an order that follows zoological classification – family by family, with the families appearing genus by genus and then the species that belong to them, in

order. The families are considered in the following order:

- Typhlopidae (blind snakes)
- Leptotyphlopidae (slender blind snakes)
- Boidae (boas)
- Colubridae (colubrids)
- Lamprophiidae (house colubrids, Montpellier snakes, sand snakes and ground vipers)
- Elapidae (cobras and sea snakes)
- Viperidae (vipers and pit vipers)

A table at the end of the book (pages 366–71) lists all the species along with the countries where they occur. This table allows readers to tick off the species they have seen in the wild, as is often done for birds. At the very end is a selected list of references giving interested readers details of where to find further information.

## The Distribution Maps

The geographic zone covered by this book corresponds almost exactly to the Western Palaearctic as defined by Cramp and Simmons for their *Handbook of the Birds of the Western Palaearctic*. It is a vast region that includes the whole of Europe (eastwards as far as the Ural Mountains in Russia and to the southeast as far as the northern side of the Caucasus chain of mountains),

North Africa (southwards as far as the 21st Parallel, near the southern edge of the Sahara) and the Middle East (to the east as far as the Caspian Sea and the borders of Iran, to the southeast as far as northern Saudi Arabia and including Kuwait). The main difference with the area covered in *Handbook of the Birds of the Western Palaearctic* is the southern limit in North Africa, where we have excluded the Tibesti range in northern Chad. All snake species that occur within this area are treated individually in the present work.

# Overview of Snakes

Snakes are one of the groups of animals that suffer most from negative prejudice. This is perhaps due to their secretive behaviour and hidden lifestyle, along with the fact that many species have a venomous bite. Many people are afraid of snakes to some degree, but many also respect or even venerate them.

Snakes do not always play a negative role in myths, fairy tales and legends; in fact, they sometimes exhibit a protective aspect.

For the ancient Greeks, especially in association with the god of medicine, Asclepius, the snake was a symbol of rejuvenation due to its regular moulting. This symbol has continued to this day in the form of Asclepius's staff, which depicts a snake coiled around a stick. Asclepius's staff is used today as a medical symbol in many countries. The name of the Aesculapian Snake recalls this association.

In Italy each year on the first Thursday in May, a strange ceremony occurs in the village of Cocullo (Abruzzes); it is designed to scare away venomous snakes so that the inhabitants will not be bitten over the next year. The Processione dei Serpari (*serpari* is a name for men who capture snakes) occurs during the Saint Dominic Festival. The statue of Saint Dominic (he provides protection against snakebite; he is not the founder of the Dominican order) is covered with jewels, banknotes, and live snakes, mainly Western Four-lined Snakes *Elaphe quatuorlineata* and Aesculapian Snakes *Zamenis longissimus*. The statue is then carried around the village by four men. During the procession, the snakes move about; some coil around the necks or shoulders of the statue bearers. After the religious service, the snakes are released back into the wild. This festival attracts thousands of tourists each year. It is not a celebration of snakes but a commemoration of Saint Dominic's Day; it was this saint who rid the village of its venomous snakes in the 11th century.

Due to the scarcity of fossils – discoveries often provide only a few vertebrae – we know little about the origins of snakes. It is, however, commonly considered that their ancestors were crawling, lizard-like animals. It was during the Jurassic (between 185 and 135 million years ago)

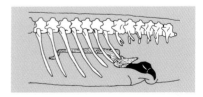

*Remnants of the pelvis in the family Boidae are proof that the ancestors of snakes had legs that have been lost during their evolution.*

*Despite their elongated body and absence of legs, neither the Slowworm* Anguis fragilis *(above) nor the Scheltopusik or European Glass Lizard* Pseudopus apodus *(right) are snakes. They are legless lizards that can be distinguished from snakes by their scale makeup, their anatomy, and their manner of locomotion. P. Geniez.*

that the phylogenetic separation from ancestral lizards began; the discovery of fossils from the Cretaceous (135 to 65 million years ago) confirms this hypothesis. The Earless Monitor Lizard of Borneo *Lanthanotus borneensis*, a lizard that inhabits Indonesian wetlands, may give us an idea of what transitional forms between ancestral snakes and modern snakes might look like.

In their long evolutionary history, ancestors of snakes lost their legs. The persistence of vestigial remnants of legs on the pelvis girdle in species belonging to primitive groups such as the Aniliidae, Typhlopidae, and Boidae is proof of this loss of legs. Such remnants are also visible externally as spurs on each side of the cloaca: they are the vestigial remnants of hind legs.

The sub-orders containing the lizards and the snakes together form the order Squamata within the class Reptilia.

## Morphology

All snakes are characterised by their very elongated form and the lack of pectoral and pelvic girdles and thus their legs. A snake's body is covered with a **scaly, keratinous skin**; their scales overlap slightly, like roof tiles.

Their **eyes** are without moveable eyelids but are covered with a transparent scale that gives them a fixed look. During the moult this transparent scale is also renewed.

The **skull** is constructed in a very articulated manner. The two mandible arcs (or upper and lower jaw) are joined simply by elastic ligaments so that there is no true articulation of the jaw. The different parts of the skull are very maneuverable, so snakes can open their mouths extremely wide when swallowing large prey. The teeth are not embedded in alveolar pits but are fixed freely on the maxillae. They serve uniquely to hold the prey, not to bite off pieces or to chew. The sternum has disappeared in snakes, which allows for the esophagus to extend greatly during the ingestion of large prey. An abundance of saliva facilitates ingestion, and very effective digestive secretions aid with digestion.

Other than the skull, the **skeleton** consists entirely of ribs and vertebrae, (between 200 and 400 on average, to 500 in some cases). The majority of the body and tail vertebrae are associated with ribs. They are attached to the vertebrae by an articulation and are free at the opposite end due to the absence of a sternum.

The **internal organs** are extremely elongated along the length of the body. In the majority of snakes, the left lung is atrophied, even absent, whereas the right lung is developed normally. At its front it has a pouch that is filled with air before a snake catches a large prey item, and it holds enough oxygen to cover the snake's needs during swallowing, since its respiratory tracts are compressed during this process. At its rear the lung is elongated by another pouch that serves as a swim bladder in aquatic snakes.

## Locomotion

Without legs, snakes move forwards by sliding over a surface using their numerous mobile ribs. Four main types of moving can be distinguished in snakes:

► "Snaking," or a lateral undulating movement, is the most common form of locomotion. Projections or plants on the ground serve as anchor points. When snakes swim, they also use an undulating movement to move forward.

► Repetition of accordion-type movements is sometimes used, especially by burrowing species. The snake compresses its body at various points in a narrow gallery or underground passage and elongates forwards (or backwards if retreating) the part of its body that is in front or behind. Outside of their underground hiding places, burrowing snakes often use the same type of action in order to move.

▶ A caterpillar-like action is mainly used by short, stout venomous snakes, particularly certain vipers. The vertebral column remains extended and straight, while the well-muscled ribs press their free ends into two grooves in the ventral surface. The snake walks on its ribs as if in a sack race. The movement, which runs in waves along the rows of ribs, resembles that of the legs of a myriapod or a crawling caterpillar.

▶ In sidewinding, or lateral movement, employed by certain snakes that live in a desert habitat, the snake's body moves sideways above the sand, the ventral surface touching the ground only at certain points. A portion of the body behind the head and another in front of the tail are lifted alternately and then lowered, in something like a stride. Snakes that move in this way can travel remarkably quickly.

## Sensory Organs

Snakes are animals of varying body temperature. They are not "cold-blooded"; their body temperature and their activity depends on the ambient temperature of their environment. How their organs perform very much depends on the ambient temperature, and cold affects them adversely.

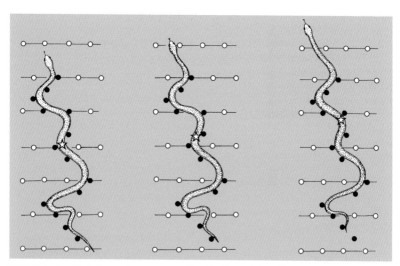

With classic "snaking", the most common way of moving forwards in snakes, projections or plants in the substrate (black dots) are used for anchorage. The star shows the movement of one part of the body.

*Among the different forms of locomotion in snakes, swimming also occurs using horizontal undulating movements. The Grass Snake – above:* Natrix (natrix) astreptophora, *in the Pyrénées-Orientales, in southwestern France – is an excellent swimmer.* Adam.

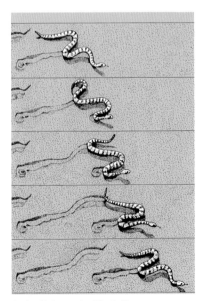

*Snakes that move by sidewinding – advancing laterally, projecting their body sideways – leave behind a distinctive trail in the sand.*

The most important sense in snakes is smell, which is often the only sense they use to find their prey.

The sense of smell functions via a sensory pit in the snake's palate called the Jacobson's organ. The snake explores its environment using its highly forked tongue, which is divided into two points; the tongue is continually darted out and in without opening the mouth by using a slit under the snout scale. It thus collects scent molecules that are delivered to the Jacobson's organ every time the tongue enters the mouth. Snakes sense their environment essentially via their tongues.

Sight has developed in very diverse ways in snakes. It is especially poor in the burrowing species that live in the ground, average in most of the vipers and excellent in certain colubrids that

*Snakes sense their environment mainly via their forked tongue, which captures scent particles that are transferred to the Jacobson's organ in the snake's palate. Here a Ladder Snake* Zamenis scalaris *photographed in southern France. P. Geniez.*

hunt by sight, such as the Montpellier snakes. Snakes react especially to moving objects.

Snakes do not have external ears, no auditory duct and no eardrum: they hear hardly anything.

Snakes exhibited by snake charmers in Cairo, Tunis or Marrakech do not move to the sound of the flute: they assume a defensive or vigilant attitude, following by sight the movements of the snake charmer's flute, giving the impression that they are moving in time to the music. In fact, they cannot hear the flute. However, the internal ear is well developed and responds to the slightest vibration, such as those from heavy footsteps of a human or large mammal.

The venomous snakes in the specialised pit viper sub-family (the Crotalinae within the family Viperidae) have developed a supplementary sense organ: the sensory pit. Situated on each side of the head between the eye and the nostril, sensory pits are very effective heat detectors. Thus, even in total darkness, pit vipers are able to detect the thermal emissions of warm-blooded animals such as mice or birds, which they follow, bite and then find later once they have succumbed to the venom. A difference in temperature of one-tenth of a degree centigrade can be detected by the sensory pits.

*Performing snakes in front of a snake charmer cannot hear the flute's music. They raise their bodies to better watch and menace their supposed aggressor, following its every move, especially that of the flute. Here, two black Moroccan Cobras* Naja haje legionis. *König.*

*Pit vipers have a heat-sensitive organ on each side of the head between the eye and the nostril. This sensory pit allows pit vipers to detect very slight changes in temperature. They are essentially found in the tropics and subtropics, although one species, the Halys Pit Viper, reaches the extreme eastern part of Europe.* Kreiner.

## Diet

Snakes feed almost exclusively on live prey. They kill their prey by suffocation or with a venomous bite and then swallow it whole without biting it into pieces or chewing it. The smallest prey items may be swallowed immediately and alive.

Swallowing takes a long time and requires much effort. When the prey is captured and immobilised, it is slowly pulled towards the back of the mouth as the skull and lower mandible close around the victim. In many vipers, the

venom fangs, which are positioned on mobile bones, fold backwards along the palate to facilitate swallowing.

Some species of snake have a specialised diet, such as the Smooth Snake *Coronella austriaca*, which feeds

*Snakes swallow their prey whole. This Aesculapian Snake is swallowing a mouse larger than its own head (right). The image below shows that the neck muscles have slowly forced this large prey item into the stomach.* Rohdich.

*Snakes are capable of ingesting prey larger than their own head, thanks to their highly extensible jaws. Here a young European Cat Snake* Telescopus fallax fallax *is preparing to swallow a young Balkan Green lizard* Lacerta trilineata *that it has previously suffocated within its strong body coils whilst inoculating it with venom. P. Geniez.*

especially on lizards and small snakes, and the Grass Snake *Natrix natrix*, which eats frogs and toads, newts and fish. However, most species with a specialised diet are found in tropical regions, including those that eat eggs, snails or snakes.

Because snakes' teeth merely serve to hold prey and do not break it up or chew it, extraordinarily active gastric juices are necessary for digestion, including of bones, feathers and hair. Venom injected into prey also helps with the start of digestion.

Snakes are well adapted to fasting; certain species are able to survive for an entire year, even two years, without eating, with no ill effects.

## The Moult

Snakes change their skin as they grow. Eliminating the outer keratinous skin, which also involves the eyes, is termed **moulting, shedding** or **ecdysis**. An imminent moult is signaled notably by the cloudy appearance of the eyes, which are normally quite clear, and an obvious dulling of the skin. The skin first breaks at the tip of the snout, and the snake then crawls out of the old skin, from the head to the tip of the tail; the skin comes off in the same way we take fingers out of a glove, by rolling it back. Moulted skins, abandoned by their owners after being shed, can be found attached to low scrub or branches in any habitats where snakes occur. When you examine a shed skin, remember that what you see is the interior of the skin before moulting occurred. Shed skins in good condition generally allow for the identification of the species concerned.

*Moult is close for this Leopard Snake* Zamensis situla, *whose eyes have become opaque.* König.

*This Nose-horned Viper* Vipera ammodytes *crawls out of its old skin in the same way as we might remove the fingers of a glove or take off a long sock. The outer part of the shed skin, shown here, is what was the inner surface before the skin was shed.* Fuchs.

*Abandoned, shed skins. Normally the species involved can be determined from the shed skin: left; the shed skin of a Grass Snake* Natrix natrix; *right; that of an Adder* Vipera berus. Cramm (left), Rohdich (right).

# Reproduction

In snakes, recognition of sexual partners occurs mainly through smell. Before copulation there is a long and varied pairing process. In certain species, when several males are in competition for a female, they may determine who is the strongest through ritual fighting.

Copulation generally lasts a long time and may continue for hours. The male copulation organs, the hemipenes – in the form of two clubs, often with spines and protrusions – are joined at the base in a common trunk. They are hidden in a vent behind the cloaca, at the base of the tail. During an erection, a single hemipenis is inverted and introduced into the female's cloaca, where it is anchored with the help of needle- or hook-shaped growths. The rough structure of the hemipenis has a particular form and shape in each species that is distinct enough to be used in the systematics and classification of snakes.

During copulation, it is always the hemipenis that is closest to the female that is everted and used.

Most snakes lay eggs (that is, they are oviparous). However, some species give birth to live young (they are ovoviviparous); such species include the Smooth Snake *Coronella austriaca* and the majority of vipers and pit vipers.

*Left: In many species of viper and colubrids, during fighting rivals intertwine and try to put down their opponent. These fights always finish without injury as the opponents never bite one another.*

*Two male Adders in a fight, which occurs without blood being spilt. The two snakes entwine together, each lifting its body and trying to put down the opponent, thus determining which of the two is the stronger.* Layer.

Mating follows a similar pattern of events in all snake species, whether a pair of Adders *Vipera berus (left)* or a pair of Asp Vipers *Vipera aspis (below)*: the male crawls onto the female, entwines her tail with his and introduces the hemipenis into the female's cloaca. Copulation can last a long time, sometimes hours. Limbrunner.

The male copulation organ consists of two appendages, called hemipenes, that are covered in spiny growths. These organs play a role in species isolation in snakes as each form of hemipenis is adapted to that of the vagina of the female of the same species. (left) Fuchs, (right) Évrard.

*The majority of snake species lay eggs. Here a Western Four-lined Snake* Elaphe quatuorlineata *that has laid a clutch of relatively large eggs in a terrarium.* Gruber.

*A Grass Snake hatching. The parchment-like eggshell is first ripped open.* Baumgart.

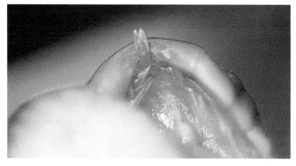

*Hatching is helped by the presence of a small, forward-pointing, horny growth – the egg tooth – that falls away as soon as the young snake hatches.* Fuchs.

*Cautiously, flicking its tongue to inspect its surroundings using its keen sense of smell, the head of a hatchling Grass Snake emerges from the egg, into which it will withdraw at the slightest sign of danger (left). Below: the young snake has finally left the egg to begin its independent life.* Baumgart.

*The Smooth Snake Coronella austriaca is a colubrid that gives birth to live young. They are ejected within a fine transparent membrane that corresponds to the envelope within the egg (above); they free themselves from the envelope as they are born by wriggling vigorously (below).* Mägdefrau.

At birth, young snakes leave the egg envelope, either within the mother's body or immediately after being expelled. The advantages of ovoviviparous incubation seem evident as the young snakes are immediately independent; they can move and hide themselves, and even defend themselves. On the contrary, clutches of eggs are exposed to the possibility of egg predation, may dehydrate if they are not in a favorable environment or may even rot if conditions are too humid. An additional advantage of being ovoviviparous is that snakes (such as the Adder) in colder climates can delay birth beyond the winter until

*Tiny, newborn snakes are independent as soon as they are born. Here, a newborn Smooth Snake. Mägdefrau.*

especially in those that normally feed on other snakes.)

## Life expectancy

Life expectancy, age of sexual maturity and growth rate are all very different depending on family, genus and species. Generally, snakes are long-lived animals, reaching at least 10 years if they avoid accidents. The average age of Adders *Vipera berus* in the wild is between 6 and 8 years, but they are known to reach 15 years. Grass Snakes *Natrix natrix* often live to 8 or 10 years and can reach 15 or 20. Nose-horned Vipers *Vipera ammodytes ammodytes* have been kept in terrariums for tens of years; it is thought that they can live to the age of 25 or 30 years in captivity, but probably a lot less in the wild.

the following spring, improving the chances of better climatic conditions and opportunities to find prey.

Females of certain species of snake, such as the Grass Snake *Natrix natrix*, sometimes gather in large numbers at favourable laying sites and lay collectively. On the edge of a sawmill, in a large pile of wood chips, we once found more than 1000 eggs.

Before leaving the egg, a young snake rips open the eggshell by using an egg tooth, a small hook positioned under the snout and pointing forwards, and then, after carefully looking around to be sure there is no danger, leaves the egg. Once the envelope is torn apart, the egg tooth is discarded.

Once the young snakes have seen daylight, their parents take no further interest in them. The young are thus totally independent and may often need to defend themselves, even against individuals of the same species. (Cannibalism exists in snakes,

Snakes increase in size throughout their life. When young, their growth is far more rapid; it progressively slows down with age but never completely stops. The largest individuals are those that have been lucky enough to avoid natural or human-induced disasters. However, within a given species, some individuals can grow more quickly than others.

## Life and Behaviour

A snake's behaviour is determined by various factors, such as habitat, relationship with their prey, breeding habits and their enemies. As they usually live quite a silent, hidden life, they are only rarely seen in the wild. Quite sedentary, they normally leave their home area only when hunting, looking for a partner of the opposite sex or, for females, finding a site to lay their eggs. The need to flee, to find prey or to answer the reproductive urge all make them move. Otherwise, they rest silently, curled up in a preferred site, mainly to control body temperature.

All snakes hunt: some hunt by following their prey, others by staying immobile, waiting for prey to pass within striking distance and taking them by surprise. Due to their colouration and patterning, many snakes are well camouflaged and are almost invisible within their surroundings. As they are very wary, rather than attack, they are more likely to flee from perceived danger or an enemy.

If they come across a rival or unexpected enemy whilst looking for a reproductive partner, they may take up a menacing and imposing attitude to encourage their adversary to flee or at least to fear them.

Most snake behaviour is common to all species, proving their common ancestry. One characteristic common behaviour is hissing as a way of advertising their presence or as a threat. Some species, such as the *Echis* vipers, emit a rasping sound by rubbing their keeled flank scales together. Certain species, such as the Viperine Snake *Natrix maura*, the Moïla Snake *Rhagerhis moilensis* and occasionally the Grass Snake *N. natrix* and the Blunt-nosed Viper *Macrovipera lebetina*, flatten the neck when threatened, raising themselves and hissing vigorously.

Other venomous species, especially the pit vipers, vibrate their tails, thus producing a rustling or buzzing sound from the ground or dead leaves. In venomous species, these warning or threatening sounds signify that if the aggressor comes too close, the snake will bite.

Snakes are rarely aggressive if not provoked. Often they need to be harassed before they decide to bite (however, many vipers and cobras assume a threatening posture and can strike suddenly). A remarkable

*As soon as a Grass snake (here* Natrix natrix helvetica*) feels threatened and it can't flee, it feigns death with belly uppermost, mouth open and tongue hanging out. It is possible that this reflex, typical of the species, serves to dissuade enemies but its efficacy is doubted.* Évrard.

phenomenon is that of feigning death, used by the Grass Snake *Natrix natrix*. Some captured animals become limp, lie wholly or partially on their backs and let the tongue hang out of the open mouth. This attitude can last 30 minutes, until the snake feels it is out of danger. If the apparently dead snake is released – for example, in water – it immediately becomes active.

## Beware of Venomous Snakes

Snakes are feared, as many species are venomous. About a fifth of the world's 3000 species are truly venomous. Their venom is often very virulent and can cause death in humans. The venom is produced by the venom glands, located in the head behind the eyes. Each venom gland is connected via a canal to the fangs located in the upper jaw.

Other than the identically shaped, smooth, non-venomous teeth of colubrids (called aglyphous dentition), there are three types of venomous teeth.

► Teeth with an inoculation groove that are located at the rear of the maxillae (opisthoglyphous dentition): a pair of enlarged teeth equipped with a groove connected to the venom gland and serving as inoculatory fangs. They are located in the rear of the mouth, which means they are usually ineffective for use on humans. With a few exceptions, their venom is

also relatively harmless. There are, however, a few opisthoglyphous colubrids whose bite can be fatal for humans – for example, the Boomslag *Dispholidus typus* of tropical Africa.

► Teeth with an inoculation groove that are located at the front of the maxillae (proteroglyphous dentition): two enlarged teeth serving as fangs solidly implanted at the front of the upper jaw and equipped with a longitudinal groove that is partially or totally enclosed and through which the venom flows when the snake bites. Certain elapids, such as the sea snakes, cobras and coral snakes, are the only species with grooved fangs located at the front of the maxillae.

► Teeth developed into tubular fangs located at the front of the mouth (solenoglyphous dentition): one large fang (sometimes more) is implanted on each half of the upper mandible, on the maxillary bone; the fangs can be retracted or positioned for biting, with the point forwards either at right angles or inclined. The fangs are tubular and in the biting process act like syringe needles. The venom passes through an outward-pointing exit above the fang tip, which allows for a better injection of venom into the bitten tissue. When not in use, the maxillary bone and the fang are folded backwards into a salivary groove. Behind the functional fangs there are replacement teeth in various stages of development. When a fang breaks or falls out, one of the reserve teeth takes its place and is only then connected to the venom duct. The vipers and pit vipers are the only snakes with forward-placed tubular fangs.

*Smooth teeth of identical shape (aglyphous dentition). Such teeth are found in Typhlopidae, Leptotyphlopidae, Boidae, and most colubrids.*

*Enlarged fangs, more or less tubular or with a groove, and located at the front of the jaw (proteroglyphous dentition: Elapidae).*

*Enlarged teeth with a groove and located at the rear of the maxillae (opisthoglyphous dentition) are characteristic of some colubrid genera, such as Telescopus and Malpolon.*

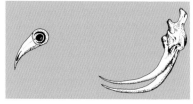

*Enlarged teeth that serve as fangs, tubular and located at the front of the jaw (solenoglyphous dentition), are found in vipers and pit vipers.*

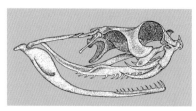

*Skull of a viper at rest (left) with the fangs folded backwards. When the snake is ready to strike, with the mouth wide open (right), the fangs are deployed forwards at right angles or angled forwards, ready to inject the venom.*

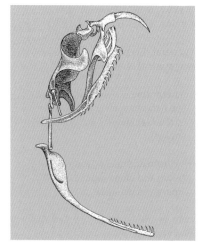

# Snakes' Venom

Here's a closer look at the different types of venom in snakes, the production of antivenins (or antivenoms, which are antitoxins to a venom), and the use of snake venoms in medicine, as well as rescue and care after a snakebite.

Snake venom has various functions:

► to kill prey;

► to prepare for digestion by liquefying some of the victim's tissues;

► as a defence against enemies.

The composition of the venomous secretion varies widely, depending on the genus or family of the snake, but generally consists of a complex mix of proteins.

In general, it is possible to distinguish two principal effects:

► a neurotoxic effect that damages the nervous system (applies mainly to the Elapidae, including sea snakes);

► a cytotoxic effect that destroys tissue and blood (applies mainly to vipers and pit vipers).

However, the two general types of venom can act simultaneously and in association with other enzymes. Indeed, snakes' venoms are made up of an extremely complex combination of different protein molecules that may have acidic or alkaline properties.

Neurotoxins paralyse the heart and respiratory or eye muscles by blocking the link between the nerve endings and the muscles and thus preventing the transmission of nerve impulses. Cytotoxic venoms act on the blood system by destroying the walls of blood vessels and red blood cells, bringing about varying degrees of haemorrhage depending on the snake species and the amount of venom injected.

Considering the complexity of venom composition, a bite from a venomous snake can result in the appearance of the following symptoms; sometimes symptoms occur together, which can aggravate the situation and mean that treatment must be even more complex:

► effects on the nervous system;

► paralysis of the respiratory muscles;

► effects on heart muscles;

► destruction of blood vessel walls (resulting in haemorrhaging);

► the breakdown of red blood corpuscles (haemolysis);

- ► necrosis of muscles and kidneys;

- ► tissue destruction;

- ► triggering of blood coagulation;

- ► prevention of blood coagulation.

The severity of a venomous bite depends on the species involved, the location of the bite, and the concentration and quality of injected venom, as well as the age, size and constitution of the victim.

A healthy adult human will recover from an Adder *Vipera berus* bite within six to eight days even without medical care. On the other hand, a bite from a Saw-scaled Viper *Echis carinatus* or Egyptian Cobra *Naja haje* usually progresses and may well cause death without medical treatment and use of an antivenin.

Furthermore, an anxiety attack may increase the effect of the bite, aggravating the general symptoms or even causing death. The victim may also collapse abruptly due to a massive allergic shock, causing the release into the body of vast amounts of antibodies that are also harmful.

The finer structure of snake venoms and their detailed compositions are still not sufficiently understood. Venoms undoubtedly evolved, however, from secretions of the digestive system.

## Serum Production

The most efficient way of treating venomous snakebites is by the administering of an antivenin. Today it is possible to counter the effects of most snake venoms by using specific or appropriate antivenin serums. Such antivenins are in principle available in two forms:

- ► monovalent serum that combats the venom of a single species;

- ► polyvalent serums that, in the form of a mixture of serums, are active against the venom of several species of the same geographic region.

Nevertheless, the use of antivenin serums for treating the bites of European vipers is increasingly controversial, and their proper conservation (antivenin serum should be kept cool), for example, whilst travelling, can be difficult. Today the preferred therapy is to treat the symptoms, placing the victim in a hospital and continuously monitoring changes in symptoms and treating them appropriately as they appear. For example, in the case of cardiac insufficiency, doctors may administer a cardiac stimulant and in the case of an allergic reaction, an antihistamine. Antivenin serum is used almost exclusively at the present time in the treatment of potentially fatal bites from exotic species.

In order to produce an antivenin, venom from a living snake is required. This "milking" of venomous snakes is mainly done at snake farms. Taking venom has to be done with caution in order to avoid harming the snake, being bitten or contaminating the venom. After collection, the venom has to be purified to isolate the toxins necessary for producing the serum and make it suitable for conservation.

To produce the serum, small doses of venom are injected into horses and sometimes sheep; the doses of venom, repeated at regular intervals, are gradually increased. The inoculated animals produce antibodies in reaction to the venom, and these increase until they ensure the animals' immunity. The antitoxin is then extracted from the blood of the immunised animal and forms the basis of the serum used against the snake venom, called antivenin serum.

## Medical Use of Snake Venom

In addition to being used in the production of antivenin, snake venom is becoming increasingly important in the production of certain medicines and in scientific research.

For example, preparations made from the venom of certain vipers are used to increase blood clotting during surgery or in medicines to inhibit blood clotting (used, for example, against phlebitis).

Snake venom is also used in the production of medicines for the treatment of epilepsy and rheumatism, in the fight against leprosy, and even in therapy for allergies. Furthermore,

some ten different venoms are used in the preparation of homeopathic dilutions.

The extent to which snake venom is used in biological research is constantly increasing – for example, in the study of the nervous system and its functions.

## What to Do in the Case of a Snakebite

### Precautions

Behaving responsibly in areas where snakes occur will generally help people avoid snakebites.

When in countryside where potentially dangerous snakes might occur, it is advisable to:

► wear closed footwear and long trousers;

► look at where you are going, especially on the ground;

► never try to scare or kill a snake;

► walk with firm steps, as snakes are quite susceptible to ground vibrations and always prefer to escape.

Few snakes rely on their camouflage and staying still; this behaviour is, however, common in vipers and pit vipers. There is always a risk, however rare, of accidently putting a hand or foot on an immobile snake and being bitten.

According to the South African Institute for Medical Research, 74.4% of bites occur on legs, 23.6% on arms or hands and only 2% on other parts of the body such as the head, neck or thorax. It is also important to be very wary of spitting cobras (one species occurs in the area covered by this book, the Egyptian Cobra *Naja haje*), which can project their venom at the eyes or face of the victim from a distance of up to 2 metres or more before deciding to bite. The only way of protecting oneself – when taking a photo, for example – is to wear glasses. However, in Europe, most snakebites occur when snakes are handled by those keeping venomous snakes in captivity or when professionals capture them in the field (sometimes when taking a photograph).

### When a Venomous Snake Bites

If a person is bitten by a venomous snake, a quick and rational reaction is usually sufficient.

There should be two main aims of this reaction:

1. To slow down the venom's dispersion in the body;

2. To neutralise the venom remaining in the body (applicable only to those with medical training).

The first thing to do is to make sure that the bite is that of a venomous snake. The mark left by the snake on the skin of the victim is a first clue. If the snake was a colubrid (most are not venomous), there will be a series of small holes arranged in a semi-circle or U shape. If a venomous snake is responsible for a bite, only the impact of the two fangs is visible in the form of two well-spaced red spots (or just one if only one fang pierced the skin of the victim). If the bite does prove to be from a venomous snake, the victim needs to stay calm, should be reassured and should drink water, but should avoid being active in a way that accelerates the venom's diffusion in the body. Obviously, rescue services should be contacted as quickly as possible in order to transport the victim to a hospital.

Inspired by traditional ideas or past experience that may be dangerous, some people who are supposedly competent or practiced in such matters may propose reducing the effect of the venom by inappropriate methods. But we explicitly discourage cutting the skin at the bite to evacuate the venom, trying to suck out the venom, or burning the bite. Cooling the area or using potassium permanganate are probably not effective and can, in certain cases, complicate medical treatment. The victim should never be given alcohol. There are devices that suck out much of the venom injected during a bite. However, to be effective, they need to be applied correctly to the bitten area within seconds of the bite happening, which is all but impossible. Also, the pain and skin damage caused by some devices may well increase the pulse of the victim, thus accelerating the spread of venom through the body.

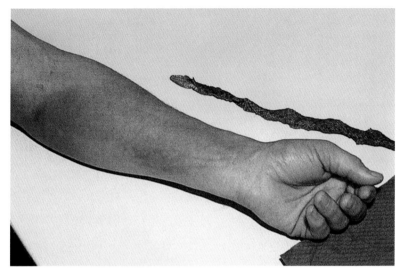

*An arm, swollen after being bitten by an Adder* Viperus berus. Synatzschke.

## Care Following a Venomous Bite

Caring for a bite victim after he or she has been given an antivenin serum, providing additional medication to help blood circulation or fight infection and treating the wound are all the work of a doctor. The serum is given in a prescribed dose as an intramuscular or intravenous injection. The doctor will decide on the appropriate medical action and monitor the effectiveness of the treatment.

Generally speaking, when this is done correctly and quickly, a venomous snakebite should not lead to serious complications and is only very rarely fatal, especially in Europe. In tropical countries and even in the southern and eastern parts of the area covered by this book, certain bites by vipers in the genera *Macrovipera*, *Daboia*, *Echis* and *Bitis* or by the Egyptian Cobra *Naja haje* can cause serious aftereffects, such as necrosis of the bitten limb, which sometimes requires amputation, even if the victim receives immediate and adequate hospital treatment.

## Immediate care following a venomous snakebite includes the following:

1. Stay calm; the victim should avoid being stressed. The victim should be reassured and, if possible, should sit or lie down in the shade. The shock and stress of a venomous bite can sometimes be even more dangerous than the direct effects of the bite itself.

2. As the result of recent research, it is now recommended not to use a tight tourniquet on the bitten limb. On the other hand, a bandage should be placed, not too tightly, on the bite to serve as a compress, and it should be removed only under medical supervision.

3. The victim should move as little as possible but be taken as soon as possible to a doctor or the nearest hospital. During travel, victims should avoid all physical effort; if possible, they should be carried or transported by car. If the bitten person has to move on their own, this should be done slowly and calmly – they should never walk quickly – to the nearest means of transport.

4. A doctor should be informed as soon as possible about the species responsible for the bite and the circumstances of the incident so that the appropriate antivenin can be found and used. The doctor will need the following information:

► place and time of the incident;

► if possible, a detailed description of the snake (size, colouring, pattern and, if possible, a photo taken with a camera or a mobile phone);

► treatment that has so far been given to the victim;

► information about the victim's past, especially whether they have been bitten before and if they have been treated in the past with serum;

► whether the victim has any allergies.

It must be remembered that venomous snakebites are often complicated to treat, and individual symptoms can vary considerably, even if the same species of snake is involved.

The priority should be taking care to avoid being bitten by a snake; this is especially important for those reading this book as it is those who are interested in snakes and in their identification who are at greatest risk of being bitten.

# Where Do Snakes Live?

Snakes occur in a multitude of biotopes and are remarkably well adapted to ambient environmental conditions. Most of the species that occur in the region covered by this book live on the ground. Their habitat will include places where the snake can lie in the sun, areas where they can find shade and areas with enough cover of protective vegetation that they can readily shelter or hide. This implies that a snake's habitat must contain varied structures: bushes, tufts of plants, open areas, tree stumps, drystone walls or piles of stones and preferably holes in the ground.

Certain species such as the sand boas (*Eryx* spp.), the worm snakes (*Xerotyphlops vermicularis* and *Letheobia* spp.), the blind snakes (*Myriopholis* spp.), the dwarf snakes (*Eirenis* spp.) and the Awl-headed Snake (*Lythtorhunchus diadema*) spend most of their lives underground and have developed partially burrowing behaviour.

Other species prefer living near water and readily swim; the principal swimming species within the area considered here are the Grass Snake *Natrix natrix*, the Dice Snake *N. tessellata* and the Viperine Snake *N. maura*.

Several species are adapted to an arboreal lifestyle, climbing into bushes and low, well-vegetated trees; these include species of the genera *Elaphe*

*The habitat of the Atlas Dwarf Viper* Vipera monticola *consists of high-elevation rocky valleys, devoid of trees or shrubs, between 2000 and 3500 metres elevation.* P. Geniez.

31

and *Zamenis*, which can be considered arboreal colubrids. The preferred habitat of these species consists of clear deciduous forest with bushes on rocky ground.

Other species that are not especially arboreal can be found lying on low bushes, particularly colubrids of the genera *Dolichophis*, *Hemorrhois* and *Hierophis* and sometimes the vipers *Macrovipera lebetina* and *Daboia mauritanica*.

The desert habitat imposes severe living conditions. Many snake species are adapted to dry, hot conditions, partly due to some very remarkable specialisations.

*The Grass Snake (here* Natrix natrix natrix) *prefers living near water; it is an excellent swimmer.* Limbrunner.

The Forskål's Sand Snake *Psammophis schokari* is extremely rapid, appearing to glide over the ground, which prevents it from burning itself on the very hot desert substrate. Snakes that advance by lateral movements, such as the Horned Viper *Cerastes cerastes* or the Sahara Sand Viper *C. vipera*, move by short "jumps" above the sand. The Sahara Sand Viper burrows into the sand until only its eyes appear above ground level.

The Egyptian Sand Boa *Eryx colubrinus* almost exclusively moves underground. Animals living in the desert without any vegetation are forced to be active at night in order not to suffer from the heat or lack of moisture whilst still avoiding their predators. Many snakes choose to live close to humans, who inadvertently often offer them a rich and varied habitat, such as hedgerows, or, in desert and arid regions, shade and humidity. Thus, for example, compost heaps are used as laying sites by the Grass Snake *Natrix natrix*, which will sometimes establish large communal nesting sites in such places. The worm snakes, whip snakes, dwarf snakes, cat snakes and many vipers will readily use drystone walls or bushes or piles of rock on the edges of fields or abandoned ruins as places to hide. Thus we humans sometimes unintentionally favour the presence of these animals that we often prefer to avoid. However, modern agricultural methods, the extension of road networks and continuing urbanisation of the countryside are a major threat to the continued existence of many species of snake.

# Snakes' Enemies

Snakes, whether venomous or not, have a large variety of predators, especially birds of prey, large wetland birds and various predatory mammals. It is widely known that storks and herons eat snakes. If very hungry, even hedgehogs will tackle vipers; this is, however, rare if unprovoked by humans. Pine and Beech Martins, Weasels and Polecats all willingly capture and eat snakes when given the chance. But the most specialised snake predator is without doubt the Short-toed Eagle *Circaetus gallicus*, a large bird of prey (1.70-m wingspan) common in Europe and around the Mediterranean. This bird, which has extremely good eyesight, often spends much time hovering over one spot, wings stretched and regularly beating, relentlessly searching the ground for snakes, which make up a high proportion of its prey. Once it has found a snake, it will dive onto it, sometimes from a height of a few hundred metres, seize it and swallow it in flight without further ado. The Short-toed Eagle is present in southern Europe from March to September, when snakes are active. In winter, this large migrant is absent from Europe and spends the European winter in sub-Saharan Africa, where snakes are active year-round.

Parasites are less conspicuous enemies of snakes. Many snakes in the wild are infested with worms, amoebae, skin infections or ticks.

However, the greatest enemy of snakes is man. Due to prejudice, superstition, and ignorance, many of these animals are simply killed, regardless of whether they are harmless or venomous. In many Mediterranean countries, snakes are actively persecuted or hunted to sell to tourists or snake charmers.

However, the largest threat to snakes (and to most living species) is habitat destruction and scarcity of prey. Semi-natural countryside is disappearing at an alarming rate. Snakes' environment is increasingly and incessantly threatened (by urbanisation, road construction, intensified agriculture, etc.), putting some species in danger of extinction, at least locally. All these threats demonstrate that we urgently need to reconsider the way we use the environment, which is also our own. The first priority should be the preservation and moderated use of the natural areas on our planet, as these are essential for the conservation of biodiversity.

# Conservation Measures

Legislation has only recently taken into account the drastic decline in numbers of most snake species. A series of laws has been introduced classifying snakes as protected species in several countries. Certainly, one of the major international treaties, the Washington Convention, includes only a small number of snake species in the region covered by this book, but most regional lists of threatened species include nearly all snake species. Today, all species of European snake are protected and may only be taken from the wild or reared in captivity with proper authorization. Obviously, protecting a species is insufficient if its habitat is not also protected. This is why nature reserves and regional and national parks have been created and biotope protection orders proclaimed.

If humans continue to ignore or to disrespect plant and animal life, despite the introduction of laws designed to protect them, including snake species, the application of such laws will have little effect. Our deepest wish is that there will come a time when people will stop stupidly killing or squashing snakes they come across, but will observe them with interest and leave them to live their lives like any other animal. This is the main aim of this book: to help redress the age-old feeling of aversion to snakes and show that they can be beautiful and fascinating, yet still mysterious, as there are so many species of snake about which we yet know very little.

*In Germany, special protection measures have been implemented for the Aesculapian Snake* Zamenis longissimus *as only a few residual populations remain; these are at the edge of the species' range and are threatened with extinction. In France, despite the species being much more widespread, it is also protected, as are all other species of naturally occurring reptiles.* P. Geniez.

# Snakes in Captivity

Snakes are not typical pets, but keeping or breeding them can be very interesting as they are difficult to observe or study in the wild. Keeping them also allows for better understanding of their physiology, behaviour, stages in the breeding cycle, etc. Good conditions for rearing snakes require a large, well-ventilated and heated terrarium with appropriate ambient humidity, as well as well-arranged plants and branches to climb into, a water container for bathing and drinking and sufficient food and drink.

Alongside the few professional zoologists who study reptiles by watching them in captivity, many amateurs look after their terrariums well and enjoy observing the lives of these distinctive and fascinating animals. The care of snakes can be time-consuming, and good snake keepers understand their responsibilities.

If correctly housed and cared for, a healthy snake will soon accept its new environment. However, there are some species that are not suitable for being reared in a terrarium, either because it is very difficult to simulate their required climate adequately (such species as the Adder or Orsini's Viper) or because their small size and underground lifestyle make it difficult to observe and feed them in a terrarium (examples are worm snakes and blind snakes). Also, it would obviously be ill-advised for a beginner to keep a very dangerous or aggressive species.

Patient and attentive observation will enable owners to quickly become familiar with their animals.

With time, many snakes in captivity lose their initial aggressiveness and shyness.

Looking after snakes in a terrarium can be instructive and rewarding. Beginners should, however, take advice from a specialist before acquiring a snake. It should also be understood that the majority of snake species are protected by law and that keeping them is only authorised, amongst other requirements, if the future owner has the appropriate license to prove their aptitude for keeping the animals; the rules for obtaining such a license vary from country to country.

In most European countries, there are many herpetological societies whose members readily provide advice and through which it is possible to meet other people with the same passion.

Reading articles in the publications produced by many of these societies, participating in events or meeting members who are passionate about snakes is a good way to improve one's knowledge about rearing snakes and to acquire solid know-how in the matter.

Many of these societies promote the protection of snakes in the wild as well as their habitats and encourage the participation of amateur herpetologists in the study of snakes in captivity.

*Male Asp Viper* Vipera aspis aspis, *Charente, western France.* Évrard.

**The Species**

# Snake Classification

Systematics is the science of classifying living groups of organisms, or taxa (singular: taxon). It allows them to be enumerated and to be classified according to kingdom, class, order, sub-order, genus, sub-genus, species, and subspecies. Originally based on the anatomical and morphological resemblance between species, systematics has recently seen spectacular changes due to the development of molecular techniques, primarily based on the study of DNA. This new form of systematics, termed phylogenetic systematics, reveals a more measureable dimension to

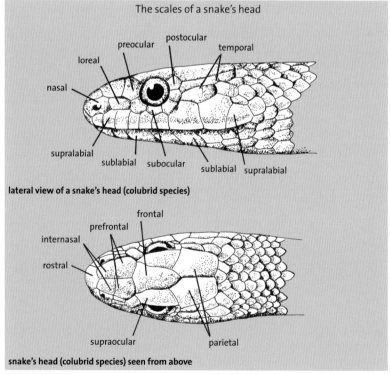

The scales of a snake's head

lateral view of a snake's head (colubrid species)

snake's head (colubrid species) seen from above

*A snake's head seen from the side and from above, showing the different scales and their names.*

evolution than traditional systematics, as it allows us to date (approximately) the moment at which a common ancestor divided into two separate lineages that have led to a present species. Phylogenetic classification now integrates in a measurable, less empirical, way the various relationships between the evolutionary lineages, and thus between orders, families, genera, sub-genera, species and distinct populations within a species.

DNA studies of snakes have shown that their phylogeny (a sort of genealogy of living beings) is a little disconcerting. In some cases, those studies have upset part of the classification of the species within the families and their respective genera. In other cases, new phylogenetic studies have confirmed what was already believed through the morphological study of species. More detailed information of reptile classification can be found in the monumental article of R. Alexander Pyron, Frank T. Burbrink and John J. Wiens, published in the journal *BMC Evolutionary Biology*, entitled "A Phylogeny and Revised Classification of Squamata, including 4161 Species of Lizards and Snakes" (13:93, 2013). As the title indicates, no less than 4161 species of squamata (lizards, worm lizards and snakes) have been studied from a molecular viewpoint and placed in an enormous phylogenetic "tree" that reconstitutes, family by family, the genealogical tree of the genera and species. Even if all species of squamata have not yet been studied, nearly all genera figure on this tree.

A large part of the results of this work (which incorporates the results of numerous other phylogenetic studies by other authors) confirms what we already thought. For example, the pit vipers and the vipers form two distinct, divergent groups but belong to the same family, the Viperidae, because they have a common ancestor, one that is not as ancient as that of the other lineages; they have retained morphological and behavioural characteristics in common that do not show recent adaptations (for example, to a certain habitat) but do show morphological traits of their ancestor.

Other results are, however, surprising. For example, sea snakes do not all belong to the same sub-family (the Hydrophiinae); they belong to various lineages of the Elapidae, certain representatives of which have adapted to a marine lifestyle. In the same way, the classification of the colubrids has been revised and contains several families (including two in the region covered by this book: the Colubridae and the Lamprophiidae).

Despite all these dramatic changes in snake classification, it is still possible to identify a species from its behavioural and morphological characteristics (shape, scale positioning, dentition, colouration). This is what we have tried to achieve in the present work, but, as the reader may well find, certain species remain very difficult to identify.

In this regard, taking into account a single morphological characteristic often leads to an incorrect identification.

On the other hand, the use of several characteristics allows for a reasonably reliable identification – one, however, that is dependent on the natural history expertise of the observer. For example, if we arrive at the combination of "small scales on the top of the head and keeled dorsal scales," the snake is obviously a viper. Most of these characters are clearly visible on photographs, and it is often possible to identify a snake later from a photograph. (But beware: photographing a venomous snake can present a real danger!)

Snake scales, based on their shape and position on the body, are given particular terms in order to characterise them, localise them and use their details in species identification. In order to interpret the species' descriptions, it is very useful to use the illustration on page 38 (two views of a colubrid snake's head), which clarifies these terms. In this book, larger more plate-like scales are called plates, but they can also be called scales.

## Sub-order Serpentes: The Snakes

The snakes are a sub-order of the order Squamata (scaly reptiles), which in turn are part of the class Reptilia (reptilian animals). The number of snake species occurring within the Western Palaearctic, the area covered by this book, is 123, according to the author's approach to the subject (of these, two have been introduced by humans); to this total must be added a few species of sea snakes that have been recorded in areas around the northern part of the Persian Gulf (Iraq, Kuwait and extreme northern Saudi Arabia). This is quite a small total compared to the 3530 species of snake in the world, but it is a lot compared

to the 13 species that occur in France and the 3 in Britain or even to the 27 species that occur in Morocco and the 51 in Turkey, the Western Palaearctic countries with the highest number of snakes. This list of 123 species has been sourced from several works, particularly genetic studies. It takes into account not only the genetic divergence between species but also their ability to maintain their morphological characteristics, even in contact zones with genetically very similar species. According to the phylogenetic classification proposed in 2013 by Pyron et al., 7 families of snake occur in the Western Palaearctic.

### Family Typhlopidae–Blind snakes
▶ Genera: *Xerotyphlops, Letheobia, Indotyphlops*
Example: Worm Snake *Xerotyphlops vermicularis*

### Family Leptotyphlopidae–Slender blind snakes
▶ Genus: *Myriopholis* (= *Leptotyphlops* of the Old World)
Example: Hook-billed Blind Snake *Myriopholis macrorhyncha*

### Family Boidae–Boas
▶ Sub-family: Erycinae (sand boas)
Genus: *Eryx*
Example: Western Sand Boa *Eryx jaculus*

### Family Colubridae
▶ Sub-family: Colubridinae (terrestrial and arboreal colubrids)
Genera: *Coronella, Dolichophis, Eirenis, Elaphe, Hemorrhois, Hierophis, Lampropeltis, Lytorhynchus, Macroprotodon, Muhtarophis, Platyceps, Rhinechis, Rhynchocalamus, Spalerosophis, Telescopus, Zamenis*
Example: Aesculapian Snake *Zamenis longissimus*
▶ Sub-family: Natricinae (semi-aquatic colubrids)
Genus: *Natrix*
Example: Grass Snake *Natrix natrix*

### Family Lamprophiidae
▶ Sub-family: Aparallactinae
Genus: *Micrelaps*
Example: Müller's Ground Viper *Micrelaps muelleri*
▶ Sub-family: Atractaspidinae
Genus: *Atractaspis*
Example: Palestine Burrowing Asp *Atractaspis engaddenis*
▶ Sub-family: Lamprophiinae
Genus: *Boaedon*
Example: Brown House Snake *Boaedon fuliginosus*
▶ Sub-family: Psammophiinae
Genera: *Malpolon, Rhagerhis, Psammophis*
Example: Western Montpellier Snake *Malpolon monspessulanus*

### Family Elapidae
▶ Genera: *Naja, Walterinnesia*
Example: Egyptian Cobra *Naja haje*

### Family Viperidae–Vipers
▶ Sub-family: Viperiinae (true vipers)
Genera: *Bitis, Cerastes, Daboia, Echis, Macrovipera, Montivipera, Pseudocerastes, Vipera*
Example: Adder *Vipera berus*
▶ Sub-family: Crotalinae (pit vipers)
Genus: *Gloydius*
Example: Halys Pit Viper *Gloydius halys*

# The Snakes of Europe, North Africa and the Middle East

This chapter, the most important of the book, is an exhaustive presentation of the 123 species of snake known to occur in the geographic area it covers, the Western Palaearctic. Each species account contains:

▶ an identification section (morphology describing the scale arrangement and colouration, criteria that differentiate it from closely related species, maximum size)

▶ an idea of the potency of its venom

▶ a succinct description of its ecological requirements

▶ information about its lifestyle and any special behavioural traits

▶ dietary preferences

▶ a short paragraph on the species' reproduction

▶ a description of its distribution

▶ a paragraph entitled "Geographic variation" that places the species within present snake classification, referring to the latest phylogenic discoveries. Any subspecies are taken into consideration.

Each species account is accompanied by one or several colour photographs, some published for the first time, as well as a distribution map that is as precise as possible. Most distribution maps have their origin in the prodigious book *The Reptiles of the Western Palaearctic* (Vol. 2: Snakes), by Roberto Sindaco, Alberto Venchi and Christina Grieco (2013). One of the strong points of this book is the inclusion of precise distribution maps, in colour, for all snake species.

# Families Typhlopidae (blind snakes) and Leptotyphlopidae (slender blind snakes)

The blind snakes Typhlopidae and the slender blind snakes Leptotyphlopidae belong to two different families, but in outward appearance they are very similar and are relatively closely related. They live a mainly underground existence; they have only minuscule eyes with which it appears that they detect only differences in the strength of light.

They are generally small (usually less than 30 cm long, but up to 95 cm for the largest species, the Zambezi Beaked Blind Snake *Afrotyphlops mucruso* of southeastern Africa) and vaguely resemble earthworms, sometimes very thin ones, in outward appearance.

Due to their lifestyle of burrowing in the soil, their skull is relatively massive and consolidated in the form of a bony capsule.

The ancient character of these snakes can be seen, among other characteristics, in their small and only slightly differentiated scales, which show little difference between those on the top of the body and those on the underside (no enlarged ventral scales), and in the remains of femur and pelvic bones in the skeleton.

The Typhlopidae (more than 250 species worldwide) and the Leptotyphlopidae (about 120 species) occur mainly in the tropical and subtropical regions of Africa, Asia, the South Sea Islands and the Americas.

From a phylogenetic perspective, these two families are placed at the base of snake evolution and, with three other families, are the most "primitive" snakes, distinct from all other species in the world.

# Family Typhlopidae (blind snakes)

## Worm Snake

*Xerotyphlops vermicularis* (Merrem, 1820) (formerly *Typhlops vermicularis*)
**Family:** Typhlopidae (blind snakes)
**Sub-family:** Asiatyphlopinae
**F.:** Typhlops vermiculaire
**G.:** Wurmschlange

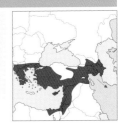

**Identification** Length 18–30 cm, to 36.5 cm, body diameter 2.5–5.5 mm. Morphology recalls that of an earthworm. A blunt head without obvious separation from the body, with minuscule eyes in the form of dark spots on the top of the head. The body is cylindrical, slightly flattened. The whole of the body is covered in small, round, smooth, shiny scales, all more or less of the same size, arranged in 22–24 rows. The short, blunt tail ends in a thorn-shaped caudal scale that can serve as a fixation point when the animal is moving forward.

Transparent brownish, pink or pale red colour; the belly is brighter, allowing the internal organs, eggs and prey remains in the body to be detected.
**Venom** Harmless.
**Habitat** At lower elevations on hillsides and mountain sides; on dry slopes covered with bushes and rocks and covered with sparse grassy vegetation. Also present in Mediterranean garigue

*The Worm Snake* Typhlops vermicularis *is morphologically reminiscent of an earthworm, but it is a snake. Right: note on the close-up the very rounded snout, the vertically elongated rostral plate and the small eyes on the top and at the rear of the head. Left: Péloponnèse (Greece); right: Azerbaijan.* P. Geniez.

45

and rural, agricultural areas, as well as on semi-arid steppe. Occurs from sea level to 1600 m elevation, rarely to 2169 m.

**Habits** Lives in a system of underground galleries; can often be found by lifting stones. Appears on the surface only rarely and then only at dusk or at night. In spring, often seen under stones. In summer, especially when it is very hot and dry, it disappears deep into the ground, only to reappear at the surface the following spring; it is then impossible to find, even by turning stones. Winter inactivity: 2–6 months, depending on region and elevation.

**Diet** Especially ants, including their eggs, larvae and nymphs; also other small insects, myriapods and small worms.

**Reproduction** Oviparous. Lays 2–6 elongated eggs that hatch in late September.

**Range** The southern Balkans from Albania and southern Bulgaria as far as Greece, including the Aegean Sea islands; farther east, the countries boarding the eastern Mediterranean from Turkey as far as northeastern Egypt and including Cyprus, Syria, Lebanon, Jordan and Israel. In central Asia, its range continues from Transcaucasia to northwestern Afghanistan.

**Geographic variation** *Xerotyphlops vermicularis* is a monotypic species belonging to a genus composed of just four species, occurring in Asia, Europe and Africa. This genus belongs to the sub-family Asiatyphlopinae.

# Israeli Worm Snake

*Letheobia simonii* (Boettger, 1879) (formerly *Rhinotyphlops simoni*)
**Family:** Typhlopidae (blind snakes)
**Sub-family:** Afrotyphlopinae
**F.:** Typhlops d'Israël
**G.:** Israelische wurmschlange

**Identification** Minuscule snake (16–24 cm in length); burrowing, wormlike; flesh pink or red-pink in colour; extremely slim (2 or 3 mm in diameter), with hardly visible, minuscule eyes hidden under a large ocular plate. Very small head, indistinct from body, is an orange or slightly yellowish colour. Remarkably conical, pointed snout, with a flattened, sharp front edge. Rarely large rostral plate extends from the first supralabial to the nostril. One large preocular (which is not as large as the nasal plate), which is in contact with the 2nd and 3rd supralabials; the prefrontal plate and the supraocular scales are larger than the body scales;

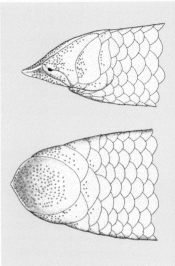

*Letheobia simonii is a minuscule, wormlike snake. Its very strange head is shaped like a flute mouthpiece. As seen in the diagram, the eyes are not visible as they are located completely under the large, vertical ocular plate situated towards the back of the head. The nostrils could be taken for eyes. Note as well the enormous frontal plate that covers nearly the whole of the head.* Talbi.

large undivided ocular plate; 2 or 3 postoculars; 3 or 4 supralabials on each side. The tail is short (less than 2% of the total length) hardly longer than wide (about 4 mm). The body scales are small and shiny, essentially all the same size, including those of the belly (no differentiated ventral plates) and arranged in 20–22 longitudinal rows; 403–488 transverse rows of scales along the body; the anal scute is undivided, and there is no caudal spine.

**Venom** Harmless.

**Habitat** Mediterranean regions. Soft and damp soils in wadis and near urban areas; also in cultivated areas and gardens.

**Habits** Almost unknown. Lives underground in soft soils, between plant roots and under stones.

**Diet** Essentially ants, including their eggs and larvae, but also other small insects and small spiders.

**Reproduction** Oviparous; the female lays a few elongated eggs.

**Range** Israel, northwestern Jordan, Lebanon and western Syria.

**Geographic variation** *Letheobia simonii* is a monotypic species. The genus *Letheobia* contains 31 species occurring principally in Africa and belongs to the sub-family Afrotyphlopinae.

# Southeastern Turkey Worm Snake

*Letheobia episcopus* (Franzen and Wallach, 2002)
(formerly *Rhinotyphlops episcopus*)
**Family:** Typhlopidae (blind snakes)
**Sub-family:** Afrotyphlopinae
**F.:** Typhlops de Turquie, Typhlops de Bischoff
**G.:** Sudosttürkische Wurmschlange

**Identification** Closely related to *Letheobia simonii*, but for the present known only from southeastern Turkey. A small wormlike snake that is a little larger than *L. simonii* (total length 25–33 cm). Eyes invisible externally. Conical snout, pointed and angular in front like that of *L. simonii*. Large, undivided nasal plate; the ocular plate is often divided into 2 or 3 narrow scales (undivided in *L. simonii*); 5 or 6 postoculars; 4 labials on each side; 20 rows of scales at mid-body. The number of transverse rows of scales around the body is larger than in *L. simonii* and diagnostic: 544–595 (compared to 403–488). No spine on the tail. The body colour is pink or orange-pink, with a slightly vermiculated pigmentation on the back; snout yellowish.

**Venom** Harmless.

**Habitat** Dry areas undisturbed by humans with grassy vegetation, many rocks and isolated shrubs (oaks, terebinth, mastic, olive). Soil often brown loam on limestone karst. Semi-arid climate with rain in winter but little or no precipitation in summer.

*Southeastern Turkey Worm Snake* Letheobia episcopus *is a tiny, blind, burrowing snake discovered in 2002.* Franzen.

**Habits** Hardly known. Lifestyle like that of the closely related *Letheobia simonii*: underground, in soft soil, between bush roots or under stones.
**Diet** Unknown; probably small insects such as ants, their eggs and their larvae, and small worms.

**Reproduction** Unknown but probably oviparous.
**Range** For the present, known from a small area in southeastern Turkey, around Halfeti, in the Upper Euphrates and near Şanlıurfa, at elevations between 500 and 640 m.

# Brahminy Worm Snake

*Indotyphlops braminus* (Daudin, 1803) (formerly *Ramphotyphlops braminus*)
**Family:** Typhlopidae (blind snakes)
**Sub-family:** Asiatyphlopinae
**F.:** Serpent aveugle des pots de fleurs or Typhlops de Brahma; **G.:** Brahmanen-Wurmschlange

**Identification** Small, wormlike snake that resembles the Worm Snake *Xerotyphlops vermicularis* but is slightly slimmer and of more variable colour, often dark violet-brown, ash-grey or even pale blue. The length when adult, 13–20.3 cm, is less than that of the Worm Snake (it is one of the smallest snakes in the world). The body scales, all almost the same size, are arranged in 20 rows around the body (22–24 in the Worm Snake) and in 292–368 transverse rows, depending on the individual. The narrow rostral plate and the 2 nasal plates overhang the mouth to form a sort of rounded shovel that allows the snake to burrow in the soil. Nasal plate completely divided by a furrow (only partially in the Worm Snake); the nasal furrow is in contact with the preocular (in contact with the 2nd supralabial in the Worm Snake).

**Venom** Harmless.
**Habitat** In Egypt, the Brahminy Worm Snake was found in newly constructed urban sites in the desert planted with tropical gardens and regularly watered lawns. In the Canary Islands, this species has been found in flowerpots. Unlike other species of Typhlopidae and species of Leptotyphlopidae of the Western Palaearctic, it avoids arid areas and desert.
**Habits** A burrowing, nocturnal species that lives in burrows dug in light, more or less humid soil; in gardens and plant containers in urban areas. It is found within clumps of roots, under stones, in heaps of vegetation or leaf litter. The best chance of finding one is to look under flowerpots, where it often shelters, or in watered lawns or gardens in hot countries; at night, look in over-watered areas.

Indotyphlops braminus. *Above: a juvenile; below: an adult darting its minuscule whitish tongue. Aguadulce (Ameria province, Spain).* Rodríguez Luque.

**Diet** Small invertebrates such as ants, termites, worms and caterpillars.

**Reproduction** The Brahminy Worm Snake is very remarkable in the fact that it is one of only two species of snakes in the world known to be parthenogenetic. Without having to mate with a male, females lay 1–18 eggs that all hatch into females. Males are unknown and probably do not exist.

**Range** The Brahminy Worm Snake is considered to be the species of terrestrial snake with the most extensive range in the world, due to the fact that it can reproduce from a single individual and the ease with which it can be accidentally transported in flowerpots. Its natural origin is uncertain, but it is probably on the Indian subcontinent and Sri Lanka. Because it is easily transported in soil (particularly in that of flowerpots), it has been introduced into sub-Saharan Africa, in other parts of Southeast Asia, northern Australia, islands in the Indian and South Pacific oceans, the southern United Sates and to Hawaii, Mexico and Guatemala. In the area covered by this book (where it has also been introduced), since 2004 it has been reported from the Arabian Peninsula; Kuwait; Egypt, around Cairo and in the Sinai; northwestern Libya; Grand Canaria (Canary Islands) and Madeira (Portugal). In 2011 and 2012 the species was found on mainland Spain at Aguadulce, Almeria province, by Francisco Rodriguez Luque. It is probable that new sites will soon be found in warmer parts of the Mediterranean region.

**Geographic variation** *Indotyphlops braminus* is a monotypic species. The genus *Indotyphlops* contains 22 species, all of Southeast Asian origin, and belongs to the sub-family Asiatyphlopinae, as does *Xerotyphlops vermicularis*.

# Family Leptotyphlopidae (slender blind snakes)

## Hook-billed Blind Snake

*Myriopholis macrorhyncha* (Jan, 1860) (formerly
*Leptotyphlops macrophynchus*)
**Family:** Leptotyphlopidae (slender blind snakes)
**F.:** Leptotyphlops à bec crochu
**G.:** Hackensnabel-Schlankblindschlange

**Identification** Minuscule, burrowing, wormlike, very thin (2–3 mm diameter) snake with a short tail (9–10% of the total length). 17–28 cm long. Minuscule eyes appear as points hidden under the very large ocular scale. The upper part of the snout is curved forward, giving the head a hooked appearance; the upper mandible is obviously longer than the lower. The mouth is on the underside. The rostral plate is concave in its lower part, and the large nasal plates are divided in their lower part, where the minuscule nostrils are located. The ocular plates are very enlarged, their lower edge located between 2 supralabials, the front one very small. The lower part of these 3 scales (rostral, nasal and ocular) is in direct contact with the mouth. The occipital plate is undivided. The body is cylindrical, slightly flattened. The body scales are smooth and, for most, identical around the entire body (no large ventral plates), arranged in 14

*The Hook-billed Blind Snake* Myriopholis macrorhyncha, *wormlike and very slim, obtains its name from the curved, hooked shape of the snout. Here, several juveniles from southeast Turkey.* Göçmen.

51

Myriopholis macrorhyncha, *in southwestern Turkey. Note the obviously prominent "beak", which is, however, less hooked than that of the closely related* M. algeriensis *(following species).* P. Geniez and A. Teynié.

rows at mid-body; 300–490 rings of scales from the head to the anal scute; 43–53 rings of scales along the tail arranged in 10 rows at mid-tail. A pale, translucent pink, red or yellowish-brown colour.

**Venom** Harmless.

**Habitat** Low-lying arid or desert regions with sparse vegetation, stones and, locally, slight humidity. Sometimes near human habitations, in oases and uncultivated gardens, but also in wilder or desertlike areas.

**Habits** Little known. Nocturnal. Lives under stones and tufts of vegetation or in the root systems of trees or bushes. Avoids direct sunlight.

**Diet** Small insects such as ants and termites.

**Reproduction** Oviparous.

**Range** Northeastern Africa, Arabian Peninsula, Israel, Jordan, Lebanon, Syria, southeastern Turkey, Iraq, western and southern Iran, to the east as far as western Pakistan and south as far as Kenya and northern Tanzania.

**Geographic variation** Since the populations in northwestern Africa have been considered to belong to another species, *Myriopholis algeriensis*, and since *M. macrorhyncha bilmaensis* is considered to be synonymous with *M. algeriensis, M. macrorhyncha* is considered to be a monotypic species (without any subspecies). However, the status of the *hamulirostris* taxon, present in Iran and also probably in Turkey, needs to be clarified as it may concern a subspecies, or maybe even a separate species.

# Algerian Hook-billed Blind Snake

*Myriopholis algeriensis* (Jacquet, 1895) (formerly *Leptotyphlops algeriensis*)
**Family:** Leptotyphlopidae (slender blind snakes)
**F.:** Leptotyphlops d'Algérie
**G.:** Algerische hackensnabel-Schlankblindschlange

**Identification** Minuscule, wormlike snake, closely related to the Hook-billed Blind Snake but with an even more pronounced hooked "beak". An inhabitant of the Sahara zone and with a more western distribution, centred on northwestern Africa. A pale pink colour.
**Venom** Harmless.

**Habitat** Low- to medium-elevation desert with sparse vegetation, stones and, sometimes but not always, damp areas here and there.
**Habits** Nocturnal and burrowing. Lives under stones and tufts of vegetation and in the root system of shrubs. Avoids direct sunlight. Sometimes comes to the surface – for example, when the ground has been watered.

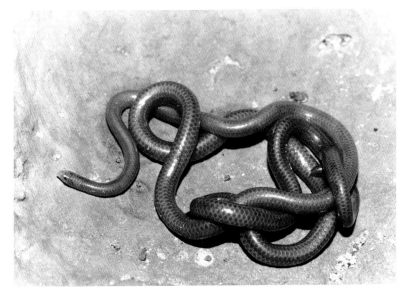

*Algerian Hook-billed bland Snake* Myriopholis algeriensis. *Note the even more hooked snout compared to that of* M. macrorhyncha, *forming a protruding "beak". Morocco.* P. Geniez.

53

**Diet** Little known, but most probably small insects, ants and termites, for example.
**Reproduction** Oviparous.
**Range** Northwestern African deserts; in the area covered by this book: southern Morocco, including the northern part of the Atlantic Sahara, desert steppe in Algeria as well as the Ajjer massif in the extreme southeastern part of the country, and southern Tunisia. Also present in Mauritania, northern Mali and northeast Niger. It probably occurs in Libya.

**Geographic variation** *Myriopholis algeriensis* is considered to be a monotypic species.

*Close-up of* Myriopholis algeriensis. *Morocco.* P. Geniez.

# Cairo Earthsnake

*Myriopholis cairi* (Duméril and Bibron, 1844) (formerly *Leptotyphlops cairi*)
**Family:** Leptotyphlopidae (slender blind snakes)
**F.:** Leptotyphlops du Caire
**G.:** Ägyptische Schlankblindschlange

**Identification** Minuscule, wormlike snake, 15–25.3 cm in total length for a diameter of 2–3 mm. The constantly darted forked tongue immediately distinguishes this snake from a worm. Rounded snout much less protruding than in *Myriopholis macrorhyncha* and *M. algeriensis*. Minuscule eyes covered by a large ocular scale. A supraocular scale located on each side. The ocular plate extends to the corner of the mouth; it is located between 2 supralabials, the front one of which is very small (less than a quarter of that of the rear supralabial). Divided occipital plate (undivided in *M.*

*macrorhyncha* and *M. algeriensis*). The body scales are smooth, translucent and shiny, all nearly the same around the body and arranged in 14 rows of scales at mid-body, and 267–364 transverse rows from the rear of the head to the tail tip. Short tail (1.2–2.0 cm) that finishes in a sharp point by the terminal scale.
Upperparts pale pink or pale purple-brown, whitish belly.
**Venom** Harmless.
**Habitat** Needs a certain amount of ground humidity. In the Nile Valley, often occurs near human habitations, in gardens and agricultural zones

*Cairo Earthsnake* Myriopholis cairi. D. Fuchs.

and at field edges. Lives under stones, between roots of bushes, in soft ground and small mammal burrows.
**Habits** Little known. Nocturnal. These slender blind snakes may be found in February, March or April at night, near the soil surface, even on the soil. Avoids direct sunlight. Actively digs in and turns over the soil. Goes into water voluntarily and swims quite well.

**Diet** Small insects.
**Reproduction** Oviparous.
**Range** Within the scope of this book, occurs only in the Nile Valley in Egypt; also in Sudan, Ethiopia, Eritrea, Somalia and Niger.
**Geographic variation** *Myriopholis cairi* is considered to be a monotypic species.

# Family Boidae (boas)

The Boidae also represents a group of very primitive snakes whose skeletons contain vestiges of hind legs and a pelvis. These vestiges are sometimes visible on each side of the cloaca and are called "cloacal spurs".

The Boidae contains, along with the Pythonidae, some of the world's largest and thickest snakes; the Anaconda *Eunectes murinus* of tropical America can grow to a length of 8.45 m and weigh to 200 kg. However, in the Western Palaearctic the only representatives of the Boidae are in the sub-family Erycinae, members of which are small and thickset, shy, discreet and burrowing, with a length of 30–50 cm, up to a maximum of 90 cm.

The Boidae kill their prey by constriction and strangulation. They are variously adapted to a multitude of habitats, anything between living in the canopy of tropical forest to on the ground in wetlands (watercourses and marshes) and even underground, especially in arid areas.

The principal area of distribution of the Boidae is centred in the tropical and subtropical zones. Depending on the various interpretations of the phylogeny of the Boidae, of those that occur within the range of this book, the sub-family Erycinae is represented by either a single genus (*Eyrx*) or by four distinct genera.

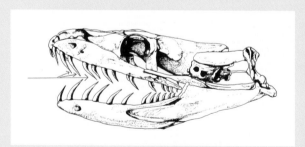

*Skull of a boa showing its sharp teeth, all of almost equal size.*

# Western Sand Boa or Javelin Sand Boa

*Eryx jaculus* (Linnaeus, 1758)
**Family:** Boidae (boas)
**Sub-family:** Erycinae (sand boas)
**F.:** Boa des sables occidental or Eryx javelot
**G.:** Westliche Sandboa

**Identification** Relatively thickset (especially when adult), smallish terrestrial snake, with small scales characteristic of the sub-family Erycinae, composed of shiny, smooth (sometimes slightly keeled at the rear of the body), lozenge-shaped scales, arranged in 40–54 rows at mid-body. Average length 30–50 cm, maximum 90 cm. In profile, the head is wedge-shaped, indistinct from the body (no obvious neck). The snout plate (rostral) is straight-sided seen from above; very thick and shovel-shaped, it helps the snake to burrow in the ground. The top of the head is covered in small, smooth, shiny scales (there are less than 10 between the eyes). Small eyes, with pupils with a vertical slit, orientated to the side; the iris golden

Eryx jaculus jaculus, *61.5-cm-long adult, found in northeastern Morocco.* García-Cardenete.

orange vermiculated with black. 1 or 2 rows of small subocular scales between the bottom of the eye and the supralabials; 10–14 supralabials on each side. The ventral plates are very narrow and quite numerous (165–200). The anal and subcordal scales are undivided. The tail is short and rounded, appearing truncated and ending with a rounded scale. There are large, irregular markings on the back, located transversely, that may be

Eryx jaculus turcicus, 51-cm-long adult. Thrace (eastern Greece). Note that the two internasal plates, located just behind the rostral, touch two scales (not three, as in E. j. jaculus). P. Geniez.

Eryx jaculus familiaris, *the easternmost-occurring subspecies and most genetically distinct. Northeastern Turkey.* P. Geniez and A. Teynié.

joined together. These markings are either brown or dark grey to green-black on a pale ground colour of near-white, beige, yellowish, sometimes even reddish. The flanks are marbled with small dark markings on a darker ground colour than that of the back. These markings can be more or less joined into an irregular dark line that runs along the bottom of the flanks. A dark, oblique line runs backwards from the eye to the corner of the mouth. The belly is pale, whitish or pale grey, but the ventral plates have a yellow-orange, bright orange or brick-red tint and are dotted with irregular, ill-defined dark spots.

**Venom** Harmless.

**Habitat** In dry lowlands, hillsides and mountains, especially in valleys of sandy or clay substrate. Dry areas with sparse vegetation, especially on soft soils. Also in Mediterranean garigue and uncultivated areas between fields. Occupies areas of steppe but avoids true desert. Generally occurs from sea

level to 1600 m elevation, but occurs in the Nagorno-Karabakh Republic (near Armenia) between 2036 and 2354 m.

**Habits** Essentially burrowing, the Western Sand Boa spends most of its time in rock crevices or in networks of small mammal burrows. Often found under flat stones or old wood. Basically active in the evening or at night, when it may be encountered on the surface. Hibernation: 3–5 months.

**Diet** Small rodents, lizards, young of ground-nesting birds, sometimes snails and insects. This constricting snake suffocates its larger more active prey items by squeezing them tightly in its powerful body coils.

**Reproduction** Ovoviviparous. Mating occurs in April or May. Gives birth to 3–12 live young in September. The newly born young are tiny, 9–14 cm long.

**Range** Essentially Mediterranean. The southern Balkan states, from southern Albania, southern Macedonia and Greece as far as southeastern Romania and southern Bulgaria. Also present on several of the Aegean Sea islands, including the Dodecanese. Father east extends through Turkey, the Caucasus region, western Iran, extreme northeastern Saudi Arabia, Syria, Jordan, Lebanon and Israel. Its range continues along the Mediterranean in North Africa, from Egypt to northeastern Morocco, the westernmost point of its vast range.

**Geographic variation** *Eryx jaculus* is quite a variable species, represented by three subspecies:

▶ *Eryx jaculus jaculus* (Linnaeus, 1758): from North Africa to Israel, Syria and extreme southeastern Turkey (Şanliurfa region). More often than not, 3 medium-sized scales are in contact with the 2 internasals, and there is just 1 row of scales between the eye and the supralabials.

▶ *Eryx jaculus turcicus* (Olivier, 1801): the Balkans and Turkey. More often than not, there are 2 rows of scales between the eye and the supralabials; however, this feature is quite variable.

▶ *Eryx jaculus familiaris* Eichwald, 1831: Caucasus region (Georgia, Armenia, Azerbaijan), eastern Turkey, northern Iraq, and northwestern Iran. Generally has a single row of scales between the eye and the supralabials. Genetically it is the most distinct of the subspecies. It has a colouration subtly different from that of the other two subspecies.

The last two subspecies, differentiated, among other characters, by the presence of 2 large transverse scales in contact with the internasals, are sometimes considered to be synonyms despite their quite distinctive colouration and their marked genetic divergence.

## Desert Sand Boa

*Eryx miliaris* (Pallas, 1773)
**Family:** Boidae (boas)
**Sub-family:** Erycinae (sand boas)
**F.:** Boa des sables or Eryx miliaire
**G.:** Wüsten-Sandboa

**Identification** With a length of 35–55 cm, this is one of the smallest sand boas. The small eyes are slightly orientated upwards (a useful character to separate it from *Eryx jaculus*, which it greatly resembles). Pupils with a vertical slit, as in all Boidae. The body

and tail form and scale makeup are quite similar to that of the Western Sand Boa (page 57). 2 rows of scales between the eye and supralabials (contrary to *E. jaculus familiaris*, the subspecies that has a distribution closest to that of *E. miliaris*, but which

*Desert Sand Boa* Eryx miliaris, *western Kazakhstan, 150 km west-northwest of Atyraou.* Crochet.

is otherwise quite distinct); 12 or 13 supralabials on each side. Ground colour between pale sandy yellow and dark brown. There is a dark temporal band, sometimes very ill-defined, on the sides of the head from the eye to the corner of the mouth. The dark markings on the back appear less geometric than those of *E. jaculus*, and can be loosely connected. Sometimes these dorsal markings are joined on the sides, and thus the surrounding pale marks on the back may be reddish with cream edges. On the tail, the dorsal markings slide progressively onto the sides and come together to form an irregular longitudinal band. Those individuals in the westernmost part of the species' range (*E. miliaris nogaiorum*, steppes of Kalmykia and

Chechnya) are especially dark; the black dorsal markings may nearly cover the whole back.
**Venom** Harmless.
**Habitat** Arid and semi-desert steppes with bushes or tufts of vegetation.

*Close-up of a Desert Sand Boa* Eryx miliaris. *Note that the eyes are slightly orientated upwards and that the internasal plates touch the four frontal scales (compared to two or three in* E. jaculus*).* Crochet.

61

*A melanistic Desert Sand Boa, in the European part of Kazakhstan, 150 km to the west-northwest of Atyraou. Colouration identical to the majority of individuals of* Eryx miliaris nogaiorum, *normally present farther west.* Crochet.

Especially sandy or silty areas, but also at times areas with stony subsoil.

**Habits** With its eyes orientated slightly upwards, it is especially well adapted to a life in desert sands. It can, when danger threatens, burrow into the sand very quickly. Commonly lives in rodent burrows. Active at dusk and at night.

**Diet** Small rodents, lizards, insects, young of ground-nesting birds.

**Reproduction** The same as the Western Sand Boa.

**Range** Steppes of central Asia and northern Transcaucasia, from southwestern Russia and western Kazakhstan (northwestern shore of the Caspian Sea) to the west as far as southeastern Kazakhstan; eastern Uzbekistan and extreme northeastern Iran.

**Geographic variation** *Eryx miliaris* is considered to be polytypic, with two subspecies; however, their validity has yet to be fully established:

► *Eryx miliaris milaris* (Pallas, 1773): the major part of its range.

► *Eryx miliaris nogaiorum* Nikol'skij, 1910: extreme western part of the species range; Kalmykia and Chechnya, countries in the northeastern Caucusus (Russian Federation).

The taxonomic status of the Tartar Sand Boa *Eryx tataricus* (Lichenstein, 1823), which nearly attains, at the northern end of the Caspian Sea, the zone covered by this book, is not clearly established. This species is part of the *E. tataricus* complex and, according to certain authors, it is a completely distinct species. According to others, *E. tataricus* and *E. miliarus* are part of the same species. Among all the *Eryx* species, *E. miliarus* and *E. tataricus* are genetically the most closely related, and it is likely that certain records of *E. tataricus* concern in fact *E. miliarus*, and vice versa.

# Arabian Sand Boa

*Eryx jayakari* Boulenger, 1888) (formerly *Pseudogongylophis jayakari*)
**Family:** Boidae (boas)
**Sub-family:** Erycinae (sand boas)
**F.:** Boa des sables d'Arabie
**G.:** Arabische Sandboa

**Identification** The smallest of the burrowing boas (38 cm maximum total length), easily recognised by its upwards-orientated (not sideways) eyes that are very close together (4 or 5 scales between the two, generally more than 5 in other *Eryx* species). Orange iris with central part almost white, highlighting the vertical pupil. In contrast with species of the *E. jaculus* group, the conically shaped (not truncated) tail terminates in a pointed scale. The dorsal scales are small, smooth and shiny, arranged in 37–51 rows at mid-body. 11 supralabials on each side. 158–184 ventral plates; the subcordals are undivided.

The top of the back is orange to mandarin-orange with many dark, irregular, transverse bands that continue onto the flanks. The flanks are pale yellow, very distinct from the orange back. There is no well-defined dark band behind the eyes. The undersides are a uniform off-white, generally unmarked.
**Venom** Harmless.
**Habitat** Very sandy deserts and desertlike steppes. Also close to oases and land devoted to crops, where it can be found by looking under various man-made shelters (corregated iron, planks of wood, sheets of cardboard).

*Arabian Sand Boa* Eryx jayakari, *juvenile. Wafrah Farms (Kuwait).* P. Geniez.

63

Eryx jayakari. *Note the small, upwards-orientated eyes. Kuwait.* P. Geniez.

top of the head, just protruding from the surface and well adapted to the situation, as in the Sahara Sand Viper *Cerastes vipera*.

**Diet** Especially small lizards and, above all, terrestrial geckos such as the *Stenodactylus* and *Bunopus* species, but also *Diplometopon zarudnyi*, an amphisbaenian (burrowing lizard) endemic to the Arabian Peninsula.

**Reproduction** Oviparous; lays a small number of large eggs that hatch after 66 days.

**Range** Occurs in the Arabian Peninsula. To the north, it reaches Kuwait, southwestern Iraq and southwestern Khuzestan, and Iran.

**Geographic variation** *Eryx jayakari* is considered to be monotypic.

**Habits** Essentially burrowing, very occasionally seen at night. Sometimes lies in wait just under the surface of the sand, with its eyes, situated on the

# Egyptian Sand Boa

*Eryx colubrinus* (Linnaeus, 1758) (formerly *Gongylophis colubrinus*)
**Family:** Boidae (boas)
**Sub-family:** Erycinae (sand boas)
**F.:** Boa des sables d'Égypte
**G.:** Ägyptische Sandboa

**Identification** A burrowing terrestrial boa, short and thickset. Length 50–77 cm. Head not differentiated from body. The small eyes are orientated to the side. The body scales are small, smooth and shiny, becoming more and more keeled towards the rear. They are arranged in 47–53 (rarely 55) rows at mid-body. Narrow ventral plates. The top of the head is covered in small, smooth scales, more numerous than in *Eryx jaculus* (11–13 scales between the eyes), 2 or 3 rows of small scales between the bottom edge of the eye and the supralabials, 12–14 supralabials on each side, 171–197 ventral plates; anal and subcordal scales are undivided. The tail is short and truncated, ending with a conical scale. Ground colour of the back is white, yellowish or yellow-

Eryx colubrinus. *Note the very wide brown markings and, above all, the strongly keeled dorsal scales, typical of this species.* Joger.

Eryx coubrinus loveridgel. *Individual in which the colour of the brown markings covers the whole body. Southern Ethiopia.* Mazuch.

grey, sometimes bright orange. There are wide dark brown or black markings on the back; a narrow, irregular white, yellowish or orange band appears between them. The dorsal markings are sometimes joined on the sides and separated in their centre, in which case an irregular pale line runs along the back; the result is rather attractive. There is a dark temporal band from the eye to the corner of the mouth. The belly is uniform whitish, yellow-grey or pale grey.

**Venom** Harmless.

**Habitat** Alluvial ground of loamy or sandy soil, with scattered stones and sparse scrub. Readily close to human habitations (e.g., oasis and agricultural land). Also found in coastal dunes on the edge of the Red Sea.

**Habits** A burrowing snake, normally active at dusk and at night. During the day rests under stones, in the root systems of bushes or in small mammal burrows. Burrows into sand and soft soil.

**Diet** Small mammals (rodents) and lizards. Extremely quick when catching prey that it suffocates by constriction within its powerful body coils.

**Reproduction** Ovoviviparous. Mating occurs in summer. Gives birth to live young in the autumn, as many as 15 young at a time. Their first moult occurs 8–10 days after birth, after which they start feeding.

**Range** Within the region covered by this book it only occurs in Egypt, along the alluvial plain of the Nile and in the Faiyum depression, north as far as the 30th Parallel. Apart from there, it occurs in the Sudan, Ethiopia and Somalia as far as northern Kenya and Tanzania and to the west as far as Niger (the Aïr massif). It is uncertain as to whether it occurs on the Arabian Peninsula.

**Geographic variation** Generally two subspecies are recognised:

▶ *Eryx colubrinus colubrinus* (Linnaeus, 1758): Egypt, from the Sudan to Somalia passing through Ethiopia, Chad and Niger.
▶ *Eryx colubrinus loveridgei* Stull, 1932: northern Kenya and northeastern Tanzania. The validity of this subspecies is questioned by certain authors who consider *Eryx colubrinus* to be monotypic.

# Families Colubridae and Lamprophiidae (colubrid snakes)

Thanks to recent advances in molecular techniques, the classification of the world's snakes has undergone much upheaval. Several phylogenies, quite congruent with each other, have been obtained by several authors. All agree that the "colubrid snakes" as previously defined, represent a polyphyletic group, meaning that they are made up of several large lineages between which are inserted other lineages belonging to families other than the Colubridae. If the phylogeny proposed by Pyron, Burbrink and Wiens, published in 2013 (the most recent and most complete at the time of writing), is followed, the colubrid snakes are in fact represented by two extensive phylogenetic groups (clades) between which, or within which, occur the family Elapidae (e.g., cobras and sea snakes) and the Atractaspididae, proving that the colubrid snakes do not constitute an homogenous group from the phylogenetic classification point of view, but two families, to which must be added that of the Elapidae. Although the Elapidae are a sister-group to one of the two large clades, they maintain their status as a separate family, whereas the Atractaspididae are demoted to sub-family status as they are situated within this same large clade. The two families within the "colubrid snakes" in the broad sense are:

► The Colubridae, which within the area covered by this book bring together numerous genera and species with a terrestrial or arboreal lifestyle that have the classic appearance of a "colubrid" snake: slightly thickset to very slim, with smooth scales, or slightly to moderately keeled in some species. An exception is the semi-aquatic species of the genus *Natrix* that are relatively thickset and have highly keeled dorsal scales, like those of vipers. The genus *Natrix* belongs to a special sub-family, the Natricinae, considered by some authors to be a family in its own right. All other Colubridae of the Western Palaearctic belong to the sub-family Colubrinae. From a morphological point of view the Colubridae are extremely diverse. Most species have smooth, shiny dorsal scales but in certain genera they are keeled as in vipers. All the species within the geographic zone considered here have round pupils, except those of the *Telescopus* (the cat snakes) and *Dasypeltis* (egg-eater snakes) genera in which the pupil has a vertical slit, a little like that of vipers. Finally, there are two types of dentition in this family: an aglyphous dentition (all the teeth are of almost the same shape and size, are not either grooved or tubular and do not connect directly with the venom glands) and an opisthoglyphous dentition (the last tooth of the upper jaw is enlarged, connected to the venom gland and equipped with a groove that allows the venom to descend when biting). A large majority of those of the Western Palaearctic Colubridae are aglyphous, and only members of the genera *Telescopus* and *Macroprotodon* have an opisthoglyphous dentition. Pupils with a vertical slit and the presence of venomous fangs are two characteristics found in other families (Lamprophiidae, Elapidae and Viperidae), proving that such special adaptations have developed independently in several families, and that it is illogical to base snake classification only on these features, although this is still seen in certain publications.

► The Lamprophiidae constitute a very heterogeneous family from not only the morphological aspect but also phylogenetically. It is divided into several sub-families very distinct from each other. Within the geographic zone covered by this book there are representatives of the following sub-families Lamprophiinae (*Boaedon fuliginosus*), Psammophiinae (genera *Malpolon*, *Rhagerhis* and *Psammophis*) and Atractaspidinae (genus *Atractaspis*). Note that the exact position of the *Micrelaps* genus does not as yet seem to be resolved. According to Pyron et al. (2013) this genus, represented in the Western Palaearctic by Muller's Ground Viper, is obviously to be

placed in the Lamprophiidae family, but it constitutes a very distinct lineage that could justify the creation of a sub-family of its own. In the absence of further research, most authors classify *Micrelaps* within the Aparallactinae, a sub-family quite closely related to the Atractaspidinae. Here also, as in the Colubridae, there are species that occur in the Western Palaearctic with a pupil with a vertical slit (*Boaedon fuliginosus*) whereas the others have a round pupil; species with an aglyphous dentition (*B. fuliginosus*) or opisthoglyphous (all members of the sub-family Psammophiinae, such as the Eastern Montpellier Snake, as well as the genus *Micrelaps*) and even proteroglyphous (*Atractaspis engaddensis*) species.

## Family Colubridae

### Colubrid snakes once placed in the genus *Coluber*

The systematics of the well-known genus *Coluber* of "colubrid" snakes has undergone much change due to molecular techniques. Genetic work has shown that the genus *Coluber*, as considered in the past, was an assemblage of "colubrid" snakes, certainly of similar appearance but belonging to completely divergent lineages. Inserted between these were other genera, e.g., the genus *Eirenis* (dwarf snakes) and the genus *Spalerosophis* (diadem snakes). The genus *Coluber*, in the strict sense, now contains about 10 species distributed in the Americas, of which the type species is the Eastern Racer *Coluber constrictor* Linnaeus, 1758, which has an extensive range in North America.

Summarizing that which concerns the Western Palaearctic:

- ► The ex-members of the genus *Coluber* genus are now placed in four genera: *Hierophis*, *Dolichophis*, *Platyceps* and *Hemorrhois*.

- ► *Eirenis* is the sister-genus of the genus *Hierophis* (that is, it is closer to *Hierophis* than the other genera *Dolichophis*, *Platyceps* and *Hemorrhois*).

- ► *Spalerosophis* is the sister-genus of the genus *Platyceps*, but other phylogenetic work places it close to the genus *Hemorrhois*, which is more likely considering their morphological resemblance.

It has been decided to present the different species in a "conventional" order within this book in order not to confuse the reader. We consider all the ex-*Coluber* species together, in the following order: the genera *Hierophis*, *Dolicophis*, *Platyceps* and *Hemorrhois*, followed by the genera *Spalerosophis* and *Eirenis*. A phylogenetic arrangement would logically have the genus *Eirenis* placed after *Hierophis* as they are the most closely related, but the absence of genetic data on another genus of dwarf snakes, the genus *Rhynchocalamus*, leads us to place this last genus just after *Eirenis*, and thus provisionally regroup all the dwarf snakes (*Eirenis*, *Rhynchocalamus*, and *Muhtarophis*) for pure convenience. The systematic position of the Sinai Whip Snake *Coluber sinai* has not been established: we have chosen to place it in the genus *Platyceps* next to *Platyceps elegantissimus* as they have similar colouration and geographic distribution.

### Genus *Hierophis*
There are only four species within the genus *Hierophis* in the world, three of which occur in the Western Palaearctic. The fourth species, *Hierophis andreanus* (F. Werner, 1917) is endemic to the southern Zagros mountains in western Iran. Finally, *Hierophis spinalis* of eastern Asia is now placed in a distinct genus, with the name *Orinetocoluber spinalis* (Peters, 1866).

# Western Whip Snake

*Hierophis viridiflavus* (Lacépède, 1789) (formerly *Coluber viridiflavus*)
**Family:** Colubridae
**Sub-family:** Colubrinae
**F.:** Couleuvre verte et jaune
**G.:** Gelbgrüne Zornnatter

**Identification** Long, slim, extremely lively and rapid snake. Average length 120–150 cm, but some may reach 180 cm. Very long tail. Narrow, elongated head, only slightly distinct from the body. Large eyes with round pupils with an orange rim; the rest of the iris is black. The body scales are smooth and quite shiny, arranged in 19 (sometimes 17) rows at mid-body. On top of the head there are 9 large plates as in most colubrid snakes. One large and high preocular scale partly inserted in a point between the prefrontal and the supraocular but does not reach the frontal (this character appears to apply to the three species in the genus *Hierophis*,

whereas in the *Platyceps* and *Hemorrhois* the preoculars and the frontal are nearly always in contact), 1 subocular at the front of the eye, 2 (rarely 1 or 3) postoculars, 8 supralabials on each side. 187–227 ventral plates, divided anal scute, subcordals divided and arranged in pairs, as in nearly all the colubrid snakes of the area covered by this book.

The top of the head is black decorated with very obvious pale yellow spots and markings. Ground colour black. On the front part of the body there is a network of transverse, more or less intertwined, very contrasting pale yellow stripes that progressively change into small yellow

*Western Whip Snake. Puy-de-Dome, southern France. Note the entirely black and yellow colouration and the stripes on the tail, typical of* Hierophis viridiflavus viridiflavus. *P. Geniez and A. Teynié.*

69

spots towards the rear and eventually into longitudinal yellow lines, seen to best effect on the tail. The belly is yellow or white, with irregular dark markings on the ventral plates on the front part of the body, then without any markings towards the rear of the body or under the tail. The adults are always black and yellow without any other colour. Juveniles are completely distinct: the head is already black with yellow patterning but the rest of the body is a pale almond-green subtly marbled with indistinct yellow-brown markings. Subadults have an intermediate colouration but are quite distinct with the black barred yellow top of the head. Individuals in southern and eastern Italy, also of Istria, Sicily and Malta are entirely shiny black and iridescent or with a bluish reflection when adult (subspecies *carbonarius*). Those on Corsica and Sardinia are either almost identical to those of continental France and northwestern Italy or with much black, and in subadults the pale areas are sometimes a turquoise-blue colour.

**Venom** Very aggressive when captured; bites ferociously to defend itself, chewing the hand of the person holding it, sometimes drawing blood, but it is not considered to be venomous.

**Habitat** A variety of habitats, more often than not dry with places to hide. Slopes with scattered bushes and stones, forest edges and open woodland, areas of maquis, scrub along roadsides, scree, ruins and undisturbed gardens, hedgerows separating cultivated areas, occasionally on the edges of wet meadows. Occurs in

mountains to 2000 m in certain areas. In France and the north of Spain, it avoids the Mediterranean zone, where it is replaced by an ecological competitor, the Western Montpellier Snake. On the other hand, where this last species is absent, as in Corsica and Sardinia, the Western Whip Snake occurs everywhere, including along the coast.

**Habits** Diurnal. More often than not on the ground but can climb well in bushes and on rocks. Very shy, rapid and agile. A long inactive period in winter.

**Diet** Quite varied: mainly lizards but also other snakes, birds and their eggs, and less commonly small mammals, frogs and toads. Young individuals feed on small lizards, beetles, grasshoppers and crickets.

**Reproduction** Oviparous. Mating occurs in April or May. During mating the male holds the female by the neck with its mouth. In June or July, 5–15 elongated eggs are laid under stones, in old tree stumps or decomposing vegetation. The young hatch from the eggs after 6–8 weeks' incubation.

**Range** From the Pyrenees (including the Spanish side), throughout central and southern France, southern Switzerland, Italy and as far as Istria (Croatia). Also on many western Mediterranean islands: Corsica, Sardinia, Elba, Sicily, Malta and a few smaller islands near the Slovenian and Croatian coasts. A species endemic to the small island of Gyaros, off the Greek coast, *Hierophis gyarosensis* (Mertens, 1968) has recently been invalidated. Genetic studies show that it is in fact a presumably

FAMILY COLUBRIDAE (COLUBRID SNAKES)

introduced population of *H. viridiflavus carbonarius*.

**Geographic variation** *Hierophis viridiflavus* is represented by two subspecies each of a distinct colour, but also quite distinct from the genetic point of view, so much so that certain authors consider the second subspecies as a separate species, *Hierophis carbonarius*:

► *Hierophis viridiflavus viridiflavus* (Lacépède, 1789): northern Spain, France, western Switzerland, northwestern and western Italy. Variegated black and yellow pattern.

The Corsican and Sardinian populations are included in this subspecies, following molecular biological data.

► *Hierophis viridiflavus carbonarius* (Bonaparte, 1833): Istria, northeast, eastern and southern Italy, Sicily and Malta. The population of Gyaros also belongs to this subspecies. *H. v. carbonarius* is essentially characterised by the entirely melanistic colouration of adults (iridescent shiny black) with the exception of the underside of the head and sometimes the sides. Juveniles are not melanistic and resemble those of the nominate subspecies but have a browner tone.

*Left*: Hierophis viridiflavus viridiflavus *juvenile. Puivert, Aude (southern France). Note the head contrasting with almost uniform body, typical of juvenile Western Whip Snakes.* Pottier.

Hierophis viridiflavus carbonarius, *subspecies characterised by its melanistic colouration. Saint Paul's Bay (Malta).* Peyre.

# Balkan Whip Snake

*Hierophis gemonensis* (Laurenti, 1768) (formerly *Coluber gemonensis, C. laurenti*)
**Family:** Colubridae
**Sub-family:** Colubrinae
**F.:** Couleuvre des Balkans
**G.:** Balkan-Zornnatter

**Identification** Relatively slim snake but less so than the Western Whip Snake; of medium size, average length 70–100 cm, rarely to 130 cm, with a long tail. Oval, quite high head little distinct from body. Large eyes with round pupils. The dorsal scales are smooth and arranged in 19 (rarely 17) rows at mid-body. 1 preocular scale, 1 subocular at the front edge of the eye, 8 supralabials on each side, 166–187 ventral plates, thus less than *Hierophis viridiflavus* that always has more than 186. Ground colour from yellowish-brown to olive, sometimes appearing quite dark. The rear of the body is finely speckled, the scales being tricolour: there is a fawn-brown central line irregularly bordered with black and white. These markings gradually disappear towards the middle of the body that becomes uniform brown or has indistinct, fawn-coloured longitudinal lines. In certain individuals with a very contrasting colouration, the rear of the body and tail are patterned with pale longitudinal lines, as in the Western Whip Snake. The yellowish or whitish ventral plates often have, at least in the front part of the body, dark markings on their outer edges. Contrary to the Western Whip Snake,

juvenile Balkan Whip Snakes have a colouration similar to that of the adults, the rear half of the body already having dark and pale mottling. In Istria (northwestern Croatia) it is possible to mistake it for the closely related Western Whip Snake *H. viridiflavus*. However the latter is larger, slimmer and in juveniles the dorsal scales are bicoloured, not tricoloured. Adult *H. viridiflavus* in this region are entirely shiny black; finally, the Western Whip Snake is absent from Mediterranean garrique, where the Balkan Whip Snake occurs.

**Venom** Very shy, but does not hesitate to bite if captured. Not venomous.

**Habitat** Various habitats with dry slopes dotted with scrub and piles of stones, drystone walls, ruins, vineyards, garigues, scree and occasionally open woodland, more often than not in areas of Mediterranean influence.

**Habits** Diurnal snake. Essentially terrestrial but can climb with agility into bushes and low trees. An inactive period in winter of 1–4 months depending on elevation.

**Diet** Lizards, small mammals, young birds, large crickets, sometimes even small snakes. Follows and catches its prey with extreme rapidity.

**Reproduction** Oviparous; lays 4–10 eggs. The young initially feed on large insects.

**Range** The Adriatic coast of the Balkans, from extreme northeastern Italy as far as Albania, continental Greece and certain islands in the Aegean Sea (e.g., Ionian islands, Euboea, Crete, Kythira and Ghayoura).

Hierophis gemonensis. *Croatia. Note the colouration, speckled at the front of the body, uniform at the rear and on the tail.* Cluchier.

*Close-up of* Hierophis gemonensis. *Crete. Note the tricoloured scales (whitish, fawn and black). In* H. viridiflavus, *there are only two colours when adult (black and yellow).* P. Geniez.

# Cyprus Whip Snake

*Hierophis cypriensis*
(Laurenti, 1768) (formerly *Coluber viridiflavus*)
**Family:** Colubridae
**Sub-family:** Colubrinae
**F.:** Couleuvre de Chypre
**G.:** Zypern-Schlanknatter

**Identification** Not very large (70–90 cm in length, maximum 116.5 cm), slender, elegant, quite similar to Western Whip Snake but more slender and darker. Elongated head quite distinct from neck. Large eyes with golden or orange rims round pupils. 1 large, high preocular plate above 1 small subocular scale at the front edge of the eye, 2 postoculars, 8 supralabials on each side, the 4th and 5th touching the eye. The body scales are smooth and shiny, arranged in just 17 rows at mid-body.

The ground colour is olive-green, grey-brown, blackish or black. Top of the head dark grey to black with a few small, pale yellow or whitish markings.

There is a vertical white line in front of the eye, another behind. On the front half of the body there are more or less well-defined, thin white or whitish transverse stripes that steadily disappear towards the middle of the body. The rear third of the body and the tail are uniformly dark. The belly is pale yellow, cream or dirty grey. The ventral plates at the front of the body have lateral marks, towards the rear of the body they join into a continuous dark line. Old individuals are black with the vestige of a few pale markings vaguely aligned transversely.

The Cyprus Whip Snake is sometimes confused with the Large Whip Snake

*Cyprus Whip Snake*
Hierophis cypriensis,
*adult*. Baier.

*Dolichophis jugularis*. The latter species is often larger and a little more thickset; it has 19 rows of dorsal scales at mid-body (compared to 17); the adults are uniform black above, whitish or reddish on the throat; juveniles are finely marbled with black and pale on a beige ground colour; their ventral plates have a black spot on the outer edge. *D. jugularis* is far more widespread in Cyprus than *Hierophis cypriensis* and often occurs in dryer areas.

**Venom** Harmless.

**Habitat** The more humid and cooler areas of Cyprus: areas with dense scrub and bushes, the ground with grass cover or open forest. More often than not near water, along rivers or near reservoirs.

**Habits** Little known. Diurnal, terrestrial snake that also climbs into bushes, very rapid and shy.

**Diet** Especially lizards and similar. Also eats amphibians and other snakes (*Natrix natrix cypriaca*) as well as large insects (crickets and grasshoppers), centipedes and small rodents.

**Reproduction** Oviparous.

**Range** Rare, endemic to Cyprus, where it was first described as recently as 1985. Known from the Troodos Massif and its foothills in the western part of the island, it occurs from 400 to 1900 m elevation. The Cyprus Whip Snake is considered to be threatened with extinction.

**Geographic variation** *Hierophis cypriensis* is monotypic.

*Close-up of a Cyprus Whip Snake* Hierophis cypriensis, *subadult*. Briola.

### Genus *Dolichophis*

There are only three species in the genus, all closely related and whose distribution ranges from the central Balkans to northeastern Iran and from the extreme south of European Russia to Sinai and southwest to Iraq. The three species – *Dolichophis jugularis*, *D. caspius* and *D. schmidti* – have steppe and Mediterranean affinities, are diurnal, shy and rapid, and will bite hard when captured.

From a phylogenetic point of view the genus *Dolichophis* is a sister-group of the *Hierophis* + *Eirenis* pairing. The large clade of *Hierophis* + *Eirenis* + *Dolichophis* is a sister-clade of all other ex-Coluber Western Palaearctic snakes and the genus *Spalerosophis*.

# Caspian Whip Snake

*Dolichophis caspius* (Gmelin, 1789) (formerly *Coluber caspius*)
**Family:** Colubridae
**Sub-family:** Colubrinae
**F.:** Couleuvre de la Caspienne
**G.:** Kaspische Springnatter

**Identification** Very large colubrid with average length of 140–170 cm but can be longer than 200 cm (individuals of 250 cm have been cited). Despite its size, this snake appears slender with a proportionally long tail and relatively small head only slightly distinct from the body. Large eyes with round pupils. Body scales smooth and quite shiny, arranged in 19 (rarely 17) rows at mid-body. 1 loral plate, 2 preoculars, the upper much larger than the lower, 2 (sometimes 3) postoculars, generally 8 (rarely 7 or 9) supralabials on each side, the 4th and 5th in contact with the lower edge of the eye. 189–211 ventral plates, slightly angular on the sides, giving the belly a squared shape. The back is pale grey, brown-grey, yellowish, olive-brown, or red-brown, generally without dark markings in the adult. The dorsal scales have a dark rim (nearly black) with a pale centre giving the Caspian Whip Snake a "fishnet" appearance typical of the species. In large adults, especially males, the head and front of the body sometimes have a reddish or orange tint. In the western and central Balkans, certain adults, even of large size, retain the black spots or bars on the front half of the back, reminiscent of the juvenile colouration. The belly is uniform yellowish-white, pale yellow or orange-yellow, without markings. The juveniles have a very distinct colouration, the head variegated with dark and the body grey or beige with small dark spots; each dorsal scale has a longitudinal band paler than the rest of the scale; the belly of juveniles is white or yellowish with a black or reddish spot on the outer edge of most of the ventral plates.

**Venom** Very shy, but readily bites if captured. Non-venomous.

**Habitat** Open countryside, garrigue, steppe-like phryganas, rocky hillsides, hillside scree, clearings in deciduous woodland, vineyards, untended gardens, agricultural lowlands where it can be common at the base of hedgerows, occasionally near water.

**Habits** Diurnal, terrestrial snake. Basks in the sun for a long time in the morning and late afternoon. Always ready to flee, which it does noisily. Cornered, this colubrid defends itself by biting vigorously. Its German name, Springnatter ("jumping snake"), comes from its ability, when on the defensive, to jump to a height of half its body length. The Kazakhs say that this snake can, by their "jumps", cause horses to flee. A 4–6 months inactive period in winter depending on region and elevation.

**Diet** Small mammals (rodents, insectivores), birds, lizards, snakes, frogs. The young feed on small lizards and large insects.

**Reproduction** Oviparous. Mates in May, just after winter inactivity, during which the male holds the female by the neck with its mouth. 5–18 eggs are laid in underground cavities, piles of leaves or under stones. Hatching occurs in September.

**Range** The Balkan countries from Albania and Greece, through Macedonia, including the Aegean islands (but absent from Attica, Peloponnese and Crete); Bulgaria, Romania, extreme eastern Croatia, Hungary (at the northwestern limit of its range); southern Ukraine and Moldavia, western and northern Turkey, edges of the Black Sea and farther east, as far as the Caspian Sea, including southwestern Russia, western Kazakhstan and extreme northeastern Azerbaijan.

**Geographic variation** *Dolichophis caspius* is considered to be a monotypic species (with no subspecies) despite a quite significant genetic diversity between the European populations and those in Asia. On certain Aegean islands (Andros, Kythnos, Tinos, Karpathos), Caspian Whip Snakes are quite small (100 cm), and adults retain traces of their juvenile colouration.

*Caspian Whip Snake* Dolichophis caspius, *adult in eastern Greece. Note the distinctive aspect of the dorsal scales, pale with black edges, resulting in a fishnet pattern. P. Geniez.*

Dolichophis caspius, *adult, which has retained some of its juvenile colouration (black bars on the back). Northwestern Greece. P. Geniez.*

*Caspian Whip Snake* Dolichophis caspius, *52-cm-long juvenile. Note a superficial resemblance to dwarf snakes of the genus* Eirenis, *as well as the dark-edged dorsal scales (which allows it to be differentiated from D.* schmidti, *in which the edges of the scales are paler than the centres). Northern Turkey. P. Geniez and A. Teynié.*

# Large Whip Snake

*Dolichophis jugularis* (Linnaeus, 1758) (formerly *Coluber jugularis*)
**Family:** Colubridae
**Sub-family:** Colubrinae
**F.:** Couleuvre flèche
**G.:** Pfeilnatter

**Identification** A uniformly shiny black, very large snake, one of the largest in the Western Palaearctic, average length 160–200 cm but can grow to 250 cm. Relatively small head compared to body, not very distinct from neck. Long, slender tail giving this species a very slender appearance overall. Large eyes with golden-yellow or orange rims round pupils. The dorsal scales are smooth, arranged in 19 (sometimes 18) rows at mid-body, 1 loreal plate, 2 preoculars, the lower one much smaller, 2 (rarely 3) postoculars. 8 (occasionally 7) supralabials, the 4th and 5th in contact with the eye's lower edge. 189–220 ventral plates.

Colouration generally shiny black, with the throat, belly and sides of the head reddish or whitish depending on the subspecies. The young and subadults are spotted, not uniform as in *Dolichophis caspius*, the dorsal scales have a longitudinal pale line in their centre, generally giving them a pattern of thin longitudinal streaks, as in the Caspian Whip Snake. The juvenile's colouration changes progressively from reddish to black after 3–4 years. The ventral surface of juveniles is very similar to that of *D. caspius* and *D. schmidti*, but

the external border of each ventral plate has a much more pronounced black or reddish spot than the other two species. It is very difficult to distinguish between the juveniles of *D. jugularis* and those of *D. caspius*. The observation of an adult at the same site greatly helps in the identification of juveniles.

**Venom** Wary and very rapid. Aggressive when cornered or when being captured. Non-venomous.

**Habitat** A species showing a clear preference for Mediterranean habitats (especially compared with the closely related *Dolichophis schmidti* that prefers steppe). Open countryside, sunny, dry hillsides with scrub and a little ground vegetation, sloping scree, meadows with abundant thorn bushes, vineyards and neglected gardens, areas with ruins or old buildings. In mountainous areas locally to 1400 m.

**Habits** Diurnal, terrestrial snake that can climb well in bushes. Basks for long periods in the sun in the morning and late afternoon, always ready to flee at the slightest sign of danger. Readily swims and follows amphibians into water. Has a relatively short inactive period in winter.

**Diet** Small mammals, birds, lizards, snakes (including individuals of its own species and the Blunt-nosed Viper *Macrovipera lebetina*), anuran amphibians and newts. Kills larger prey by constricting within its powerful coils until suffocated.

**Reproduction** Oviparous. Lays 5–15 eggs in a rodent's burrow, under a rock or in tree roots.

**Range** Southern and southwestern Turkey to the west as far as the edge of Izmir and to the northeast as far as Karakaya, in Muş province, many of the Greek islands opposite the Turkish coast including Rhodes, Cyprus, northern and western Syria, Lebanon, western Jordan and Israel. Farther west, a few records from Iraq and western Iran.

**Geographic variation** *Dolichopsis jugularis* has either a red, whitish or yellow throat that has allowed for it to be separated into three or four subspecies:

▶ *Dolichophis jugularis jugularis* (Linnaeus, 1758): a subspecies characterised among other things by the throat and the front of the undersides being red in adults. Occurs in Turkey, north and central Syria. Records from Iraq and Iran also concern this subspecies.

▶ *Dolichophis jugularis asianus* (Boettger, 1880): characterised by the throat and front part of undersides being whitish, even in adults. Southern Syria, Jordan, Lebanon, Israel.

▶ *Dolichophis jugularis zinneri* Cattaneo, 2012: island subspecies characterised by having a yellow throat and less intense black back. The Greek islands of Rhodes, Symi, Halki and Tilos (Dodecanese). The sub-specific status of the Cyprus populations is not completely clear, probably due to various releases of subspecies from the continent. Certain authors recognise an endemic subspecies *D. j. cypriacus* (Zinner, 1972), characterised by much of the underparts being black but the throat and front part of the belly are whitish marbled with red.

*Large Whip Snake, here* Dolichophis jugularis jugularis, *southern Turkey, distinguished from the Caspian Whip Snake* D. caspius *by its black colouration contrasting with its red underparts.* Mebert.

Dolichophis jugularis jugularis, *110-cm-long subadult, southern Turkey. As in* D. caspius, *young individuals are highly spotted.* P. Geniez and A. Teynié.

Dolichophis jugularis asianus, *adult from Jordan showing its whitish throat, typical of this subspecies.* P. Geniez.

## Schmidt's Whip Snake

*Dolichophis schmidti* (Nikol'skij, 1909) (formerly *Coluber schmidti*)
**Family:** Colubridae
**Sub-family:** Colubrinae
**F.:** Couleuvre de Schmidt
**G.:** Schmidts Pfeilnatter

**Identification** Large snake of rust-red to brick-red colour that resembles the Caspian Whip Snake *Dolichophis caspius* (see page 76). Unlike this latter species the centres of the dorsal scales are darker than their edges, they have a pale edge. Length 140–160 cm. The dorsal scales are smooth and shiny, arranged in 19 rows at mid-body. 1 loreal, 1 or 2 preoculars, 2 postoculars, 8 (rarely 7 or 9) supralabials, the 4th and 5th in contact with the lower edge of the eye. 195–212 ventral plates.

The adults generally have a red-brown to brick-red colouration. The pattern on the body scales (back and flanks) is characteristic: dark red centres with light upper and lower edges: this results in very beautiful pattern of longitudinal, alternating red and white lines. The belly is a light coral-red to cherry red colour with a bright pink shine. The young have dark spots and confusingly very much resemble the young of *D. caspius* and *D. jugularis* but the middle part of the scales is darker than the edges (except for the dark markings on the back and flanks). This juvenile pattern can persist for a long time (until they reach a length of around 140 cm). Certain young adults are beige-grey without a trace of red, but the dorsal scales have light edges.

**Venom** Non-venomous. Rapid and wary, bites when caught.

**Habitat** Dry hillsides with grass and bushes, stony steppe with little vegetation, untended gardens, sites with ruins. Shows a preference for quite dry regions of steppe, more so than its two congeners; it does however occur in the more humid banks of rivers and lakes. Occurs to at least 1900 m elevation.

**Habits** A wary, diurnal terrestrial snake, always ready to flee. During hot periods it is only active in the morning and late afternoon. Sometimes seen moving on warm nights. Has an inactive winter period of several months (from mid November to mid March in Armenia).

**Diet** Small mammals (rodents, insectivores), ground-nesting birds. Probably also takes other snakes and maybe amphibians.

**Reproduction** Oviparous. Mating in April or May, lays 7–12 eggs in July or early August that hatch from mid September.

**Range** Most of Turkey's steppe (absent from the Mediterranean belt and the damper parts of the northern shores of the Caspian Sea), southern Armenia, east of the Caucasus region, south of Russia (southeastern Dagestan), Azerbaijan, as far north as Iran and Turkmenistan. Farther south, there has been recent evidence of its presence in Syria and the extreme north of Jordan. Its precise distribution is still uncertain due to confusion with the two other *Dolichophis* species.

**Geographic variation** *Dolichophis schmidti* is considered to be a monotypic species.

*Schmidt's Whip Snake* Dolichophis schmidti, *adult on Mount Ararat (northeastern Turkey).* P. Geniez and A. Teynié.

Dolichophis schmidti, *juvenile from Turkey. Note the mottled colouration, similar to that of other* Dolichophis *species, but in Schmidt's Whip Snake the dorsal scales are pale-edged, not dark-edged.* Teynié.

### Genus *Platyceps*

The genus *Platyceps* includes quite a large number of species that occur from the Adriatic coast as far as central Asia and in East Africa. Most of these medium-sized, slender, wary colubrid snakes are agile and rapid.

The *Platyceps* are placed in a large clade that also includes the ex-*Coluber* snakes of the genera *Hemorrhois* and *Spalerosophis*. Within the genus *Platyceps*, the phylogeny is disparate and does not reflect the morphological similarities between the different species, to such a point that it is impossible with the phylogeny proposed today to dissociate the different groups.

Using morphological characteristics, six groups can be distinguished for the region covered by this book:

► the *Platyceps najadum* group (*P. najadum* and *P. collaris*);

► the *Platyceps rhodorachis* group (for several species of which the systematics are still debated and controversial);

► the *Platyceps ventromaculatus* group which should probably be attached to the *Platyceps rhodorachis* group;

► the *Platyceps florulentus* group of East Africa;

► the *Platyceps elegantissimus* group, with just one species;

► the *Platyceps sinai* group. Its exact placing within the genus is still unknown; it could belong to another genus.

## Dahl's Whip Snake

*Platyceps najadum* (Eichwald, 1831) (formerly *Coluber najadum*)
**Family:** Colubridae
**Sub-family:** Colubrinae
**F.:** Couleuvre-fouet à cou tacheté
**G.:** Schlanknatter

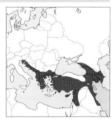

**Identification** An elegant, very slender snake with a line of brown or black marks with white rims (looking like eyes) on the sides of the neck. The marks diminish towards the rear of the body to disappear completely after the first quarter. The first pair of marks directly behind the head are sometimes fused, forming a transverse band (but not a pronounced band on the nape as in *Platyceps collaris*). There is a dark band on the side of the head from the nostril to the corner of the mouth; this band is interrupted by a vertical white or yellowish-white bar in front of the eye and another just behind (contrary to *P. collaris*, which can resemble this species, in which the black band is continuous). In the western part of the species' range (*P. najadum dahlii*), the head and neck are grey, bluish-grey or olive-grey, passing progressively to uniform red-brown or brick-red towards its rear. Farther east (*P. n. najadum*), this difference of colour between the front and the rear is less pronounced and the markings at the front of the body are less well-defined, giving them a more or less marbled appearance. The belly is an unmarked yellowish-white. Average length to 100 cm (*P. n. dahlii*), but to 140 cm in *P. n. najadum*. The tail is very long and tapering, more than a quarter of the snake's total length. The head is narrow and elongated, but quite distinct from the thin neck. Remarkably large eyes, with a round pupil with a golden-red rim. Smooth and quite shiny dorsal scales, arranged in 19 (sometimes 17) rows at mid-body. One high preocular scale is well developed, inserting in a point between the prefrontal and the supraocular, sometimes as far as touching the frontal (but not in those *P. n. dahlii* examined); 1 small subocular at the front edge of the eye; 8 (rarely 7 or 9) supralabials on each side. 205–230 ventral plates, angular on the sides; a divided anal scute; and a large number (104–138) of pairs of subcordals.

**Venom** Non-venomous. Very shy. Some individuals can be quite aggressive when captured, whereas others try to free themselves from the hand that's holding them by wriggling wildly.

**Habitat** Quite varied, from lowlands to mountains, above 2100 m. Occupies Mediterranean valleys, hillsides of garigue-type vegetation with bushes and grasses, neglected gardens, forest clearings, vineyards, sites with ruins, meadows with bushes; also on wetland dykes. In mountains and on plateaus, especially in the east, in rocky, steppe-like areas with scree and bushes. Thick vegetation, scrub, piles of stones or drystone walls provide ideal hiding sites for Dahl's Whip Snake.

**Habits** Terrestrial, diurnal snake that will climb into bushes. Very shy and

*Two different* Platyceps najadum najadum *from northeastern Turkey. Note as characteristic of the species: little difference in colour between the head and the body and the presence of many small dark spots following on from the large marks on the neck. The two white bars before and behind the eye are typical of this species.* P. Geniez and A. Teynié.

rapid. Active from March to October; inactive winter period of 3–6 months, depending on elevation and region.

**Diet** Especially lizards; in periods of lizard paucity, also large insects, young rodents and young birds in the nest. Hunts its prey with incredible speed.

**Reproduction** Oviparous. Lays 3–16 very elongated eggs in cavities in the ground or under stones. Hatching occurs in late August or September; the newly hatched young feed mainly on young lizards (e.g., skinks) and crickets.

**Range** The Balkan countries, coastal areas of the Adriatic, Greece, some of the Dodecanese islands, southern Macedonia, southern Bulgaria, a large part of Turkey, the Caucasus region, east as far as Iran, extreme northern Iraq and southwestern Turkmenistan. To the south, extends along the eastern Mediterranean as far as Syria and Lebanon. Recorded from Cyprus in 1910 by herpetologist G. A. Boulenger, it has been observed there only once since, in 1996.

**Geographic variation** *Platyceps najadum* is a quite variable species across its vast range. At least seven subspecies are currently recognized by several authors, whereas others, wrongly, consider that it is certainly a very variable species but that these variations are without geographic structure. It is, however, probable that certain subspecies (e.g., *P. n. dahlii*) are, in fact, separate species and that *P. najadum* represents a species complex.

▶ *Platyceps najadum najadum* (Eichwald, 1831): eastern Turkey as far as the Caucasus and northern Iran.
▶ *Platyceps najadum dahlii* (Fitzinger, 1826): the Balkans, the Adriatic coast, southern Macedonia, Greece, some of the Dodecanese islands, southern Bulgaria, western and southwestern Turkey.

*Close-up of* Platyceps najadum dahlii. *The Peloponnese (Greece).* P. Geniez.

Platyceps najadum dahlii. *Thrace (eastern Greece). Note the bluish head, the brick-red body and the large spots on the neck, followed by only a few small spots; the rest of the body is completely uniform.* P. Geniez.

▶ *Platyceps najadum kalymnensis* (Schneider, 1979): Kalymnos island in the Aegean Sea. A micro-insular melanistic subspecies, not recognised by all authors.

▶ *Platyceps najadum albitemporalis* (Darevsky and Orlov, 1994): southeastern Azerbaijan and northwestern Iran. Among other features, characterised by the lower part of the sides of the head, including the lower temples, being entirely yellowish-white (dark in *P. n. najdum* and *P. n. dahlii*).

▶ *Platyceps najadum schmidtleri* (Schätti and McCarthy, 2001): the southern Zagros mountains in southwestern Iran.

▶ *Platyceps najadum atayevi* (Tuniyev and Shammakov, 1993): Turkmenistan (Kopet Dag massif) and probably northeastern Iran.

The sub-specific status of the populations of Rhodes (Greece), Syria, Lebanon and Iraq has yet to be established.

## Red Whip Snake

*Platyceps collaris* (Müller, 1878) (formerly *Coluber rubiceps*)
**Family:** Colubridae
**Sub-family:** Colubrinae
**F.:** Couleuvre-fouet rougeâtre
**G.:** Rötliche Zornnatter

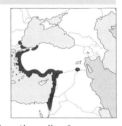

**Identification** Small to medium-sized (total length of about 70 cm for Western Palaearctic and western Turkey, often more than 100 cm in the Levant countries, to 112 cm in Jordan). A very slender, elegant snake, even more so than the similar Dahl's Whip Snake, with a very long tail. General colour a uniform reddish-grey to red-brown. On the nape, a transverse black or dark brown band with white edges forms a collar. At the front of the body, there are a few small, white-edged dark markings, well spaced one from the other, that dwindle to eventually disappear at mid-body. There is often another transverse band on the

neck, smaller than the collar. Some individuals have a brick-red top to the head, whereas others have a grey top to the head and neck, as in *Platyceps najadum*. A horizontal black band runs from the rostral plate to join the collar, separating the contrasting brown head above from the white underparts. On the upper preocular scale, there is a small white spot, sometimes subdued, just in front of the eye. The belly is a bright yellowish-white.

The head is long and quite small, little distinct from the body. The large eyes have a round pupil and a wide golden-red rim. The dorsal scales are smooth, quite shiny and arranged in

85

19 rows at mid-body; 2 or 3 preocular plates, the upper much larger and higher than the others insert in a point between the prefrontal and the supraocular as far as to touch the frontal plate; 2 postoculars, 8 or 9 (occasionally 7) supralabials on each side. 186–200 ventral plates in the western part of the species range, 200–223 in the east; the paired subcordals are on average less numerous than in *P. najadum*, 79–128.

The closely related Dahl's Whip Snake is easily distinguished by the white around the eyes, which cuts across the black horizontal band, and by the presence of large white-rimmed dark spots along the sides of the neck, not below.

**Venom** Non-venomous. A quite wary colubrid that if caught tries harder to free itself by twisting incessantly than by biting.

**Habitat** Plains near the coast, as far as the first foothills near the shore with a salt spray influence. Arid, garigue-type terrain with scrub, stones and rocks, open oak woods on rocky substrate, edges of vineyards and gardens.

**Habits** A terrestrial, diurnal snake that can be slightly active at dusk on warm nights. Extremely rapid, even more so than Dahl's Whip Snake; it straightens in a flash to escape. in winter, there is an inactive period of several months in the northern part of its range; in the south, it is probably more or less active throughout the year.

**Diet** Feeds mainly on lizards and large insects, sometimes also small snakes. Prey is swallowed live. It has been said that Red Whip Snakes have been seen to attack lizards that are far too large for them to swallow, with the result that the lizard's tail breaks (this happens easily in lizards); the tail is then immediately swallowed.

**Reproduction** Oviparous. 2–6 very long, cylindrical eggs are laid under a large stone or in a rodent burrow in July. They hatch after 6–8 weeks.

**Range** Isolated populations along the Bulgarian and Turkish coast as far as the area west of the Black Sea, then along the Turkish eastern Mediterannean coast as far as Israel, passing through western Syria, Lebanon and westernf Jordan. Quite

Platyceps collaris. *Northwestern Turkey, to the east of Istanbul. Note the continuous black "bridle" bar unbroken by the vertical white bars. There are no large black spots on the side of the neck, only a few well-spaced sparse smudges (contrary to* P. najadum dahlii, *which it resembles).* P. Geniez and A. Teynié.

rare in the west, more common in the Levant countries (e.g., Lebanon).

**Geographic variation** *Platyceps collaris* is considered to be a monotypic species, despite the large variation in size and scale makeup between eastern and western populations.

---

**The *Platyceps rhodorachis* group**

*Platyceps rhodorachis* represents a complex whose systematics and relationships with closely related groups is contested to such an extent that for certain species it is undecided whether they belong to this group or not, due to a lack of genetic data on all populations. The *P. rhodorachis* complex has a vast range that extends from the southeastern Sahara and the Horn of Africa as far as the steppes of central Asia, passing via the southern Levant countries and the Arabian Peninsula. The "real" Jan's Cliff Racer *P. rhodorachis* (Jan, 1863) is an eastern species with a distribution from Iran to northwestern India. It reaches the Western Palaearctic in northeastern Iran, within the geographic range covered by this book. This complex includes two other species: the Saharan Whip Snake *P. tessellatus* and the Israeli Whip Snake "*Platyceps* cf. *ladacensis*".

---

# Jan's Cliff Racer

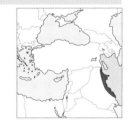

*Platyceps rhodorachis* (Jan, 1863) (formerly *Coluber rhodorachis*)
**Family:** Colubridae
**Sub-family:** Colubrinae
**F.:** Couleuvre-fouet de Jan
**G.:** Jans Pfeilnatter

**Identification** A medium-sized slender colubrid snake; appears very long with a long head that is, however, quite distinct from the neck. Adults normally 100–110 cm in length, to 138 cm. Eyes quite large; round pupils have a narrow golden or orange rim. Smooth dorsal scales generally arranged in 19 (rarely 21) rows at mid-body. 1 well developed and high preocular plate, below which is a small subocular scale.

2 (sometimes 3) postoculars. Generally 9 (sometimes 10) supralabials; the 5th and 6th normally touch the lower edge of the eye, sometimes the 6th and 7th, sometimes only the 5th. 204–244 ventral plates; anal scute normally divided; divided subcordals are arranged in pairs, but sometimes the 2nd and 3rd subcordals are undivided.

Two very distinct types of colouration. More often, an overall quite drab grey or brown-grey colouration. There are generally small, dark grey markings on the front half of the back, arranged transversely and alternating, sometimes forming transverse bands across the neck and even occasionally on the front half of the back; this colouration gradually disappears on the rear part of the back, which becomes uniform grey or beige, as does the tail. Quite a high proportion of the population have a spectacular and unique colouration; the whole body is uniform grey, without markings but with a thin red vertebral line that slowly disappears towards the rear of the body. Others have an intermediate colouration, with markings on the back but not along the vertebral column, where they are replaced by a narrow grey line. The top of the head is grey or brownish, either uniform or with small, dark, ill-defined spots. On the sides of the head are 3 pale or sulphur yellow, sometimes orange, vertical bands, 1 in front of the eye, the 2nd behind the eye and the 3rd, sometimes indistinct, on the temple. This colouration is also present in uniform individuals with a red band. A short, dark band under the eye does not continue back to the corner of the mouth. The belly is ivory or yellowish, in certain individuals becoming progressively pinkish towards the rear of the body and under the tail; there are small dark spots on the trailing edge of the ventral plates that disappear towards the rear of the body, though not under the tail.

**Venom** Harmless.

**Habitat** A large variety of relatively arid or rocky habitats, from sea level to 3000 m elevation in Afghanistan. Occupies canyons, rocky hillsides, steppe and also the sides of watercourses (e.g., areas with tamarisks), scrub, abandoned farmland, orchards, gardens and edges of construction sites. It enters houses on occasion, searching for small mammals, a hiding place or even a hibernation site. In Asia, it is replaced by *Platyceps karelini* in near-desert or sandy environments.

**Habits** Jan's Cliff Racer is an excellent climber. Very agile, it is capable of swimming across open water. Very wary, with good eyesight, it hides in a tree at the slightest alert. In mountainous areas, it has a long inactive period in winter, but at low elevations it may be active even in mid-winter during sunny weather.

**Diet** Mainly lizards and small mammals (rodents and shrews) and to a lesser extent birds and sometimes their eggs, and other snakes (*Spalerosophis diadema*, *Myriopholis macrorhyncha*). It is possible that Jan's Cliff Racer also takes amphibians and fish. Juveniles feed mainly on crickets, grasshoppers and small lizards.

**Reproduction** Oviparous. The female lays 4–9 very elongated eggs in June or July. Hatching occurs in August or September.

**Range** Vast distribution from northeastern Iraq (Kurdistan), Iran and the eastern side of the Caspian Sea as far as northern India (the region of Uttarakhand) and extreme western Nepal. In Iraq (thus in the Western Palaearctic), it is known only from a few specimens in museum collections and a few cited observations.

**Geographic variation** *Platyceps rhodorhachis* is closely related to *P. karelini* (Brandt, 1838), and hybrids have been found in these species' contact zone. Many subspecies are recognised, but their systematic order is confused. There are populations that have distinct morphological characteristics but have not been formerly described, whereas others already described belong to distinct species or do not, depending on the author (e.g., *P. r. ladacensis*, from Ladakh).

In Iraqi Kurdistan (Arbil and As-Sulaymaniyah regions) and western Iran (to the south of Lake Urmia) a new subspecies, provisionally named Kurdish Cliff Racer, is suspected by Beat Schätti *et al.* In this form, the dorsal markings are joined in large, transverse, well-defined very dark grey bands, separated from each other by narrow yellow transverse bands. As in *P. r. rhodorachis*, this patterning gradually fades in the rear half of the back and tail, which are a uniform sandy yellow (greyer in *P. r. rhodorachis*). The top of the head is densely covered with dark markings that are separated from each other by complex yellowish patterning. The first transverse neck band forms a collar, with a central rounded point as far as the rear of the parietal plates. There are more ventral plates than in the nominative subspecies: 242–249 compared to 204–244.

Platyceps rodorachis, *of the lined form, with its magnificent red band in the centre of the back. Iran, Hormozgan province. This form, well known in Iran, has not been recorded from Iraq.* Heidari.

# Saharan Whip Snake, Saharan Racer

*Platyceps tessellatus* (F. Werner, 1909) (formerly *Platyceps saharicus*)
**Family:** Colubridae
**Sub-family:** Colubrinae
**F.:** Couleuvre-fouet du Sahara
**G.:** Saharische Pfeilnatter

**Identification** A slender, elegant colubrid of medium size: less than 100 cm in Jordan but to 135.9 cm in Egypt and 160 in southern Algeria (personal observation). Narrow head slightly distinct from body. Large eyes; the round pupils have a silver-grey or golden-orange rim. Smooth dorsal scales arranged in 19 rows at mid-body. 1 high, well-developed preocular scale above a small subocular scale. 2 (sometimes 3) postoculars; 9 supralabials, the 5th and 6th touching the lower edge of the eye. Very many ventral plates (223–264). Undivided anal scute.

The back is pale grey, brown-grey or greenish-grey, gradually becoming more reddish-brown towards the rear, and on the tail it has a series of dark spots or transverse bands that alternate with the flank's markings. At the front of the body, these transverse marks are separated by yellowish-white scales. Starting at the middle of the body, the dorsal markings gradually fade and disappear at the rear and on the tail, areas that are uniformly coloured. Older individuals have a less contrasting colouration than juveniles; the colours become quite dull. Juveniles are brightly coloured. The top of the head is dark grey or dull brown with small, yellow, arabesque markings that disappear in adults. There are 3 pale yellow or orange vertical bands on the side of the head, 1 in front of the eye, the 2nd behind the eye and the 3rd, sometimes indistinct, on the temple. There is a short dark bar under the eye that does not reach the corner of the mouth. The sides of the neck sometimes show an orange tint. The belly is ivory or yellowish, becoming progressively more reddish towards the rear and under the tail, with blackish markings on the rear edge of the ventral plates that disappear towards the rear of the body and on the tail.

**Venom** Harmless; wary and unaggressive.

**Habitat** Rocky and stony areas in desert or semi-desert: often in gorges or steep-sided valleys, sometimes close to man, in abandoned gardens, on the edges of fields and in oases. Also in sandy areas with acacias. Appears not to venture far from water, which ensures its survival in desert areas where it occurs.

**Habits** Normally crepuscular and nocturnal, but can be seen in the sun in the morning, more so than truly nocturnal species. Readily hides in deep rock crevices during hot weather.

**Diet** Especially lizards and similar, but sometimes other snakes and amphibians.

**Reproduction** Oviparous.

**Range** Jordan's western and southwestern boarders, the southern half of Israel, Sinai, Arabian Peninsula, eastern Egypt and the mountain ranges of the southern Sahara (extreme southeastern Libya, northern Chad and southeastern Algeria, as far as the Tassili n'Ajjer and Hoggar ranges).

**Geographic variation** *Platyceps tessellatus* is considered to be a monotypic species.

*Saharan Whip Snake* Platyceps tessellatus, *a large adult. Tassili n'Ajjer (southern Algeria).* Peyre.

Platyceps tessellatus, *subadult. Rum Wadi (Jordan). Note the lack of intermediate bands between the dark dorsal bars; also the rear half of the body and the tail are without markings.* M. Geniez.

# Israeli Whip Snake

"Platyceps sp. *incertae sedis* Schätti and McCarthy, 2004" (formerly "*Platyceps ladacensis* sensu Perry, 2012")
**Family:** Colubridae
**Sub-family:** Colubrinae
**F.:** Couleuvre-fouet d'Israël
**G.:** Israelische Pfeilnatter

A strange colubrid was recently found in the Mediterranean part of the southern Levant countries (southern Israel and western Jordan). Named "*Platyceps* sp. *incertae sedis*" by Schätti and McCarthy (2004), it is considered by Gad Perry (2012) as belonging to *P. ladacensis* (Anderson, 1871), whose range extends discontinuously as far as Israel. We consider that the name *P. rhodorachis* should be applied uniquely to *P. rhodorachis* with a red band and that other *P. rhodorachis*, with markings and without the red band, belong to a different species, *P. ladacensis*. We prefer, in the absence of further information, especially genetic, to follow the classification proposed by Schätti, Tillack and Kucharzewski (2014) in which *ladacensis* is a subspecies of *P. rhodorachis*, those snakes with a red vertebral band being a morphotype that occurs pretty much everywhere in the range of *P. rhodorachis*, mixed with marked individuals without any red.

**Identification**. A slender, elegant colubrid snake of medium size (about 100 cm total length) with an appearance similar to that of Saharan Whip Snake *Platyceps tessellatus*. It is differentiated by the smaller number of ventral plates (209–245, average 228, compared to 223–264, average 244) and a slightly different colouration. The transverse markings on the back are often more black-grey: on the pale ground colour that separates them there are other much less dark, very narrow transverse marks. There is little contrast in the pattern on the top of the head. As in *P. tessellatus*, the markings on the back and the flanks progressively disappear towards the rear of the body and tail, which are less clearly pink than in the other species.

**Venom** Harmless.

**Habitat** Less arid regions than Saharan Whip Snake: as opposed to the latter, it occupies Mediterranean and sub-Mediterranean habitats (e.g., open garigue).

**Habits** Essentially the same as *P. tessellatus*.

**Diet** Probably lizards and similar.

**Reproduction** Oviparous.

**Range** Mediterranean and sub-Mediterranean countries from the northern half of Israel and northwestern Jordan. An ancient record from Cairo in Egypt is probably erroneous. It is possible that populations of *P. rhodorachis* that occur in the mountains of the southern Arabian Peninsula – e.g., in Yemen – actually are of this taxon.

**Geographic variation** No geographic variation known for this taxon at present.

*Israeli Whip Snake. Note the thin darkish lines between the principal bars and the markings on the back, which continue well beyond the middle of the body.* (above) Shacham, (right) Haimovitch.

The Glossy-bellied Racer *Platyceps
ventromaculatus* (Gray, 1834;
formerly *Coluber ventromaculatus*)
belongs to a complex of species
whose classification is debated
and controversial. The "real" *P.
ventromaculatus* is an oriental species
whose range covers southern Pakistan
and northwestern India. Very similar
colubrids found in Iran are provisionally
named "*Platyceps* cf. *ventromaculatus*".
Other species, such as *P. karelini* from
central Asia, may well belong to this
complex. Two species found within the
area covered by this book, *P. rogersi* and
*P. chesneii*, also have a controversial
taxonomic status and may be the
same species. Certain authors consider
these two taxa as subspecies of *P.
karelini* despite contradictory genetic
data (perhaps based on errors in the
identification of analysed specimens):
in effect, *P. karelini* and *P. rogersi* are
placed in distinct sub-clades (see
Pyron et al., 2013) and have very
distinct geographic distributions and
colourations, demonstrating how little
known these taxa still are.

# Anderson's Whip Snake, Spotted Racer

*Platyceps rogersi* (Anderson, 1893) (formerly *Coluber
rogersi*)
**Family:** Colubridae
**Sub-family:** Colubrinae
**F.:** Couleuvre-fouet d'Anderson
**G.:** Anderson Pfeilnatter

**Identification** A medium-sized
colubrid (80–100 cm total length)
characterised by a series of large dark
grey to reddish brown elongated,
transverse markings, very close
together and separated by a white or
pale yellow line (giving the impression
of thin, pale, transverse bars on the
back). The whitish flanks also have
round markings that alternate with
the back markings. All these markings
gradually fade towards the rear of
the body and on the tail, which is
uniform beige, with occasionally a
vestige of markings in the form of a
diffuse median line. The 1st mark, on
the nape, is very elongated with pale
edges. The top of the head is the same
colour as the dorsal markings, dark
grey, brown-grey or reddish grey, often
with a few small, pale marks on the
supraocular scales and the frontal and
exterior edge of the parietals. Some
individuals have an obscure, transverse
dark band from the eye to the top of
the head. There are 3 white or pale
yellow, oblique, vertical bands on the
side of the head, 1 in front of the eye, 1

Platyceps rogersi. *Negev desert (Israel). Note the very elongated mark on the nape and the dorsal markings, which are very wide and close to each other.* Crochet.

just behind the eye and the 3rd on the rear part of the temple. Under the eye, there is a short, dark, oblique band. The belly is a creamy-white with dark markings on the front half of the body and dark blotches on the outer edge of some of the ventral plates.

A narrow head, distinct from the body. Large eyes with round, orange-rimmed pupils. The dorsal scales are smooth and quite matte, arranged in 9 rows at mid-body; 1 large preocular plate above 1 or 2 small suboculars, 2 postoculars, 9 (sometimes 10) supralabials, the 5th and 6th (sometimes the 6th and 7th) touch the lower edge of the eye. The ventral plates have angular lateral edges that form a lateral ridge along the sides of the belly. Divided anal scute.

**Venom** Harmless.

**Habitat** Arid semi-desert and steppe regions with stones and scrub, occasionally near watercourses.

**Habits** Diurnal and nocturnal; during the hottest parts of the year, mainly active at night. Very wary and extremely rapid.

**Diet** Mainly small mammals and lizards; perhaps also amphibians.

**Reproduction** Oviparous. Lays remarkably large, long eggs, up to 5 in a clutch.

**Range** Southern Syria, Lebanon, Jordan, Israel, northwestern Saudi Arabia, Sinai, northern Egypt and northern Libya, to the west as far as Cyrenaica.

**Geographic variation** *Platyceps rogersi* is a monotypic species. Sometimes considered to be a subspecies of *P. chesneii* or even *P. karelini*, a central Asian species that it does not resemble at all.

*Close-up of* Platyceps rogersi, *Shawbak (Jordan)*. Cluchier.

# Western Spotted Whip Snake

*Platyceps chesneii* (Martin, 1838)
**Family:** Colubridae
**Sub-family:** Colubrinae
**F.:** Couleuvre-fouet de Chesne
**G.:** Chesnes Zornnatter

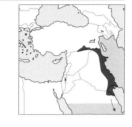

**Identification** Medium-sized colubrid snake (total length 90–110 cm), closely related to *Platyceps rogersi*, this last taxon sometimes considered a subspecies of *P. chesneii*. It is distinguished, apart from its eastern distribution, by its somewhat distinct colour. The top of the head is uniform grey without any pale markings; there is a black or dark brown transverse band from eye to eye (when it exists in *P. rogersi*, it is indistinct). The markings on the neck and back are generally less extensive and are widely separated one from the other (2–4 pale scales between marks, compared to 1 in *P. rogersi*). The mark on the nape is even longer and slimmer. In some individuals, the markings are quite wide and brown with yellow edges, a similar colouration to *P. rogersi*. Others, on the contrary, have quite narrow black transverse markings and thus superficially resemble Algerian Whip Snake *Hemorrhois algirus* (see page 109). As in *P. rogersi*, the dorsal and flank markings gradually fade towards the rear from mid-body and are completely missing on the tail. The belly is a bright yellowish-white; some individuals have small black marks on the external edges of one ventral plate in two or three, whereas other individuals have no markings.

**Venom** Harmless. Some individuals can be aggressive when captured, whereas others remain docile.
**Habitat** Desert and steppe regions at low elevation with grass and scattered bushes. Also encountered in oases and the edges of human habitations.
**Habits** Little known. Especially active at dusk and at night, but can be found in the sun in the early morning. An essentially terrestrial colubrid that is, however, at ease climbing in bushes or on rocks. Wary and quite rapid.
**Diet** Principally lizards and similar; sometimes also small mammals.
**Reproduction** Oviparous.
**Range** From southeastern Turkey across the Euphrates and Tigris valleys as far as Iran (western foothills of the Zagros mountains), as well as the northern Arabian Peninsula: Iraq, Kuwait and northeastern Saudi Arabia.
**Geographic variation** Here we consider *Platyceps chesneii* to be a monotypic species. The question as to whether *P. chesneii* and *P. rogersi* represent two distinct species or simply two subspecies of *P. chesnii* is not yet clearly established.

*Western Spotted Whip Snake (Platyceps chesneii). Kuwait. Note the very long dark stripe on the neck and, compared to P. rogersi, the black bar between the eyes and the more widely spaced dorsal spots (separated by two or three rows of pale scales). (left) P.-A. Crochet, (right) P. Geniez.*

# Egyptian Whip Snake

*Platyceps florulentus* (Geoffroy Saint-Hilaire, 1827)
(formerly *Coluber florulentus*)
**Family:** Colubridae
**Sub-family:** Colubrinae
**F.:** Couleuvre-fouet d'Egypte
**G.:** Ägyptische Zornnatter

**Identification** A medium-sized, quite slim colubrid (total length 80–110 cm), with relatively dull colours compared to other members of the genus *Platyceps*. Generally grey to olive-brown with some yellowish mottling, becoming progressively reddish towards the rear of the body and tail. Pattern on back composed of slightly contrasting transverse greyish or reddish marks. These marks break into smaller spots and gradually fade and disappear towards the rear of the back and on the tail. Spots smaller and less distinct on the flanks. On the head, there are blurred spots on the parietal and supraocular scales, sometimes also on the frontal scale. There is a blurred vertical pale bar just in front of the eye, another just behind. Belly between cream and whitish-grey, sometimes also reddish; the ventral plates are dark-spotted on the sides. Young have a more contrasting design, whereas older individuals are almost a uniform grey or brown colour.

Slim head, distinct from body. Medium-sized eyes with a quite dark iris. 21 (rarely 20 or 23) rows of smooth dorsal scales at mid-body, 1 loreal scale, 1 very high preocular, 2 postoculars, 1 (or 2) small suboculars above the 4th upper labial that is less high than the others, 9 supralabials (rarely 10 or 11), the 5th and 6th in contact with the lower edge of the eye. 201–228 ventral plates, with angular sides, forming a keel along each side of the belly. Preanal scale undivided.

**Venom**. Non-venomous.

**Habitat** Within the region under consideration (Nile Valley in Egypt), found mainly near wetlands in oases and farmland. Frequents canal banks, ruins, sides of cultivated areas on desert edge.

**Habits** A terrestrial snake active during the day and at dusk. Very wary and rapid.

**Diet** Small mammals, lizards, ground-nesting birds, amphibians. Kills prey larger than itself by suffocating it within its body coils.

**Reproduction** Oviparous.

**Range** Within the considered area, found only in Egypt, along the Nile Valley, to the north as far as the Delta and from there eastwards as far as the northern Sinai. Also occurs farther south, in East Africa: Sudan, Eritrea, Ethiopia, Somalia, Uganda, Kenya, Nigeria and Cameroon.

**Geographic variation** *Platyceps florulentus* has three subspecies. Only the nominate subspecies penetrates into the Western Palaearctic, via the Nile Valley.

*Egyptian Whip Snake*
Platyceps florulentus,
*photographed in Egypt.* Baha El Din.

# Elegant Whip Snake

*Platyceps elegantissimus* (Günther, 1878) (formerly *Coluber elegantissimus*)
**Family:** Colubridae
**Sub-family:** Colubrinae
**F.:** Couleuvre-fouet élégante
**G.:** Pracht-Zornnatter

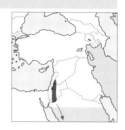

**Identification** A magnificent, slender snake with a distinctive dorsal pattern formed of black rings that are wider on the back than on the flanks and, though not always, enhanced by an orange or red vertebral line. Ground colour varies between whitish-grey, bright yellow and pale orange or olive. There are between 19 and 28 black rings on the back (counting from the nape to the base of the tail) and 7–13 on the tail. The rings on the front half of the body are wide above (5 or 6 rows of scales), coming to a triangle on the flanks; on the back, they are separated by 5–9 rows of pale scales. There are two transverse black bands on the head, the front one going from eye to eye; the hind one is larger and crosses the parietal scales. There is a vertical dark stripe under the eye. The belly is yellowish-white.

Average length 50–70 cm, rarely a little more. The head is flat, quite short and hardly distinct from the body. Medium-sized eyes with round pupils and dark iris. Quite pointed snout. Medium-sized eyes with round pupils and dark iris. Smooth dorsal scales in 19 rows at mid-body. An almost square loreal scale; 2 preoculars, the upper in a wedge between the prefrontal and the supraocular, touching the frontal scale; often 1 or 2 suboculars

Platyceps elegantissimus, *a marvelous colubrid snake distinguished by its well-spaced wide black bands.* Talbi.

99

between these and the supralabials; 2 postoculars, 8 supralabials on each side. The 197–200 ventral plates are slightly keeled on the sides; divided preanal scale.

*Platyceps elegantissimus* might be confused with another beautiful and even rarer species: *P. sinai*. In the latter species, the rings on the back are narrower (2 or 3 scales wide at the front, 1 at the rear of the body), and it has 17 rows of dorsal scales at mid-body instead of 19; it has a larger number of subcordal scales (91–100 instead of 78–84).

**Venom** Harmless.

**Habitat** Arid and pre-desert zones. At the foot of rocky slopes or in stony, dry riverbeds. Also on the edge of oases.

**Habits** Little known. Very discreet ground-dwelling snake, especially active in morning and afternoon (on the contrary, *Platyceps sinai* is active at dusk and at night), but also in the early nighttime hours.

**Diet** Small lizards, in particular nocturnal geckos.

**Reproduction** Oviparous.

**Range** Southern Israel (Arada Wadi), southern and southwestern Jordan (Aqaba, Rum Wadi and lower part of Al-Mujib Wadi), western and central Saudi Arabia. Apparently quite a rare species.

**Geographic variation** *Platyceps elegantissimus* is monotypic.

## Sinai Whip Snake

*Platyceps sinai* (Schmidt and Marx, 1956) (formerly *Coluber sinai*)
**Family:** Colubridae
**Sub-family:** Colubrinae
**F.:** Couleuvre-fouet du Sinaï
**G.:** Sinai-Zornnatter

**Identification** One of the most brightly coloured snakes of the Western Palaearctic. A small colubrid snake (known maximum length 53 cm in total), quite slim but with a relatively large, somewhat flattened head with a slightly angular snout, as in *Platyceps elegantissimus*. Medium-sized eyes, dark iris, round pupil with thin dark brown surrounding. Head and body sandy-grey, almost white. A thin black bar from eye to eye passes on to the 4th supralabial: another, larger bar crosses the back of the head, descending onto the sides as far as the corner of the mouth. 42–51 black rings across the body and 20–28 on the tail. These rings are narrower than in *P. elegantissimus*, between 1 and 3 rows wide, becoming narrower towards the back of the body and on the tail. A pale red vertebral line extends from behind the head to the tip of the tail. Belly white.

Two preocular scales sometimes partly fused, the upper forming a wedge between the prefrontal and the supraocular, just touching the frontal shield; 2 postoculars, 8 upper labials on each side, the 4th and 5th in contact with the eye. Smooth dorsal scales aligned in 17 rows at mid-body, 171–181 ventral scales, 91–100 pairs of subcordal scales.

**Venom** Harmless.

**Habitat** In full desert, rocky, dry riverbeds with sand and gravel. Also occurs on the edges of oases and in abandoned cultivated areas.

**Habits** Little known. This rare and local ground-dwelling snake, very discreet, is active at dusk and at night, rarely during the day.

**Diet** Small lizards (e.g., the skink *Ablepharus rueppellii*).

**Reproduction** Oviparous.

**Range** Endemic to a very small area where it is rare and seldom recorded: southern Sinai (Egypt), in particular at the Sainte-Catherine hermitage and in Feiran Wadi; the Negev desert

*Sinai Whip Snake* Platyceps sinai. *Rum Wadi (Jordan). Note that compared with* Platyceps elegantissimus *the black rings are narrow, narrower than those that separate them.* Sindaco.

Platyceps sinai. *Israel or Jordan.* Shacham.

(southern Israel) and southwestern Jordan (Rum Wadi, Ghoer Wadi and the lower part of Al-Mujib Wadi). May occur at 1500 m elevation at Sainte-Catherine.

**Geographic variation** *Platyceps sinai* is a monotypic species. Its place in the genus *Platyceps* needs confirmation by genetic analysis.

### Genus *Hermorrhois*

This genus contains only four species: two in southwestern Asia, *Hemorrhois ravergieri* and *H. nummifer*, are very similar, and two in North Africa, *H. hippocrepis* and *H. algirus*, are equally very similar to each other but quite distinct from the *ravergieri-nummifer* group. The genus *Hemorrhois* is considered to be genetically part of the *Platyceps* + *Spalerosophis* group. The name "Hemorrhois" makes reference to the often reddish colour of the belly of the Horseshoe Whip Snake *H. hippocrepis*.

## Ravergier's Whip Snake

*Hemorrhois ravergieri* (Ménétriés, 1832) (formerly *Coluber ravergieri*)
**Family:** Colubridae
**Sub-family:** Colubrinae
**F.:** Couleuvre de Ravergier
**G.:** Ravergiers Zornnatter

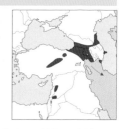

**Identification** Relatively robust typical snake of medium size (adults average 100 cm, but up to 150 cm in the easternmost part of its range). A relatively large head, quite distinct from the neck and slightly triangular, especially when in a threat posture. The eyes are relatively small, with round pupils encircled with silver-gold or orange-gold. Dorsal scales are slightly keeled, generally with 21 rows at mid-body. Very large supraocular scale; 1 loreal; 2 preoculars, the upper forming a wedge between the prefrontal and the supraocular and touching the frontal (a character that allows differentiation between the genera *Hemorrhois* and *Hierophis*); 2 postoculars, 1 small subocular scale at the leading edge of the eye; normally 9 supralabials on each side. 199–207 ventral plates, divided anal scute, 70–99 pairs of subcordals.

Quite variable in colour. Dorsal pattern composed of dark rhomboids, sometimes with pale borders, often reduced to transverse bars or, on the contrary, sometimes very large and interconnected. These dorsal patches join at the rear to become an undulating band and a longitudinal dark line on the tail. This dorsal pattern is sometimes poorly marked. A row of dark spots or vertical bars along the flanks, on the rear third of the body and the tail, also form a dark longitudinal line. The colour of the dorsal and flank markings varies between grey to black and can be brown or even orange. There are 2 arced dark bands forming a V at the back of the head; 2 small dark spots between the eyes; on the side of the head a dark band from the eye to the corner of the mouth and a shorter one just below and behind the eye. Belly is dirty white to grayish-white, with pale mottled markings in the adult, with much brown colouring in the juvenile. There are dark spots on the external edges of certain ventral scales that on the rear third of the body, as far as the anal scale, also come together on each side to form a longitudinal line. In the eastern part of its range (e.g.,

Ouzbekistan), some individuals have black heads.

**Venom** Non-venomous, but bites vigorously in self-defense.

**Habitat** Semi-deserts; southerly exposed areas of steppe or mountain slopes with many bushes. Also in rocky valleys. In south of range, only occurs in mountainous areas to 2600 m elevation; lower down and farther south it is replaced by the closely allied Coin Snake.

**Habits** Relatively rapid, agile snake that will climb walls and small cliffs. Diurnal; encountered in the morning or late afternoon sunning itself next to a bush or rock pile. When feeling menaced it can very much flatten the head, which becomes triangular and well distinct from the body (this can cause confusion with vipers). It has an

Hemorrhois ravergieri, *Osmandere (Turkey). Note the dorsal pattern in large zigzag form, which becomes a straight band on the tail. The orange-brown colour is reminiscent of that of* Montivipera bulgardaghica albizona *(page 319), with which it cohabits in the area.* P. Geniez and A. Teynié.

*Juvenile* Hemorrhois ravergieri. *Near Göle (northeastern Turkey). Note the completely different dorsal pattern composed of thin, well-spaced transverse bars, reminiscent of* Vipera ammodytes transcaucasiana *(page 312), which occurs in the area. This individual when adult will remain almost the same colour.* P. Geniez and A. Teynié.

inactive winter period that lasts 2–6 months, depending on elevation.

**Diet** Mice and other small rodents, lizards, birds at the nest. Suffocates larger prey in its strong body coils before swallowing it; smaller prey is swallowed alive.

**Reproduction** Oviparous. In June or July, lays 4–10 eggs under a pile of rocks or stones. Eggs hatch in September.

**Range** From eastern Turkey to northwestern China (Xinjiang) and from Afganistan and western Pakistan through Transcaucasia and the steppes of central Asia (Iran, northern Iraq, Uzbekistan, Kazakstan). There are isolated mountain populations in Syria (Mount Hermon), extreme northern Jordan (Jabal Druz) and Lebanon (Mount Lebanon).

**Geographic variation** *Hemorrhois ravergieri* is considered monotypic, but genetic data suggests a strong geographic variation within this species.

# Coin Snake

*Hemorrhois nummifer* (Reuss, 1834) (formerly *Coluber nummifer*)
**Family:** Colubridae
**Sub-family:** Colubrinae
**F.:** Couleuvre à monnaies or Couleuvre nummulaire
**G.:** Münzennatter

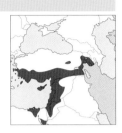

**Identification** A medium-sized, sturdy colubrid snake, often 100–110 cm long, but may attain 140 cm. The large triangular head is quite distinct from the body. The relatively small eyes have round pupils with golden rims; the golden speckled iris does not appear to be very dark. Obviously keeled matte or only slightly shiny dorsal scales in 23 or 25 rows at mid-body.

Of very variable colour. Along the back is a line of large, dark, coin-shaped spots outlined or not with a black line, close together; not an undulating joined band as in the closely related Ravergier's Whip Snake. This dorsal line of spots continues along the tail as a dark longitudinal line with undulated edges (not straight as in *Hemorrhois ravergieri*). It has a line of smaller spots on the flanks alternating with those on the back. An either indistinct or very marked dark band crosses the top of the head from eye to eye, and there are irregular spots on the back of the head. There is a dark line from the eye to the corner of the mouth and a shorter one from under the eye to the mouth. The whitish-grey belly often has fine dark mottling. There are dark spots on the sides of the ventral scales. Most adults are light grey or matte beige, with only slightly contrasting dorsal markings

that give the species a resemblance to the Blunt-nosed Viper *Macrovipera lebetina*, with which it often coexists. On the contrary, other individuals have much more contrasting colours, especially in the Mediterranean part of its range.

Some Coin Snakes are almost impossible to separate from Ravergier's Whip Snake by colouration alone. It is necessary to look at scale details for certain identification. In comparison to *Hemorrhois ravergieri*, *H. nummifer* has a greater number of rows of dorsal scales (23–25 in *H. nummifer*, only 21 in *H. ravergieri*), of ventral plates (males 195–212 compared to 189–207, females 208–230 compared to 199–206) and pairs of subcordals (males 82–107 compared to 77–99, females 79–107 compared to 70–94). In male *H. nummifer*, there are more basal spines on the hemipenis than in *H. ravergieri*, but this character can only be examined in dead specimens – for example, road casualties.

**Venom** Non-venomous. Not particularly shy, relying on its mimetic colouring to go unnoticed. If cornered, it hisses loudly and faces up to its aggressor; its bite is painful and may cause bleeding.

**Habitat** Arid south-facing slopes with a few small bushes, rocks and scree. A low-elevation species, often below 900 m, occasionally to just over 1000 m (by comparaison *Hemorrhois ravergieri* occurs to an elevation of 2600 m in mountains).

**Habits** Very similar to Ravergier's Whip Snake. Diurnal. Lies in the sun for long periods when the temperature is not very high. When in danger, flattens its head and hisses loudly (a habit similar to that of Blunt-nosed Viper, with which it may be confused at first glance).

**Diet** Small mammals, lizards, small birds. Kills prey by suffocation due to constriction in its coils.

**Reproduction** Oviparous. In June and July, lays 4–10 eggs in the ground under stones, scree or rocks. Eggs hatch in September.

**Range** The Mediterranean Basin: some of the Dodecanese islands, Rhodes, Cyprus, southern and southeastern Anatolia as far as Transcaucasia, northwestern Iran, extreme northern Iraq, western Syria, western Jordan, Lebanon, Israel, northeastern Egypt as far as the Nile. It has been recorded as far east as Tajikistan, Kyrgyzstan, and southern Kazakhstan.

**Geographic variation** *Hemorrhois nummifer* is considered monotypic.

Coin Snake Hemorrhois nummifer. *Gülek region (southeastern Turkey). Not very well marked grey individuals, like this one, look very similar to the* Blunt-nosed Viper Macrovipera lebetina *at first glance; they can be found in the same area.* P. Geniez and A. Teynié.

*A particularly well-marked individual of* Hemorrhois nummifer. *Dortyol region (southern Turkey).* P. Geniez and A. Teynié.

# Horseshoe Whip Snake

*Hemorrhois hippocrepis* (Linnaeus, 1758) (formerly *Coluber hippocrepis*)
**Family:** Colubridae
**Sub-family:** Colubrinae
**F.:** Couleuvre fer-à-cheval
**G.:** Hufeisennatter

**Identification** Active and often brightly coloured snake, medium or large, most often 100–140 cm (rarely to 185 cm in Morocco; personal observation of the author). Males are on average larger than females. The head is less slender than in the closely related Algerian Whip Snake and quite distinct from the body, especially in adults. The eyes are quite large, with round pupils and silvery-white to golden-orange rims. Smooth, shiny dorsal scales, more numerous than in *Hemorrhois algirus*, 25–29 rows at mid-body. There is a row of subocular scales between the lower edge of the eye and the supralabials, so that no labial is in contact with the eye (a rare phenomenon in colubrid snakes, but there are rare cases, as seen in Spain, where a supralabial touches the eye). 214–258 ventral plates; divided anal scale.

Yellowish, greyish-white or orange ground colour, sometimes with a lot of dark colouring that can make the snake appear blackish. On the top of the head, there is a very dark transverse bar with light edges in front and behind from eye to eye, and a dark horseshoe-shaped mark at the back of the head. In the centre of the back, there is a longitudinal row of large round or rhomboidal black or reddish spots that contrast with a fine pale border; these spots tend to fuse towards the tail, finishing as a continuous median line; the whole is quite beautiful. There are smaller spots on the flanks, their position alternating with those on the back. The belly is yellow, orange or pearly-red, with dark spotting on the sides. Juveniles usually show more contrasting colours than older individuals.

**Venom** Harmless, but large individuals can inflict a nasty bite if handled.

**Habitat** Arid hillsides with bushes, stones or scree in steppe or Mediterranean habitats. Also present at the top of beaches and even cultivated areas as long as they have refuge, such as hedgerows, drystone walls or rocky areas. They occur around inhabited areas and in unkept gardens. In mountain areas, they occur in elevations to 2430 m in the Betic sierras in southern Spain and at 2660 m in the Moroccan High Atlases.

**Habits** This terrestrial snake can climb into low bushes and over rocky terrain. Diurnal during winter and spring, but avoids very hot or sunny weather. Often active in the early morning and again in later afternoon, sometimes in the early nighttime hours. Very shy, it flees rapidly. Inactive in winter for 4 or 5 months in mountainous areas, much less near the coast and in the southern part of its range, where it can sometimes be seen sunning itself during clement weather in the heart of winter.

**Diet** Small mammals, birds and reptiles, including snakes. Young animals may also sometimes eat large insects. Very occasionally it will eat amphibians.

**Reproduction** Oviparous. Mating occurs during May and the first half of June, as early as April in the south. Lays 4–11 eggs between late June and mid-July; they hatch from mid-August to mid-September.

**Range** The southern two-thirds of the Iberian Peninsula north as far as about Girona; northwestern Africa (Morocco, northern Algeria and northern Tunisia), Pantelleria island (Italy) and Sardinia, where it is local and considered to have been probably introduced by man.

**Geographic variation** *Hemorrhois hippocrepis nigrescens* (Cattaneo, 1985), endemic to Pantelleria island (between Sicily and Tunisia), has recently been described but its validity is doubtful. Thus *Hemorrhois hippocrepis* is probably monotypic (with no subspecies).

*Young Horseshoe Whip Snake* Hemorrhois hippocrepis. *Marrakech region (Morocco).* P. Geniez.

# Algerian Whip Snake

*Hemorrhois algirus* (Jan, 1863)
(formerly *Coluber algirus*)
**Family:** Colubridae
**Sub-family:** Colubrinae
**F.:** Couleuvre d'Algérie or Couleuvre algire
**G.:** Algerische Zornnatter

**Identification** To about 100–140 cm, but often shorter. Slender, very rapid snake. The narrow head is hardly distinct from the body. Large eyes and round pupils with a silver-grey to orange rim. Smooth dorsal scales, not very shiny, in 23–25 rows at mid-body. Unlike the closely related Horseshoe Whip Snake *Hemorrhois hippocrepis*, 1 or 2 labial scales always touch the eye. 1 elongated loreal scale, 1 large, high preocular above 1 or 2 small suboculars, 2 or 3 postoculars, 9 (sometimes 8 or 10) supralabials on each side. 209–237 ventral scales.

The back is pale grey, pale brown or even yellow, with well-spaced dark transverse bars, or round or rhomboid spots close to each other. These bars or spots sometimes have pale edges. There are 1 or 2 rows of alternating round spots on the flanks. In individuals in the southern and eastern parts of the species' range *H. algirus algirus*, the top of the head and front of the neck are often uniform dark grey, sometimes black. On the other hand, individuals in Morocco and northern Algeria *H. algirus intermedius*, and also sometimes in those of the Atlantic Sahara and Mauritania, the head is not dark but has a more or less well-defined pattern similar to that of the Horseshoe Whip Snake: a dark transverse band, with pale borders, joining both eyes, with another in the form of a horseshoe at the back of the head; the first dorsal mark forms a triangular point on the nape. The belly varies from pearl-grey to pearly-red and can be yellowish or orange. There is often a dark quadrangular spot on certain ventral scales, corresponding to a prolongation of the bars on the flanks. Young individuals have highly contrasting and bright colours, whereas adults are normally much duller.

**Venom** A harmless, shy snake. If caught, it may bite.

**Habitat** Steppe-type landscape, arid and stony, often with spiny bushes; waste land between cultivated areas with bushes and stones; scree, ruins, overgrown gardens, patches of vegetation along the coast on desert edge, desert areas. Avoids the Mediterranean zone, where it is replaced by the Horseshoe Whip Snake.

**Habits** There is a marked rhythm to its daily activity: at the start of morning, it basks in the sun, then retreats to a shaded hiding place; in the late afternoon, just before dusk, it is active again, hunting. When it is very hot, this snake is only active at dusk and the early nighttime hours.

**Diet** Lizards, small mammals, young birds, large insects (e.g., grasshoppers)

**Reproduction** Oviparous.

**Range** Desert and pre-desert areas of North Africa: from northern Mauritania to northwestern Egypt along the northern edge of the Sahara; also present in the Hoggar range (southern Algeria). Introduced in Malta, where it is local.

**Geographic variation** *Hemorrhois algirus* is quite a variable species. Two subspecies are generally recognized; here we propose a third one:

► *Hemorrhois algirus algirus* (Jan, 1863): eastern part of the range (eastern Algeria, Tunisia, northern Libya and northwestern Egypt). Head and neck grey or black, dorsal marks form narrow transverse bands, anal scales often divided.

► *Hemorrhois algirus intermedius* (Werner, 1929): western part of the range (Morocco and northwestern Algeria). Appearance and colour are intermediate between *H. hippocrepis* and *H. a. algirus*. Head rarely dark, generally with the horseshoe-shaped mark of its close relative; dorsal marks very variable, often lozenge-shaped or round; general colouration more "desert"-like than that of *H. hippocrepis* (more sandy-beige or pinkish-brown). To distinguish it from the Horseshoe Whip Snake, it is often necessary to examine scale details: at least 1 upper labial scale in contact with the eye on both sides, and always less than 26 rows of dorsal scales. It seems quite possible that *H. a. intermedius* is the result of an introgression between the Horseshoe and Algerian Whip Snakes and thus represents a hybrid zone between the two.

► *Hemorrhois algirus villiersi* (Bons, 1962): Atlantic Sahara (in southwestern Morocco), and northwestern Mauritania (Bons, 1962): Atlantic Sahara (southwestern Morocco) and northwestern Mauritania (Bir Moghrein, Zouirat and Nouadhibou). This subspecies has for a long time been considered synonymous with *Hemorrhois algirus intermedius*.

Hermorrhois algirus villiersi, *subadult. Atlantic Sahara, southwestern Morocco.* Cluchier.

Hemorrhois algirus algirus. *The top of the head and neck are darker than the rest of the animal, quite a typical colouration in the type subspecies.* Kreiner.

Hemorrhois algirus intermedius, *near Imitek, in southwestern Morocco. It is the western North Africa subspecies that can be very different from* H. a. algirus. *Particularly, the head is very rarely dark and the dorsal bars are often replaced by large spots, giving the appearance of a Horseshoe Whip Snake* H. hippocrepis. *To differentiate it from that species, it is necessary to count the number of rows of dorsal scales (23–25 compared to more than 25 in general) and to verify that at least 1 supralabial touches the eye.* Martínez del Mármol Marín.

However, it is very slender and is coloured like the type *algirus algirus*, with the head often black or grey, the back often yellowish-grey or brownish-yellow, well-spaced dorsal bars (sometimes well-spaced round markings) and dorsal scales normally arranged in 23 rows (more often 25 in *intermedius*). Its general appearance does not resemble that of *H. hippocrepis*, and, in particular, it usually has the peculiarity of having an undivided anal scale. Here we propose reevaluating *Coluber algirus villiersi* Bons, 1962 with the name *H. algirus villiersi* (Bons, 1962). Affiliation with one or other of these two subspecies is not clearly established for the population in Adrar Atar (Mauritania) and the Hoggar range (southern Algeria).

### Genus *Spalerosophis*

This genus is represented by some six species of arid habitats, steppe or desert, distributed in North Africa, Somalia, the Arabian Peninsula and central Asia as far as India and Pakistan. They are quite thickset colubrid snakes when adult, with far more numerous scales than other Palaearctic Colubridae. In particular, certain head scales (e.g., the loreal) are split, and the frontal scale is often separated from the prefrontals by very small scales; the eye and the supralabials are always separated by a complete row of subocular scales.

From a phylogenetic point of view, *Spalerosophis* is placed between the genera *Platyceps* and *Hemorrhois*.

# Diadem Snake

*Spalerosophis diadema* (Schlegel, 1837)
**Family:** Colubridae
**Sub-family:** Colubrinae
**F.:** Couleuvre à diadème
**G.:** Östliche Diademnatter

**Identification** A medium to large snake (120–140 cm, to 154.5 cm in Egypt, 200 cm in central Asia), relatively slender. Head quite triangular and distinct from the neck, especially when it adapts an aggressive attitude. Medium-sized eyes with round pupils that are slightly vertically elliptical in strong light; the pupil, which has a diffuse white rim, contrasts with the very pale sandy-beige to bright orange iris. The numerous, very slightly keeled body scales are arranged in 25–31 rows at mid-body. The frontal plates are sub-divided into smaller scales (whereas in most colubrid snakes there are 9 large plates on the head, there are always more in the genus *Spalerosophis*). The loreal is sub-divided into 2–6 smaller scales, 2–4 preoculars, 3 or 4 postoculars; there is a complete row of small suboculars between the bottom edge of the eye and the upper labials (the eye is completely surrounded with small scales except for the upper edge, which is in contact with the supraocular); 10–14 supralabials on each side. 205–254 ventral plates. The anal scale is undivided and the subcordals are divided and occur in pairs.

Quite variable in colour. Ground colour grey, sandy, yellowish, ochre or reddish. On the back, a series of generally contrasting, transverse and quite large (but sometimes quite indistinct) spots, often in a lozenge form, extended transversely, sometimes forming well-separated transverse bars. They are quite variable in colour, from greenish-brown to reddish-brown; each dorsal mark has a pale edge in front and behind but not on the sides. 2 lines of darks spots on the flanks alternate with the dorsal markings. The 2 first dorsal marks are sometimes fused into one very long mark, reminiscent of Western Spotted Whip Snake's *Platyceps chesneii* dorsal pattern. It is the same for the first flank spots on the side of the neck, which fuse to form a dark longitudinal band. The underside of the head is often finely mottled or vermiculated with white on a dark background; a dark transverse band – sometimes very distinct, sometimes indistinct – goes from eye to eye; this band is sometimes broken in the middle. There is a dark temporal band from the eye to the corner of the mouth. The belly is white, yellowish or reddish, sometimes with dark mottling on the sides.

**Venom** It is said to be slightly venomous. Cases of poisoning have been recorded, and its saliva is toxic. Quite slow-moving when active at night. When disturbed at its daytime retreat, it can be aggressive and bite violently, even causing bleeding.

**Habitat** Real desert regions; in the north, in desert-edge steppe. Essentially in flat sandy areas with stones. Also present on oasis edges and on the edges of cultivated areas, and in coastal dunes or rocky ravines of wadis.

*Clifford's Diadem Snake* Spalerosophis diadema cliffordii. *Atlantic Sahara, between Dakhla and Aoussert. Note the very distinctive pattern compared with the Kuwait specimen on the next page.* Rufray.

Spalerosophis
diadema cliffordii,
*subadult. Kuwait, near
Ratqa.* P. Geniez.

By preference this snake occurs in rock crevices, under stones and in small mammal tunnels. Occurs from lowland plains to 2000 m in mountain areas in the eastern part of its range.

**Habits** Ground-dwelling colubrid snake that is active during the day in spring and at night during the hottest part of the year.

**Diet** Lizards and allies, snakes, small mammals (rodents) and birds.

**Reproduction** Oviparous. Mates in the spring. Lays 3–16 eggs that are stuck together.

**Range** Vast geographic distribution: almost the whole Sahara, from the Atlantic Sahara (southwestern Morocco), Mauritania, Mali and Niger as far as Egypt and Sudan, through Algeria, southern Tunisia, and Libya; Sinai, Israel, Jordan, Syria and extreme southeastern Turkey; the whole of the Arabian Peninsula and much farther east across Asia Minor as far as Pakistan and northern India.

**Geographic variation** A quite variable species, recognized as polytypic. Three subspecies are recognized by most authors:

▶ *Spalerosophis diadema cliffordii* (Schlegel, 1837): the whole of the Sahara, the Arabian Peninsula and desert countries to northern Saudi Arabia as far as southwestern Iran. It is the only subspecies present in the area covered by this book. It is sometimes considered to be a separate species, but molecular analysis contradicts this separation.

▶ *Spalerosophis diadema diadema* (Schlegel, 1837): Pakistan and northern India.

▶ *Spalerosophis diadema schirazianus* (Jan, 1865): from Iran and Turkmenistan as far as Uzbekistan.

# Werner's Diadem Snake

*Spalerosophis dolichospilus* (F. Werner, 1923)
**Family:** Colubridae
**Sub-family:** Colubrinae
**F.:** Couleuvre à diadème du Maghreb
**G.:** Westliche Diademnatter

**Identification** One of the most beautiful snakes of North Africa; at first glance, it resembles a miniature Ball Python. Average length 120–130 cm, rarely more. Elongated head, wider at rear, well distinct from the neck. Medium-sized eyes, round or slightly elliptical pupils with thin whitish rim, iris pale grey to orangish. Dorsal scales obtusely keeled, numerous, are arranged in 31–33 rows at mid-body (thus more than in *Spalerosophis diadema*). The frontal plates are subdivided into smaller scales; the loreal split into 5–9 small scales and 8–12 small scales around the eye, which thus is not in contact with the supralabials; 10–14 upper labials on each side. Yellowish, pinkish or almost white ground colour. A series of splendid, regularly shaped, large round or oval marks along the back (in *S. diadema* the marks are irregular), with thin dark and pale edges around the entire mark (pale edges only on the sides in *S. diadema);* they are quite often greenish-grey, thus contrasting with the pinkish-beige ground colour of the back, but can also be dark mahogany-brown. In some individuals the dorsal marks are less regularly shaped and thus can appear to be intermediate between *S. dolichospilus* and *S. diadema*. There are smaller elongated marks on the flanks of the same colour and alternating with the back markings. As in *S. diadema*, the top of the head has fine and dense

Spalerosophis dolichospilus, *adult. Near Assa (southwestern Morocco).* García-Cardenete.

115

Spalerosophis dolichospilus, subadult. Morocco. Note, at the front of the head, the divided prefrontals, characteristic of the genus Spalerosophis. Aymerich.

dark mottling on a whitish background and a dark transverse band that, often broken, goes from eye to eye and continues on each side of the head from the eye to the back corner of the mouth. Every scale on the side of the head is half dark, half white. The first back and flank markings are often fused to form large longitudinal bands. Pale yellow to yellowish-white belly with more or less dark marbling or spotting.

**Venom** Slightly venomous. A quite slow-moving snake when seen, active at night. If disturbed in its daytime retreat, it is often aggressive, adopting an intimidation posture; it can bite a hand of an aggressor and draw blood, sometimes holding on to inject reputedly slightly venomous saliva.

**Habitat** Dry, stony steppe, with or without bushes. Avoids true desert, where it is replaced locally by

*S. diadema.* Occurs as high as 1450 m in the Moroccan Anti-Atlas.

**Habits** Ground-dwelling colubrid snake. More often than not active at night particularly during the hottest months, when it can be found, for example, on roads. Can also be encountered at dawn or dusk. The rest of the time it hides in a hole, in a stone wall or under a rock.

**Diet** Lizards and allies, smaller snakes, small mammals (rodents), birds.

**Reproduction** Probably the same as the Diadem Snake. Oviparous. Lays 3–15 eggs.

**Range** Endemic to North Africa: Morocco, northern Algeria and northern Tunisia.

**Geographic variation** *Spalerosophis dolichospilus* is a monotypic species.

## Genus *Eirenis*

Represented by 18 species, 14 in the Western Palaearctic. The centre of diversification of this genus is southeastern Turkey. Apart from Turkey, the genus occurs in the Levant countries, the Arabian Peninsula, Transcaucasia, Iran and farther east as far as Pakistan, with one species in East Africa, *Eirenis africana* (Boulenger, 1914); genetically its systematic position is unknown. Although the monophyly of the genus *Eirenis* has been confirmed by molecular analysis (that is, that all *Eirenis* species, except maybe

*E. africana*, have a common ancestor and their grouping is justified), the phylogeny among the genus does not appear to be clear and does not always reflect morphological affinities.

Here we propose placing them in two groups (for practical reasons that have no phylogenetic value), those that have 17 rows of dorsal scales and those that have only 15.

# Ring-headed Dwarf Snake, Asia Minor Dwarf Snake

*Eirenis modestus* (Martin,1838)
**Family:** Colubridae
**Sub-family:** Colubrinae
**F.:** Couleuvre naine modeste
**G.:** Kopfbinden-Zwergnatter

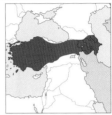

**Identification** A small, generally slender snake, head hardly distinct from the body, average length of 50 cm, but may rarely reach more than 60 cm. Eyes with round pupils, as in all *Eirenis* species, rimmed reddish golden-brown, especially at the top of the very dark iris. Smooth, shiny body scales, dorsal scales arranged in 17 (rarely 18) rows at mid-body. As in nearly all members of the genus *Eirenis*, there is 1 loreal scale, 2 postoculars, 7

supralabials on each side, the 3rd and 4th in contact with the eye, 171–190 ventral scales, a divided anal scale, 61–81 pairs of subcordals. Head with a dark mark in 3 parts separated by narrow transverse yellow or orange bands: a wide dark band between the eyes, a large half-moon-shaped mark on the parietal scales and a wide neck collar. The collar is curved forwards and bordered behind by a pale line that does not extend under the neck; in a

Eirenis modestus modestus, *juvenile. Near Ispir (Turkey). Note the very contrasting colours compared with those of the adult, shown on the next page.* P. Geniez and A. Teynié.

good number of individuals, it extends with a pointed centre as far as the dark mark on the parietals, dividing in half the pale mark in front of the collar. The head pattern is very contrasting in juveniles; it becomes duller in adults. The supralabials are whitish, marked with more or less well-defined small dark bars at their edge. The back is generally a uniform grey, yellowish-grey, sandy-brown or reddish-brown when seen from a distance, each dorsal scale crossed by a longitudinal orange-brown bar (hardly visible in juveniles) that is reminiscent of the dorsal scales of young *Dolichophis caspius* and *D. jugularis*. In the subspecies *semimaculatus*, as its name suggests, a proportion of individuals is more or less well marked with black spots on the back (sometimes even transverse bars) and black streaks aligned longitudinally that fade towards the rear of the body. The shiny ventral plates are white to pale grey.

**Venom** Quite a slow-moving harmless snake; occasionally bites when handled.

**Habitat** Dry slopes with little vegetation, garigue, edges of wetter areas, areas of stony steppe. In mountainous areas as high as 2000 m. Hides under stones and in rock crevices.

**Habits** Mainly diurnal but prefers shaded areas. The species rarely comes into the open, but after rain it may be seen drying itself, lying in the sun. Quite slow and not scared when discovered. Quite common in spring under stones, it totally disappears in summer, when it goes underground. Totally inactive for several months during winter, it may become active for a period in February or March if the weather allows.

**Diet** Centipedes, scorpions, insects and on occasions small lizards and similar.

**Reproduction** Oviparous. Lays 3–6 relatively large, longish eggs. Freshly hatched young are extremely small, only 8–10 cm long.

**Range** Turkey (from the northwest as far as the Marmara Sea coast, certain Dodecanese islands facing the Turkish coast, the Caucasus region,

Eirenis modestus modestus, *adult. Tavçancil, near Istanbul (Turkey).* P. Geniez and A. Teynié.

Eirenis modestus semimaculatus, *adult. Lesbos island (Greece). Note the numerous small black streaks on the body, typical of this subspecies.* P. Geniez.

Eirenis modestus cilicius, *old adult with faded pattern. Southern Turkey.* Teynié.

Armenia and Azerbaijan. Replaced in Cyprus and countries along the eastern Mediterranean coast by *Eirenis levantinus*. Its presence in Iraq and northern Iran is uncertain.

**Geographic variation** Three subspecies are presently recognized:

▶ *Eirenis modestus cilicius* Schmidtler, 1993: southern Turkey, along the coast between Anamur and Mersin. Uniform back, often grey with little contrast.
▶ *Eirenis modestus semimaculatus* (Boettger, 1876): western and southwestern Turkey; also certain islands in the Aegean Sea, including the Dodecanese (e.g., it is abundant on Lesbos, Greece). Some individuals have black spotting on the back alongside others with a uniform back.
▶ *Eirenis modestus modestus* (Martin, 1838): the rest of the species' range. In principle, the back is always uniform. The status of *E. m. werneri* (Wettstein, 1937), endemic to Alazonisi (Dodecanese islands) is not clear, but it is probably the same subspecies as *E. m. semimaculatus*.

# Golden-lined Dwarf Snake

*Eirenis aurolineatus* (Venzmer, 1919)
**Family:** Colubridae
**Sub-family:** Colubrinae
**F.:** Couleuvre naine à lignes d'or
**G.:** Goldlinien-Zwergnatter

**Identification** A dwarf colubrid snake morphologically close to *Eirenis modestus* but with a distinctive and characteristic head and neck pattern; dark collar (black in young individuals) on the neck, bordered by a yellow line with a black edge present on the typical form of this species; it is straight across the neck, then suddenly turns to become an oblique line continuing forwards as far as the last supralabial. A pale yellow line in front of the dark collar surrounds the parietal scales, forming a narrow oblique straight band that joins the 5th upper labial, just behind the eye – when seen from above, a straight-sided horseshoe shape. A thin yellow transverse band on the top of the head continues on the sides to just behind the eye. This patterning, which is contrasting yellow and black in juveniles, becomes less marked but persists in adults. The back is a uniform beige-brown or reddish-brown seen from a distance; each dorsal scale has a longitudinal orange-brown line, as in *E. modestus*.

Eirenis aurolineatus, *adult. Near Akçatekir (southern Turkey).* P. Geniez and A. Teynié.

Eirenis aurolineatus, *juvenile*. Schmidtler.

Average length 45 cm. Head slightly distinct from body. Very dark iris, darker than in *E. modestus*. Smooth, quite shiny body scales, arranged in 17 rows at mid-body. 149–168 ventral plates, 71–84 pairs of subcordals in males, 60–70 in females.
**Venom** Harmless.
**Habitat** Quite similar to that of Ring-headed Dwarf Snake; on southern slopes in the Bolkar range; in the Mediterranean and subalpine stages, to about 1800 m.
**Habits, Diet and Reproduction** Similar to that of Ring-headed Dwarf Snake.

Probably active during the day and at dusk; probably eats insects, centipedes, scorpions and sometimes small lizards and similar. Almost certainly oviparous.
**Range** Endemic to a small area in southern Turkey, the Bolkar range.
**Geographic variation** *Eirenis aurolineatus* is a monotypic species. Genetically, it is the species closest to *E. modestus*.

## Levantine Dwarf Snake

*Eirenis levantinus* Schmidtler, 1993
**Family:** Colubridae
**Sub-family:** Colubrinae
**F.:** Couleuvre naine du Levant
**G.:** Levante-Zwergnatter

**Identification** Small colubrid snake (35–40 cm) that closely resembles *Eirenis modestus*. Slender head, hardly distinct from body. Black iris, such that the pupil is hardly visible. There is a dark bar across the top of the head (black in juveniles) and between the eyes, and another very large bar that covers most of the parietal scales. This mark is generally separated from the large dark collar bar by a yellow band. The central black band sometimes finishes in a prolongation at the rear that may touch the collar, such that the yellow band is separated in its centre (as in *E. modestus*). The dark collar curves forwards to touch the back supralabial scales. Curiously, in individuals in Turkey the back of the collar is relatively straight, descending low on the sides of the neck, quite similar to its form in *E. barani*, whereas in Lebanon and Israel it is separated in two on the sides by the pale band that edges the back of the collar. There are sometimes a few small black spots at the edge of the rear of the collar. The back appears uniform yellowish-brown to reddish-brown, each dorsal scale crossed with

a longitudinal orange band, as in *E. modestus*. The belly is whitish. The dorsal scales are smooth and shiny, arranged in 17 rows at mid-body. 1 preocular scale, 2 postoculars, 7 supralabials on each side. Fewer ventral plates than in *E. modestus*: 139–166 compared to 171–190; 59–76 subcordal plates in males, 55–72 in females. Distinguished from *E. modestus* by being on average smaller, having fewer ventral plates, an entirely black iris (the pupil has a brown rim in *E. modestus*), larger and more pronounced spots on the labials and often a more orange or yellow overall colour.

Above all, distinguished from *E. barani* by the head and neck pattern and the back scales never having dark centres but being crossed by a pale line; unlike in *E. barani*, the collar does not extend onto the neck; it is less square-cut, more or less curving forwards.

Eirenis levantinus, *adult. Near Dortyol (southern Turkey).* P. Geniez and A. Teynié.

**Venom** Harmless.

**Habitat** Dry stony slopes with sparse vegetation of bushes and grass, often in Mediterranean garigue. In hilly areas, it may occur to an elevation of 1100 m.

**Habits** Hardly known; likely the same as that of other dwarf snakes. Lives in such a way that it is always hidden, usually under rocks.

**Diet** As yet unknown. Probably like other dwarf snakes: insects, centipedes, scorpions, occasionally small lizards and similar.

**Reproduction** Unknown, but almost certainly oviparous.

**Range** Extreme north of Israel, Lebanon, western Syria (Mount Hermon) and southeastern Turkey (principally in Hatay province).

**Geographic variation** *Eirenis lavantinus* is considered to be monotypic.

Eirenis levantinus, *subadult. Lebanon.* Cluchier.

# Baran's Dwarf Snake

*Eirenis barani* Schmidtler, 1988
**Family:** Colubridae
**Sub-family:** Colubrinae
**F.:** Couleuvre naine de Baran
**G.:** Barans Zwergnatter

**Identification** Quite slim colubrid snake, smaller than *Eirenis modestus* (less than 40 cm total length). Narrow head hardly distinguished from body. 3 wide dark grey (black in juveniles) transverse bars on head and neck, separated from each other by a broken yellow line. These bands have a straight edge in front and behind (whereas they are curved in a half-moon in *E. modestus* and *E. levantinus*). The 1st band passes the eyes on the sides and terminates as a triangle; the 2nd stops abruptly on the 6th supralabial scale, and the last one descends in a point very low on the sides of the neck, forming a half collar with ends visible only

123

*Baran's Dwarf Snakes* Eirensis barani barani, *adults, southern Turkey.* Göçmen.

from underneath (this last trait characteristic of the species). The back is generally yellowish, pale brown or grey, the dorsal scales slightly paler on the sides than in their centre (it is the opposite in *E. modestus* and *E. levantinus*). Certain individuals, particularly those of subspecies *E. barani bischofforum* have 2 rows of large, dark, very contrasting spots on the back and small dark spots on the flanks, more or less longitudinally aligned. The belly is uniform white.

The dorsal scales are smooth and shiny, arranged in 17 rows at mid-body. 1 loreal scale, 2 postoculars, 7 supralabials on each side. There are fewer ventral plates than in *E. modestus*: 138–156 in males, 156–165 in females (compared, respectively, to 171–180 and 173–190 in *E. modestus*). There are also fewer subcordal plates: 60–68 in males and 54–62 in females

(compared to 61–81 in males and 63–72 in females in *E. modestus*).

**Venom** Harmless.

**Habitat** Dry and stony slopes with sparse bushes or grass, rocky riverbeds, meadows with remains of open woodland. In mountainous areas, to 1100 m, rarely as high as 1700 m.

**Habits** Little known, probably similar to other dwarf snakes. Quite slow. Lives mainly in cover; if found, more often than not under stones. Nothing is known about its hibernation.

**Diet** As yet unknown, but almost certainly as in other dwarf snakes: mainly insects, centipedes and scorpions.

**Reproduction** Oviparous. Number of eggs unknown.

**Range** Very restricted: southern Turkey, principally the Gulf of Aden, the Anti-Taurus range and extreme northern Hatay: also known from extreme northwestern Syria.

Eirenis barani barani,
*subadult. Osmaniye
(southern Turkey).
Note the clear-cut
collar that descends
in a point under the
neck, as well as the
dorsal scales, which
are paler on their
edges, characteristics
typical of this species.*
M. Geniez.

**Geographic variation** There are two recognized subspecies of *Eirenis barani*:

► *Eirenis barani barani* Schmidtler, 1988: southern Anatolia, northwestern Syria.
► *Eirenis barani bischofforum* Schmidtler, 1997: limited to the foothills of the northern part of the Anti-Taurus range. Characterised by the presence of dark spots on the front part of the back, at least in some individuals.

# Armenian Dwarf Snake

*Eirenis punctatolineatus* (Boettger, 1892)
**Family:** Colubridae
**Sub-family:** Colubrinae
**F.:** Couleuvre naine d'Arménie
**G.:** Armenische Zwergnatter

**Identification** Shy dwarf snake, relatively rapid for an *Eirenis*, average length 30–50 cm. Slender; grey to reddish brown-grey with 8 longitudinal rows of marks arranged two by two on the front part of the body that form continuous lines on the rear body and tail (some individuals are almost completely lacking in markings and can easily be confused with *E. modestus*, which, however, always has traces of a neck band). Narrow head is hardly distinct from the body. Top of the head has small, dark arabesque markings that tend to fade in the adult. No clearly defined collar or neck band. The supralabials are whitish to yellowish with dark borders. The belly is pale whitish-grey, without markings. The dorsal scales are shiny and smooth, disposed in 17 rows at mid-body. 1 loreal scale, 1 preocular, 2 postoculars, 7 supralabials on each side; 156–175 ventral plates.

**Venom** Harmless.

**Habitat** Dry slopes in steppe-type environments with short or cushion-like vegetation. In mountainous areas, usually at mid-elevation but can occur as high as 2000 m in various areas.

**Habits** Little known, but likely resemble those of other dwarf snakes. Rapid and shy, more so than *Eirenis modestus*. Lies in the sun in the morning more readily than most other *Eirenis* species. Hides under stones, in rock crevices and shrub roots. Active between April and October, but difficult to find during the hottest summer months, except after rain. Inactive during winter for several months.

**Diet** Essentially arthropods: grasshoppers and crickets, beetles, caterpillars, myriapods, small scorpions and spiders. Probably occasionally takes young lizards and similar.

**Reproduction** Oviparous; in late June, lays 3–8 eggs that hatch in late summer or early autumn.

**Range** Extreme southern Turkey, southern Armenia, southern Azerbaijan

Eirenis punctatolineatus punctatolineatus, *juvenile. Muradiye (eastern Turkey).*
P. Geniez and A. Teynié.

(Nakhchivan), northwestern and western Iran. May be found in northeastern Iraq.

**Geographic variation** Two or three subspecies are recognized by most authors.

► *Eirenis punctatolineatus punctatolineatus* (Boettger, 1892): eastern Turkey, southern Armenia, southern Azerbaijan and northeastern Iran.

► *Eirenis punctatolineatus condoni* (Boulenger, 1892): southwestern Iran.
► *Eirenis punctatolineatus kumerloevei* Eiselt, 1970: an insular subspecies restricted to Akdamar island (in Lake Van, eastern Tukey), described on the basis of one melanistic specimen and considered by most authors to be synonymous with the type subspecies.

# Narrow-striped Dwarf Snake

*Eirenis decemlineatus* (Duméril, Bibron and Duméril, 1854)
**Family:** Colubridae
**Sub-family:** Colubrinae
**F.:** Couleuvre naine à dix lignes
**G.:** Zehnstreifen-Zwergnatter

**Identification** Smallish slender snake with 2 thin double longitudinal dark lines along the back, which are the width of half a scale, on a grey brown to caramel-brown background, to pleasant effect. In young individuals, there are 2 or 3 additional longitudinal dark lines on each flank. It has neither a collar nor a dark neck band. The upper head is dark brown, with small, pale arabesque markings reminiscent of the markings on young snakes of the genus *Dolichophis* (*D. jugularis*, *D. caspius* and *D. schmidti*). Belly is uniform creamy-white or yellowish. The largest of the genus *Eirenis*: 50–65 cm in length, largest specimens to 75 cm. Slender head slightly

separated from the body. The dorsal scales are smooth and quite shiny, in 17 rows at mid-body, each with a pale longitudinal line in its centre that accentuates the linear aspect of this species. One loreal scale that is longer than it is wide, 1 preocular, 2 postoculars, 7 upper labials on each side, each with a thin reddish-brown border; the 3rd and 4th touch the eyes' lower edge. 136–183 ventral plates, a divided anal scale, divided subcordals in pairs.

**Venom** Non-venomous.

**Habitat** Dry parts of Mediterranean zone and steppe. Dry slopes with scattered brush and stones, stony areas on the edge of crops, also open wooded

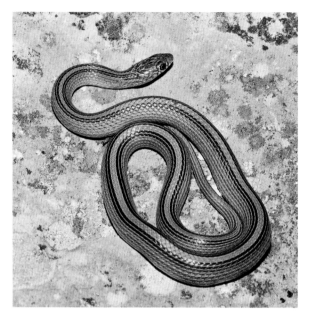

Eirenis decemlineatus, *young adult. Halfeti (southeastern Turkey).* P. Geniez and A. Teynié.

areas. More often than not, hidden under stones or holes in the ground.
**Habits** Little known, but probably similar to those of other dwarf snakes. Mainly diurnal, but with little activity during hottest period of the day. Moves slowly but usually hidden. No winter period of inactivity.

**Diet** Similar to that of other dwarf snakes: large insects (especially crickets and grasshoppers), centipedes, scorpions and spiders. Probably sometimes takes small lizards and similar, but as yet this is unconfirmed.
**Reproduction** Oviparous.
**Range** Israel, Lebanon, northwestern Jordan, northern and western Syria, southeastern Turkey. Ancient records from Iraq and Iran probably are of other species.
**Geographic variation** *Eirenis decemlineatus* is monotypic (there are no subspecies).

*Close-up of* Eirenis decemlineatus, *Lebanon.* Cluchier.

# Striped Dwarf Snake

*Eirenis lineomaculatus* Schmidt, 1939
**Family:** Colubridae
**Sub-family:** Colubrinae
**F.:** Couleuvre naine rayée
**G.:** Längsgepunktete Zwergnatter

**Identification** Very small (20–30 cm in length) and thickset (the most thickset of the dwarf snakes). Quite closely related to the Crowned Dwarf Snakes (*Eirenis coronella* and *E. coronelloides*). Slim head, high and hardly distinct from body. Very rounded snout. No loreal scale, which is rare in Palaearctic Colubridae. 1 preocular plate, only 1 (rarely 2) postoculars, 7 supralabials, the 3rd and 4th in contact with the eye. Scales smooth and quite shiny, in 17 rows at mid-body (15 in *E. coronella*

and *E. coronelloides*). 110–145 ventral plates, divided anal and subcordal scales. The back is beige-grey, olive-brown or yellow-brown with irregular transverse caramel-brown spots, each with a black border on the upper edge. There are 2 or 3 rows of identical but smaller spots on the flanks arranged in alternation with the dorsal spots. The top of the head is mottled, with a dark longitudinal line on each side of the frontal plate, becoming less visible and faded in adults. There is a wide,

Eirenis lineomaculatus, *southern Turkey. Note the thickset appearance compared to other members of the genus* Eirenis. Schmidtler.

*Close-up of* Eirenis lineomaculatus. *Near Antakya (southern Turkey). Note the absence of a loreal scale (but in this individual the preocular is divided, a rare occurrence!) and a particularly thick neck.* M. Geniez.

oblique black bar under the eye. A wide dark neck band curves forwards on the sides to form a complete V-shaped collar under the chin. A dark spot often present in front of the collar reaches the rear of the parietal plates. The pupils are round, with golden-brown to orange-brown rims. The belly is whitish, densely spotted with contrasting large black spots, isolated and regularly spaced.

**Venom** Harmless.

**Habitat** Dry, stony slopes with sparse vegetation in Mediterranean-type habitat, from plains to hills. Rarely seen; more often than not found under stones. Locally found above 1000 m.

**Habits** Little known, but probably like those of other dwarf snakes. Active by day, mainly in the morning and evening until dusk. A slow-moving and relatively slow species that shows no sign of aggression, even when handled. Lives a hidden lifestyle; sometimes found under stones in spring.

**Diet** As for other dwarf snakes: arthropods (large insects such as crickets), but also small lizards.

**Reproduction** Oviparous. Number of eggs as in other dwarf snakes (3–8).

**Range** Rarely observed; endemic to the Levant countries: Israel, Lebanon, northwestern Jordan, western Syria, southeastern Turkey.

**Geographic variation** *Eirenis lineomaculatus* is a monotypic species.

# Crowned Dwarf Snake

*Eirenis coronella* (Schlegel, 1837)
**Family:** Colubridae
**Sub-family:** Colubrinae
**F.:** Couleuvre naine couronnée
**G.:** Krönchen-Zwergnatter

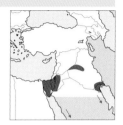

**Identification** Small, quite thickset snake that goes unnoticed and has a very hidden lifestyle. 25–35 cm in length. Narrow head hardly separated from the body; thick neck. 1 small loreal scale, sometimes absent; 1 preocular; 2 (rarely 1) postoculars; 7 (rarely 6) supralabials on each side; 15 rows of dorsal scales at mid-body. 141–163 ventral plates; a divided anal scale; 48–67 pairs of subcordals.

Back from pinkish-grey to beige with about 50 (32–65) narrow, transverse red-brown to blackish bands with irregular edges, very obvious in juveniles; they sometimes break up in older individuals and become almost invisible. A complete neat dark collar behind the head continues onto the neck. Unlike the closely related *Eirenis coronelloides*, the underside of the head is pale,

*Crowned Dwarf Snake* Eirenis coronella coronella, *adult from Israel. Note the absence of a "crown" (the trapezoid dark mark on the top of the head), despite its name.* Crochet.

*Young Crowned Dwarf Snake* Eirenis coronella coronella. *Israel. Note the absence of a crown, even in the juvenile.* Martínez del Mármol Marín.

sometimes with a few small, dark marks; it normally does not have the large dark rectangular mark characteristic of the other species. There is a dark band with irregular edges between the eyes in young individuals. All individuals have a dark rectangular mark under the eye that continues under the side of the head. Round pupils circled with orange. The belly is uniform yellowish to reddish. Often mistaken for *E. lineomaculatus*, which has 17 rows of dorsal scales, not 15, and in which the dorsal bands are irregular and do not descend onto the flanks in a continuous way; *E. lineomaculatus* has the belly entirely covered with large dark spots.

**Venom** Harmless.

**Habitat** The *Eirenis* species best adapted to arid and semi-desert conditions. Occurs especially in extensive steppe with scattered brush and many rocks. In spring, usually found under rocks. In arid mountains, occurs to 1600 m elevation.

**Habits** Little known, but probably similar to other dwarf snakes. Slow-moving and unaggressive; active by day principally in the morning and late afternoon until dusk. Leads a hidden lifestyle.

**Diet** Mainly insects and other arthropods such as myriopods, large spiders and scorpions; maybe also small lizards and similar, but needs confirming.

**Reproduction** Oviparous. Number of eggs unknown (but a female found dead and dissected contained 5 elongated eggs).

**Range** It is the most southerly occurring of the dwarf snakes: principally western and northwestern Arabian Peninsula and, farther north, Israel, western Jordan, southern Syria and northwestern Iraq. Its geographic distribution is poorly known due to confusion with *Eirenis coronelloides*.

**Geographic variation** *Eirenis coronella* is a polytypic species:

► *Eirenis coronella coronella* (Schlegel, 1837): Israel, Jordan, mid Syria, northeastern Saudi Arabia, and from there to northwestern Iraq.
► *Eirenis coronella fenelli* Arnold, 1982: mountainous areas along the southwestern border of Saudi Arabia.

► *Eirenis coronella ibrahimi* Sivan and Y. L. Werner, 2003: an isolated subspecies in the Sinai.

# Sinai Crowned Dwarf Snake

*Eirenis coronelloides* (Jan, 1862)
**Family:** Colubridae
**Sub-family:** Colubrinae
**F.:** Couleuvre naine couronnée de Turquie
**G.:** Türkische Krönchen-Zwergnatter

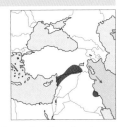

**Identification** Closely related to and difficult to separate from *Eirenis coronella*. The top of the head is covered with a large, more or less trapezoid dark mark (called the "crown" by some authors) that extends from the supraocular and frontal plates to the dark collar. There are 2 dark longitudinal bands along the belly; well marked in juveniles, they tend to fade with age. There are far fewer subcordal plates: fewer than 49 pairs in males, fewer than 41 in females. And *E. coronelloides* is smaller; maximum length never exceeds 26 cm (to 35 cm in *E. coronella*).
**Venom** Harmless.

Eirenis coronelloides, *Turkey. Note the extensive dark mark on the top of the head that joins the collar, visible especially in juveniles.* Göçmen.

**Habitat** Open and relatively dry environments, from plains to hills. Penetrates the Mediterranean zone, unlike *Eirenis coronella*, which prefers more desertlike habitats.
**Habits, Diet and Reproduction** Probably identical to those of *Eirenis coronella*.

**Range** Southwestern Turkey, western Syria, Lebanon, northwestern Jordan, western Iraq.
**Geographic variation** *Eirenis coronelloides* is a monotypic species. Until recently, it was considered synonymous with *E. coronella*.

## Lake Van Dwarf Snake

*Eirenis thospitis* Schmidtler and Lanza, 1990
**Family:** Colubridae
**Sub-family:** Colubrinae
**F.:** Couleuvre naine du lac de Van
**G.:** Thospites-Zwergnatter

**Identification** Small and slender; 40–56 cm in length. Head hardly distinct from body. Resembles *Eirenis modestus* but has 15 rows of dorsal scales at mid-body instead of 17; collar absent or very indistinct; no large mark on top of head. Parietal scales sometimes have fine dark border, as have the upper labials. Back a uniform pale olive-grey to yellowish brown-grey, with series of small dark spots at the front of the body that may form thin dark longitudinal lines; they become fainter towards the rear of the body. In *E. thospitis hakkariensis* the back is uniform yellowish-beige without spots or marks. The dorsal scales have a tawny or orangey

Eirenis thospitis thospitis. *Eastern Turkey.* Schmidtler.

longitudinal band, as in *E. modestus*. The belly is a uniform yellowish-white. The rostral scale is nearly 2 times wider than it is high and is very visible from above. 1 preocular scale, 2 postoculars, 7 upper labials on each side. 169–190 ventral plates, 48–56 pairs of subcordals.

**Venom** Harmless.

**Habitat** Plateaus and mountains between 1500 and 2000 m elevation. Arid, stony hillsides with a low vegetation sward of herbaceous plants.

**Habits** Hardly known; probably very similar to other dwarf snakes of the genus *Eirenis*. Apparently active during the day and at dusk. Hides under stones and in rock crevices.

**Diet** Probably arthropods such as crickets, grasshoppers, beetles, spiders and centipedes; perhaps also very small lizards.

**Reproduction** Unknown but probably oviparous, like other dwarf snakes.

**Range** In eastern Turkey (the Lake Van region) and southeastern Turkey (Hakkari province). The German name *Thospites* is derived from an ancient Latinized name for Lake Van.

**Geographic variation** According to genetic analysis, *Eirenis thospitis* is closely related to another species, *E. hakkariensis* (Schmidtler and Eiselt, 1991. Other authors, referring to distinguishing morphological features, consider that *E. hakkariensis* represents a species in its own right. Until further research has been carried out, here we adopt an intermediate position and consider the two as subspecies of *E. thospitis*.

▶ *Eirenis thospitis thospitis* Schmidtler and Lanza, 1990: on the east of Lake Van. Small dark spots and marks on the head and on the front of the back.

▶ *Eirenis thospitis hakkariensis* Schmidtler and Eiselt, 1991: southeastern Turkey, in Hakkari province, between 1500 and 1900 m. Back normally a uniform colour, without spots or marks.

# Collared Dwarf Snake

*Eirenis collaris* (Ménétriés, 1832)
**Family:** Colubridae
**Sub-family:** Colubrinae
**F.:** Couleuvre naine à collier
**G.:** Halsband-Zwergnatter

**Identification** Small snake, the diameter of a pencil; uniform tawny-brown with a wide dark black collar, normally with blurred edges when adult. 23–35 cm in length, occasionally 40 cm. Small head, hardly separated from the body. Smooth, shiny dorsal scales in 15 rows at mid-body. 1 loreal plate, 1 preocular, 2 postoculars, 7 upper labials on each side, 152–172 ventral plates. Ground colour yellow-brown, sand-yellow or olive-grey. Each dorsal scale appears paler in its centre, unlike species in the *Eirenis modestus* group. A wide black straight collar on the neck does not bend towards the front, but descends onto the sides of the neck but not onto the belly. This collar has precise edges in juveniles that become much less so in adults. It is preceded by a wide pale yellow band that joins the cheeks, which are the same yellow colour. This pale collar fades with age, especially in males. In juveniles and young adults the cheek scales and labial plates have dark rims. There is a well-marked black spot on the front part of each parietal scale in juveniles that disappears almost totally with age, especially in males. The iris is entirely black such that the pupil is hardly visible. The underside is a uniform pearly or creamy-white.
**Venom** Harmless.

*Collared Dwarf Snake* Eirenis collaris. *Azerbaijan. Note the blurred-edged collar and thick neck, typical of this species. Another way to differentiate from* E. modestus, *present in the same area, is to check that the number of rows of dorsal scales at the mid-body is 15. Right: close-up of a juvenile.* P. Geniez.

**Habitat** Steppe environments. Dry slopes with brush and tufts of grass. More often than not, it is hidden under stones. Found in mountainous areas to an elevation of 2000 m.

**Habits** A slow-moving snake with a hidden lifestyle. Rarely flees when found. An inactive winter period of several months can be interrupted in February or March during sunny days.

**Diet** Scorpions, centipedes, insects and spiders. May incidentally catch small lizards, but this has not been proved.

**Reproduction** Oviparous. Lays 4–8 elongated eggs that hatch in August and September. Newly hatched snakes are minuscule: a total length of 12–13 cm and a width of 5 mm.

**Range** Northeastern Turkey, southern Armenia, Transcaucasia (Dagestan, Azerbaijan), northeastern Iraq and northwestern Iran.

**Geographic variation** Certain authors recognize two subspecies:

▶ *Eirenis collaris macrospilotus* (F. Werner, 1903): southeastern Turkey, in the region of Kars.
▶ *Eirenis collaris collaris* (Ménétriés, 1832): the rest of the species' range.

# Eiselt's Dwarf Snake

*Eirenis eiselti* Schmidtler and Schmidtler, 1978
**Family:** Colubridae
**Sub-family:** Colubrinae
**F.:** Couleuvre naine d'Eiselt
**G.:** Eiselts Zwergnatter

**Identification** Small and quite slender (although older individuals can be more thickset); 30–36 cm total length, rarely to 40 cm. Head hardly distinct from body. 15 rows of dorsal scales at mid-body. 1 preocular scale, 2 postoculars, 7 supralabials each side. 140–180 ventral plates. Back is olive-grey, beige, yellow-brown or dark brown, more often than not without markings, occasionally with small black spots sometimes aligned into 2 or 4 longitudinal bands that finish in

black lines at the rear of the body and fade on the tail. In adults, the centre of the dorsal scales is slightly paler than the sides (as in *Eirenis collaris*); this is hardly visible in juveniles. The head often has 3 dark transverse bands: 1 not so contrasting on the snout, 1 between and around the eyes that continues slightly downwards and 1 on the parietal scales that continues onto the sides, where it finishes abruptly above or on the upper part of the supralabial plates. At the back

Eirenis eiselti, *adult. Akçadag, Malatya region (Turkey)*. P. Geniez and A. Teynié.

of the head there is a dark neck band in the form of a half-collar, 4–6 dorsal scales wide, and remarkably short on the sides (the shortest of the genus), descending hardly lower than the corner of the mouth; it is thus not visible from below. The iris is very dark; the pupil sometimes has a silver-grey rim. The belly is pale yellow to brownish-white.

**Venom** Harmless.

**Habitat** On plains and hillsides. Open dry countryside with brush, isolated trees and an abundance of stones; occasionally in open oak woodland. More often than not, found under stones. In mountain areas occurs to about 1900 m elevation.

**Habits** Little known; probably very similar to other dwarf snakes.

*Close-up of* Eirenis eiselti. *Halfeti (southeastern Turkey). Note that the marks on the head stop at the upper part of the supralabial plates, whereas the collar descends lower on the neck; associated with 15 rows of dorsal scales. This patterning is enough to identify this species.* P. Geniez and A. Teynié.

Essentially active in the morning and evening. Slow-moving, with a hidden lifestyle. Inactive for several months in winter.

**Diet** Essentially insects, but perhaps also small lizards.

**Reproduction** Oviparous. Number of eggs unknown but probably the same as in other dwarf snakes (3–8 per clutch).

**Range** Known only from southeastern Turkey, from Adana in the west to Lake Van to the northeast. It is probably present in extreme northern Syria.

**Geographic variation** *Eirenis eiselti* is a monotypic species. Genetically it is the closest relative of *E. collaris*.

# Roth's Dwarf Snake

*Eirenis rothii* Jan, 1863
**Family:** Colubridae
**Sub-family:** Colubrinae
**F.:** Couleuvre naine de Roth
**G.:** Roths Zwergnatter

**Identification** Small snake (total length 25–35 cm); moderately slender. Head small and slender, little distinct from the body, flattened, less high than

the neck or the body, which gives it a characteristic shape when seen from the side (like *Eirenis occidentalis*, with which it can be confused). 15 rows

Eirenis rothii, *adult.*
*Near Altinozu*
*(southern Turkey).* P.
Geniez and A. Teynié.

*Close-up of* Eirenis rothii. *Note the resemblance to* E. occidentalis. *In* E. rothii *the dorsal scales have pale edges and a black spot, and the wide prefrontal plate appears to be inserted between the nasal and preocular, which is divided in its lower part, this last part representing the loreal (unfortunately a character hardly visible in this photo). In* E. occidentalis *there is usually no loreal.* P. Geniez and A. Teynié.

of dorsal scales at mid-body. Loreal scale present more often than not, very small. 1 preocular, 2 postoculars, 133–200 ventral plates. Uniform back; brown-grey, reddish-brown to yellowish-brown. Dorsal scales with pale yellow or pinkish edges and with a small spot on their bottom edge. Top of head and neck with 3 or 4 wide black bands separated one from the other by a yellow transverse line. The second black band, often fused with the first, covers the eyes and the 2nd, 3rd and 4th supralabials. The 3rd band stops abruptly level with the 6th supralabial as in *E. eiselti*. The last very large black band extends as far as under the neck, forming an incomplete collar; it also has a yellow surround. The black band crossing the head is less marked in certain individuals, in which case it is the yellow that dominates. The iris is entirely black, such that the pupil does not stand out. The belly is grey to yellowish-white.

**Venom** Harmless.
**Habitat** Arid slopes with some brush and stones, from plains to hillsides, in Mediterranean or steppe-type environments. Voluntarily hides under stones or in small mammal burrows.
**Habits** Poorly known, probably similar to other dwarf snakes. Supposed active during the day, in the morning and evening, but also in the early nighttime hours. Relatively slow-moving with a hidden lifestyle. Locally abundant; it is possible to find several individuals under the same stone in the spring.
**Diet** Essentially large insects such as crickets, grasshoppers and beetles, and also myriapods; not known to eat lizards.
**Reproduction** Oviparous. Number of eggs unknown.
**Range** Coastal areas of the Mediterranean's western shore: Israel, western Jordan, Lebanon, western Syria and southeastern Turkey.
**Geographic variation** *Eirenis rothii* is monotypic.

# Western Persian Dwarf Snake

*Eirenis occidentalis* Rajabizadeh, Nagy, Adriaens, Avcı, Masroor, Schmidtler, Nazarov, Esmaeili and Christiaens, 2016
**Family:** Colubridae
**Sub-family:** Colubrinae
**F.:** Couleuvre naine de Perse occidentale
**G.:** Westliche Persische Zwergnatter

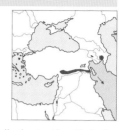

*Eirenis persicus* (J. Anderson, 1872) was the only species representing the subgenus *Pseudocyclophis*. Recently, using morphological characters and mitochondrial DNA markers, Rajabizadeh and others showed that *E. persicus* belongs in a species complex with six different species, occurring from southeastern Turkey to northwestern India through Iran. In the Western Palaearctic, only one of them is present, reaching southeastern Turkey: *E. occidentalis*, the westernmost species of the sub-genus *Pseudocyclophis*.

**Identification** Very small (30–40 cm), very slender with a very small head hardly distinct from the body, flattened like that of *Eirenis rothii*, with which it may be confused. 15 rows of dorsal scales at mid-body. Very small eyes are entirely black (pupils cannot be distinguished). The scale on the snout (rostral) is obviously larger than it is high; the internasals are very wide and in contact with the 2nd supralabial; usually no loreal scale, only 1 preocular, only 1 postocular, 1 temporal plate within the 1st or 2nd row of scales that succeed the postocular, 7 (rarely 8) supralabials, the 3rd and 4th in contact with the lower edge of the eye. 186–213 ventral plates in males, 204–224 in females. Back uniform yellowish-, reddish- or pinkish-brown.

Tail proportionally shorter than that of the true *E. persicus* occurring eastward (13.4–19.5% of the body length against 22–29% for *E. persicus* in males, 13.1–19.0% against 19.0–22.6% for *E. persicus* in females) and with a lower number of pair of subcaudals (46–59 against

*Eirenis occidentalis.* Josef Friedrich Schmidtler.

141

*Close-up of* Eirenis occidentalis. *Halfeti (southeastern Turkey). The dorsal scales are almost uniformly coloured without a black spot (contrary to* E. rothii). P. Geniez and Alexandre Teynié.

64–83 for *E. persicus* in males, 42–58 against 54–77 for *E. persicus* in females). The dorsal scales are almost uniformly coloured (not bicoloured as in *E. rothii*) and without a dark base (contrary to *E. persicus*). The upper head is black, sometimes with a pale snout and a paler transverse band just behind the eyes. Normally has a wide black collar, separated from the black of the head by a pale band that widens on the cheeks. Belly a uniform yellowish-white or pale reddish-grey.

**Venom** Harmless.

**Habitat** Arid and rocky hillsides with sparse bushy vegetation; sometimes also in open semi-arid forest. Plateau steppe at low elevations.

**Habits** Poorly known. Mainly active at dusk, occasionally at night. Quickly hides, readily concealing itself under stones or in rock crevices. A short period of inactivity in winter.

**Diet** Large insects (crickets and grasshoppers) and other arthropods.

**Reproduction** Oviparous.

**Range** Southeastern Turkey, extreme southeastern Armenia (a single specimen found in 1985), up to Kermanshah province in western Iran. Has been reported from southern Azerbaijan, but confirmation is needed.

**Geographic variation** *Eirenis occidentalis* is a monotypic species. It is the sister species of *E. persicus*, the latter occurring farther east. It belongs to a distinct subgenus, *Pseudocyclophis*. Genetic studies have shown that members of this subgenus are correctly placed with the other species of the genus *Eirenis*.

### Genus *Rhynchocalamus*

There are only three species in this genus, with distributions from southeastern Turkey to western Iran, southwards to the Levant and into the Sinai and Yemen. They are minuscule, burrow-living snakes similar to dwarf snakes of the genus *Eirenis*. The genus *Rhynchocalamus* is the sister-group of genus *Lytorhynchus*; these two genera occur in a large clade next to *Eirenis*, *Macroprotodon* and *Spalerosophis* as well as the ex-*Coluber* genera *Hierophis*, *Dolichophis*, *Platyceps* and *Hemorrhois*.

# Palestine Black-headed Dwarf Snake

*Rhynchocalamus malanocephalus* (Jan, 1862)
**Family:** Colubridae
**Sub-family:** Colubrinae
**F.:** Couleuvre naine à tête noire
**G.:** Schwarzkopf-Zergnatter

**Identification** Small (total length about 35 cm, to 49.5 cm), slender snake with a long, narrow head, indistinct from the body; top of head, nape and often the neck shiny black, except for the upper edges of the rostral and bottom of the labials, which are pure white. The black on the nape extends into a collar on the sides as far as the ventral plates. The back is a uniform beige-grey, yellowish-beige or bright reddish-brown. The sides of the neck are sometimes whitish. The belly is a uniform shiny porcelain or pinkish-white.

The dorsal scales are smooth and shiny, arranged in 15 rows at mid-body. The large rostral scale advances over the muzzle, its corner inserted between the internasals; very large nasal plates, lengthened towards the rear and at times touching the preoculars; 1 small more or less triangular loreal scale, 1 preocular, 1 or 2 postoculars, only 6 upper labials, the 3rd and 4th in contact with the eye's lower edge, 7 lower labials. 179–231 ventral plates, divided anal scale, 53–70 pairs of subcordals.

*Left:* Rhynchocalamus melanocephalus. *Western Galilee (Israel). Note the resemblance to certain dwarf snakes of the genus* Eirenis. Shacham.
*Right:* close-up of Rhynchocalamus melanocephalus. *Nizzanim (Israel). Its markings – the white edge to the rostral plate and all the white labials with black upper parts forming a straight line – does not occur in genus* Eirenis. Crochet.

143

This species is sometimes confused with the Persian Dwarf Snake *Eirenis occidentalis*, which has an even smaller head, no loreal plate and a clear zone separating the black on the head and black collar.

**Venom** Harmless.

**Habitat** Hilly or mountainous areas with a Mediterranean influence. Arid slopes with scree and sparse vegetation, often in valleys near watercourses or in wadis with sparse vegetation. Also known from steppe environments – for example, near Azraq in Jordan, where its presence is a relict of the past. In the Sinai occurs to 1000 m.

**Habits** Mainly nocturnal during the hot season. Often hides under stones and in rock crevices. Sometimes seen moving at night.

**Diet** Various arthropods such as grasshoppers, crickets, ants or myriopods; perhaps small lizards and similar.

**Reproduction** Oviparous.

**Range** Hatay province in southern Turkey, western Syria and Jordan, Lebanon, Isreal and southern Sinai (Egypt).

**Geographic variation** Two recognized subspecies: *Rhynchocalamus melanocephalus melanocephalus* and *R. m. satunini*. They are so different in general appearance, colouration, scale arrangement and even ecology that here we treat them as two distinct species. The elevation of *R. satunini* to a separate species was recently validated in 2015 by Avcı and workers on the basis of their large genetic divergence.

## Satunin's Dwarf Snake

*Rhynchocalamus satunini* (Nikol'skij, 1899)
**Family:** Colubridae
**Sub-family:** Colubrinae
**F.:** Couleuvre naine de Satunin
**G.:** Zergnatter

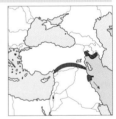

**Identification** Closely related to *Rhynchocalamus melanocephalus*, but even more slender in appearance. 7 upper labials on each side instead of 6, 8 lower labials instead of 7; a distinctive colouration: the tiny head is whitish, with a black mark on the rostral plate, a large black mark that encompasses the eyes and reaches the labials, and another that covers the parietal scales and descends to the top of the temples; there is a wide black collar on the nape and start of back that on the sides advances to just behind the temples. The body is brilliant brick-red, and the bottom of the flanks is pure white.

**Venom** Harmless.

**Habitat** Little known. Steppe with sage or with dwarf spiny bushes, dry rocky slopes with rocky outcrops covered in xerophytic vegetation. To an elevation of 1200 m in Armenia.

Rhynchocalamus satunini. *Khosrow, near Erevan (Armenia).* Orlov.

**Habits** Very little known. Discreet snake usually found under stones.
**Diet** Invertebrates; among other things, larvae and pupae, small woodlice of the genus *Oniscus*.
**Reproduction** Oviparous.

**Range** Southeastern Turkey, southern Armenia, Nakhchivan (Azerbaijan), northwestern Iraq and western Iran.
**Geographic variation** *Rhynchocalamus satunini* is a monotypic species; it is still considered to be a subspecies of *R. melanocephalus* by most authors.

## Baran's Black-headed Dwarf Snake

*Muhtarophis barani* (Olgun, Avcı, Ilgaz, Üzüm and Yılmaz, 2007) (formerly *Rhynchocalamus barani*)
**Family:** Colubridae
**Sub-family:** Colubrinae
**F.:** Couleuvre naine à tête noire de Baran
**G.:** Barans Schwarzkopf-Zergnatter

**Identification** Very slender small snake; in general appearance and colour, reminiscent of dwarf snakes of the genus *Eirenis*. Length 34–40 cm. Short head hardly distinct from body.

Eyes with round pupils. Smooth, shiny dorsal scales arranged in 17 rows at mid-body (compared to 15 in the genus *Rhynchocalamus*). Large rostral scale mainly undivided and above

145

*Muhtarophis barani. The pearl grey head, backed by a wide black collar contrasting with the rest of the body, gives this snake a colour pattern unique within the Western Palaearctic. Ilgaz.*

the muzzle. 1 large preocular and 2 postoculars on each side of the head, 5 upper labials, only the 3rd touching the eye's lower edge (it touches the 3rd and 4th in members of the genus *Rhynchocalamus*); 65–75 pairs of subcordals (compared to 53–70 in *R. melanocephalus*).

Head ash-grey, uniform or with a little black mottling, followed by a wide black collar contrasting with the uniform reddish-brown or brick-red body without spots or a pattern; the lateral borders of the dorsal scales are paler than the centre. The sides of the head are whitish, with a triangular black "tear" mark under the eye. Belly and underside of tail bright white without markings.

**Venom** Harmless.

**Habitat** Arid slopes with maquis-type vegetation at 50 to 600 m elevation; open forest between 350 and 1300 m. The climate where it occurs within its range is characteristically hot and dry in summer with abundant rain in winter.

**Habits** Small, terrestrial, shy snake about which we know little. Like the Palestine Black-headed Dwarf Snake, it is seldom in the open and is normally hidden under stones or in rock crevices. Its biology is barely known. This species was first described in 2007. Of the species covered in this field guide, it is one of the most recently discovered.

**Diet** Probably the same as the Palestine Black-headed Dwarf Snake: arthropods (such as crickets and myriopods) and maybe small lizards.

**Reproduction** At present totally unknown. Probably oviparous.

**Range** Known from a few sites in southern Turkey, in the Anamos range (Hatay province).

**Geographic variation** *Muhtarophis barani* is a monotypic species.

---

### Genus *Macroprotodon*, the false smooth snakes

The genus *Macroprotodon* regroups small opisthoglyphous snakes with a geographic distribution limited to the Iberian Peninsula and North Africa and as far as southern Israel to the east. The genus *Macroprotodon* is, along with *Bamanophis dorri* (ex-*Coluber dorri* of West Africa) very closely related to a set regrouping the genera *Hemerophis, Hierophis, Dolichophis, Platyceps, Spalerosophis* and *Eirenis*. The False Smooth Snake *Macroprotodon cucullatus* represents a species complex difficult to identify using morphology; its systematics are thus still controversial and under discussion. Molecular and morphological studies tend to agree in recognizing three or four groups, probably or certainly corresponding to four different species, according to the obtained genetic separation: *M. cucullatus, M. mauritanicus, M. brevis* and *M. abubakeri*. Although the recognition of *M. brevis* and *M. abubakeri* is well supported by molecular data, the exact status of *M. cucullatus* is poorly understood, as is its relationship to *M. mauritanicus*; they appear to be very close genetically, but this is on the basis of the study of just one specimen of *M. cucullatus*.

**Identification** All false smooth snakes are small (generally 40–50 cm, to 67 cm for the longest individuals); quite slow-moving with smooth, shiny scales that can be confused at first sight with smooth snakes (*Coronella austriaca* and *C. girondica*). Their eyes are small and slightly inclined upwards (they are easily visible from above). The pupil is very slightly vertically elliptical when the snake is in bright sunlight but appears round most of the time. The iris is pale orange, brighter in the upper half. The head is somewhat flattened, oval, hardly distinct from the neck except when in a threatening attitude. There are generally 8 supralabials (rarely 9); the 6th (at the back of the eye) is especially high, very close to or even touching the parietal plates, depending on species; 1 preocular, generally 2 postoculars, 1 loreal. An opisthoglyphous dentition with a series of 3 to 5 maxillary teeth in front of the venomous grooved fangs. Quite variable in colour; discreet, normally with small mottles or a series of small spots more or less longitudinally aligned depending on the species, on a background colour between grey and pale beige, yellowish and reddish. The flanks have no clear markings. A dark oblique line runs from the eye and more often than not joins the corner of the mouth, passing by the bottom of the rear labial plates (in contrary to true smooth snakes, in which this line crosses the top and middle of the rear labials). There is often a wide dark collar on the nape that continues forwards in a point onto the back of the top of the head (the "hood") and that extends onto the sides of the neck as far as the first ventral plates. The top of the head is entirely black in some individuals; others have no collar or black on the head and thus resemble the Southern Smooth Snake *C. girondica*.

**Venom** Slightly venomous. Opisthoglyphous dentition with grooved venom fangs located at the back of the maxillae. Slow-moving snakes, not particularly aggressive, although some individuals inflict a definite bite when handled and try to inoculate their venom if allowed. The symptoms of poisoning in humans are mild: reddening and swelling of the bitten area that disappear after a few hours.

**Habitat** Dry areas at low and mid elevations with scree and bushes, open forest, drystone walls, cultivated areas with ruined buildings, arid and near-desert steppe.

**Habits** Essentially active at dusk and at night. In the day, hides under stones, in holes or in the ground. Moves slowly but will sometimes attack if threatened. Inactive during winter for 3–6 months, depending on region and elevation.

**Diet** Small lizards and geckos and probably small snakes. Bites its prey and then clings to it until the rear-located fangs penetrate the prey. Prey is thus paralysed and dies rapidly.

**Reproduction** Oviparous. Lays 5–7 eggs that hatch in late summer.

## Eastern False Smooth Snake

*Macroprotodon cucullatus* (Geoffroy Saint-Hilaire, 1827)
**Family:** Colubridae
**Sub-family:** Colubrinae
**F.:** Couleuvre à capuchon orientale
**G.:** Östliche Kaputzennatter

**Identification** Little-known taxon. All examined individuals had a beige-grey to reddish-brown back with ill-defined dorsal markings, separated by fine pale speckling. All had jet black or dark brown tops to their heads and, more often than not, a horseshoe-shaped pale band between the eyes; the base

Macroprotodon cucullatus. *Nizzanim (Israel)*. Boaz Shacham.

of the rostral scale and the labials are extremely pale except for a large oblique black mark under the eye. The belly is a uniform pale colour. 19 rows of dorsal scales at mid-body. Other characteristics given in the literature mainly concern another taxon *M. brevis textilis* (see page 155), which has a much more western distribution.

**Range** Poorly known: southern Israel, Mediterranean shores of Egypt and Libya; to the west, probably as far as southeastern Tunisia.

Macroprotodon cucullatus. *Al Haniyah (Cyrenaica, Libya)*. Peyre.

149

# Algerian False Smooth Snake

*Macroprotodon mauritanicus* Guichenot, 1850
**Family:** Colubridae
**Sub-family:** Colubrinae
**F.:** Couleuvre à capuchon d'Algérie
**G.:** Algerische Kaputzennatter

**Identification** It has a series of 3 maxillary teeth in front of the fangs (compared to 4 or 5 in other members of the genus *Macroprotodon*). 19 row of dorsal scales at mid-body. In Algeria, the 6th upper labial generally touches the parietal. On the other hand, in nearly all individuals examined from the Balearic Islands, these scales did not touch. It is larger than the other species: to 67.2 cm in the Balearics, probably as long in North Africa. The neck collar is generally separated into 3 parts; the central part has a more or less lozenge form with, at its front, a

point reaching the back of the parietal plates. A black line behind the eye is either complete or broken and reduced to a well-defined black mark that does not reach the corner of the mouth or the more or less V-shaped mark on the head. The top of the head is rarely black (never in the Balearic Islands). The dorsal pattern is more or less aligned longitudinally, and is made up of a series of elongated brown, blackish or grey marks along the vertebrae, with another series of elongated marks along each side but smaller. The general colour is most often

Macroprotodon mauritanicus, *subadult. Majorca, Balearic Islands. Note the aligned reddish markings on the back, quite typical of this species*. P. Geniez.

*Right: close-up of* Macroprotodon mauritanicus. *Between El Kseur and Azazga (Algeria). The three-part reddish collar with an elongated middle mark is characteristic of the species. Note also that the sixth supralabial plate, behind the eye, is very high and touches the parietal (a character that is considered typical of* M. mauritanicus *but that is very rare in specimens from the Balearic Islands).* Crochet.

yellowish or yellow-reddish brown, but sometimes completely grey. The rostral plate and supralabials are often yellow (more whitish, greyish or pinkish in other *Macroprotodon* taxons), but there are some *M. mauritanicus* with white labials. The belly is a uniform yellowish, sometimes with small geometric black spots or large black checkered marks.

**Range** The eastern half of northern Algeria and northern half of Tunisia (not reaching the near-desert areas). Also present on Majorca and Minorca in the Balearic Islands (Spain), where it is considered probably to have been introduced by the Romans during 2nd century BC, and on Lampedusa Island (Italy).

## Aboubakeur's False Smooth Snake

*Macroprotodon abubakeri* Wade, 2001
**Family:** Colubridae
**Sub-family:** Colubrinae
**F.:** Couleuvre à capuchon d'Aboubakeur
**G.:** Aboubakeurs Kaputzennatter

**Identification** A series of 4 maxillary teeth in front of the fangs (contrary to 3 in *Macroprotodon mauritanicus*, the species it most resembles). 19 rows of dorsal scales at mid-body. 6th supralabial in contact or not with the parietal, depending on the individual. The neck collar is often complete, sometimes divided into 3 parts. The black line behind the eye is normally complete, reaching the corner of the mouth and generally touching the more or less V-shaped mark on the top of the head. Top of head rarely black. The back pattern is usually linear along the length of

Macroprotodon abubakeri. *Bousfer-Plage (northwestern Algeria). Note the complete collar (not divided in three), the black band on the cheek that joins the bottom of the collar and the colour of the almost unlined back.* Peyre.

the snake, as in *M. mauritanicus*, but less marked, some individuals only having ill-defined small markings. General colour more often than not grey, sometimes reddish, but without a yellow tint. The rostral plate and supralabials are white, greyish or pinkish. Large, black quadrilateral marks on the belly give the underside a checkered appearance.

**Range** Extreme northeastern Morocco and northwestern Algeria, where its occurrence sometimes overlaps with that of *Macroprotodon brevis textilis*, in areas with a Mediterranean climate rather than in arid steppes.

## Western False Smooth Snake

*Macroprotodon brevis* (Günther, 1862)
**Family:** Colubridae
**Sub-family:** Colubrinae
**F.:** Couleuvre à capuchon occidentale
**G.:** Westliche Kaputzennatter

**Identification** A series of 4 or 5 maxillary teeth in front of the fangs. 19–23 rows of dorsal scales at mid-body. The 6th supralabial does not normally touch the parietal or, if so, only occasionally. An extremely variable species with three very distinct subspecies that we prefer to treat separately.

***Macroprotodon brevis brevis*** (Günther, 1862) 21–23 rows of dorsal scales at

Macroprotodon brevis brevis. *Ras Oumlil (southwestern Morocco). Resembles* M. cucullatus *to the west, but it has, in principle, more than 21 rows of dorsal scales.* García-Cardenete.

mid-body. The 6th supralabial usually touches, slightly, the parietal plate, which is sometimes divided. The neck collar is generally complete, but is sometimes absent or reduced to a lozenge mark on the neck. The top of the head is black in some individuals. A line behind the eye is often complete, joining the corner of the mouth but rarely touching the more or less V-shaped mark on the top of the head. Diffuse dorsal pattern, more often than not consisting of indistinct marks separated by pale scales. In some individuals, the back is quite uniform, with small elongated black spots arranged along the area of the vertebrae, but they are more spaced than in *M. mauritanicus* and *M. abubakeri*, which, furthermore, have 19 rows of dorsal scales at mid-body and occur farther west. Very variable general colour, from greyish to pinkish or reddish, but without a clear yellow tint. The rostral plate and supralabials are white, greyish or pinkish. The belly is generally marked with large black quadrilaterals, giving a checkered appearance.

**Range** Only in the western half of Morocco, except for the Rif and the Tingitana Peninsula (near Tangier), where it is replaced by *Macroprotodon brevis ibericus* in a large range of habitats that are neither too desert nor mountainous. The false smooth snakes of the Atlantic Sahara (southwestern Morocco), which have 19 rows of dorsal scales and a uniformly pale belly, appear genetically to be close to *M. b. brevis* despite their colour and a number of rows of dorsal scales similar to that of *M. b. textilis* (see page 155).

**Macroprotodon brevis ibericus** Busack and McCoy, 1990 Morphologically a less variable taxon than other *Macroprotodon* species. 21 rows of dorsal scales at mid-body. Normally the 6th supralabial slightly touches the parietal plate and is sometimes split. Neck collar always present and undivided, coming to a point at the rear edge of the parietal plates. Black-headed individuals are very rare (observed in the region of Cadiz, southern Spain, and in the Rif, Morocco). Black line behind the eye often complete, reaching the corner of the mouth but only rarely touching the more or less V-shaped mark on top of the head. Back pattern more or less diffuse, normally made up of small spots separated by pale scales, sometimes aligned along the vertebrae (they are more spaced in *M. mauritanicus* and *M. abubakeri*). General colour varies little: often grey or brown with blackish or lead-grey markings. White, greyish or pinkish rostral plate and supralabials. The belly is generally marked with large black quadrilaterals, giving a checkered appearance.

**Range** Iberian Peninsula (southern half of Spain, southeastern Portugal) and northern Morocco (Tingitane Peninsula and the Rif range, to the east as far as Melilla, to the southeast probably as far as the Tazekka massif); in more or less warm and dry habitats in the Iberian Peninsula, rather cool and more mountainous habitats in Morocco.

Macroprotodon brevis ibericus. *Near* Uleila del Campo *(Spain). Note the complete collar, the general grey-brown colour (without yellow tint) and small back spots, aligned but well separated one from the other.* Nicolas.

*Close-up of* Macroprotodon brevis ibericus. *Between Ketama and Bab-Berret (the Rif, northern Morocco). Note that on the one hand this individual shows a scale abnormality (it happens quite often in reptiles), with three supralabials in front of the eye; on the other hand, it has a complete collar, such that the band on the cheek does not reach the collar and the fifth supralabial (which should be the sixth), higher than the others, does not quite reach the edge of the parietal.* P. Geniez.

***Macroprotodon brevis textilis*** (Duméril, Bibron and Duméril, 1854) The *textilis* taxon is generally considered to be a particular morphotype of *Macroprotodon cucullatus*, despite its obvious phylogenetic affinity with *M. brevis*.

19 rows of dorsal scales at mid-body. 6th supralabial plate nearly always separated from the parietal. More often than not, no collar; if present, it is large and black, prefiguring individuals with black heads. Some individuals have black tops to their heads. Those without a black head have a neat inverted V-shaped mark towards the back of the head that starts at the frontal plate, crosses the parietals and continues to the very back of the head; a very elongated neck mark, pointed in front, fits into the V. A black band behind the eye is often complete and touches the corner of the mouth but rarely the V-shaped mark on the top of the head. General colour is pale beige, pale brown or pinkish-grey (a little like *Coronella girondica*). Dorsal pattern identical to that of *M. cucullatus*, sometimes consisting of well-defined markings separated by pale scales, which gives it a latticed appearance (like a piece of "textile"; hence its Latin name); these marks are never aligned along the vertebral column (contrary to *M. mauritanicus* and *M. abubakeri*). Rostral plate and supralabials are white, greyish or pinkish. Belly white, yellowish or pinkish, nearly always void of black checkered markings.

**Range** Arid steppe in the High Plateaux area in eastern Morocco and northwestern Algeria. Also an isolated population in the Hoggar range (southern Algeria).

Macroprotodon brevis textilis. *In the local context (south of Matarka, eastern Morocco), note the forked or V-shaped mark pointing backwards on the head as well as the non-aligned pattern, which is, on the contrary, composed of blurred marks alternating with whitish ones. This false smooth snake has 19 rows of dorsal scales and a pale belly with dark markings.* García-Cardenete.
*Right: close-up of* Macroprotodon brevis textilis, *without a collar. 21 km southeast of El Ateuf (northeastern Morocco). Note that the sixth supralabial is not in contact with the parietal, which is nearly always the case in this subspecies.* P. Geniez.

### Genus *Lytorhynchus*
Small genus of just five species. Small, discreet snakes adapted to a life in sand, characterised by their very developed rostral plate, flattened into a shovel used for digging in the sand.

The genus *Lytorhynchus* is closely related to a clade regrouping the species of the former genus *Coluber* of the Old World, and the genera *Macroprotodon* and *Eirenis*.

## Awl-headed Snake

*Lytorhynchus diadema* (Duméril, Bibron and Duméril, 1854)
**Family:** Colubridae
**Sub-family:** Colubrinae
**F.:** Couleuvre fouisseuse à diadème
**G.:** Gekrönte Schnauzennatter

**Identification** Small snake (35–45 cm total length); elegant and colourful, with a slender body and relatively short tail for a member of the

Colubridae. Elongated, pointed head, bevelled when seen in profile, indistinct from body. Eyes quite large, with round or slightly vertically elliptical pupils, as in the false smooth snakes (genus *Macroprotodon*). The pupil is small compared to the size of the iris (orange-tan), which gives the species a distinctive "expression". A spatula-shaped snout appears square from above; the mouth is on the underside. The snout plate (rostral), much enlarged, covers the whole mussel, slotting in a point between the internasals, and serves as a shovel to dig in sand, its exclusive habitat. Frontal plate short and wide, 1 large loreal scale, 2 preoculars generally above a small subocular, 2 postoculars, 7 (rarely 8) supralabials on each side, the 4th and 5th in contact with the eye's lower edge. Smooth, shiny dorsal scales, arranged in 19 rows at mid-body, 152–195 ventral plates, divided anal plate, subcordals divided and

arranged in pairs. Back is pale brown, yellowish-brown, pinkish or even nearly white, with a series of very contrasting rectangular or lozenge-shaped large dark spots, often with pale edges. Flanks with smaller marks alternate with the dorsal marks. Top of the head vividly coloured with a dark transverse band at eye level, attached by a wide elongated mark to another in the form of a diadem on the nape. There is a large dark temporal band from the eye to the corner of the mouth and another just under the eye. Pure white belly, sometimes slightly cream-coloured.

**Venom** Harmless, as it is quite lethargic, it does not try to escape or bite if handled.

**Habitat** One of the snakes most dependent on desert sand. Very arid

*Awl-headed snake* Lytorhynchus diadema. *Atlantic Sahara, southwestern Morocco.*
Cluchier.

Lytorhynchus diadema diadema. *Rum Wadi, Jordan. Note the very well developed rostral plate that goes above the top of the snout and the small size of the pupil compared to that of the iris.* M. Geniez.

Saharan steppe with sand banks, even if small. Also present at the base of large aeolian dunes, locally cohabiting with vipers of the genus *Cerastes*. Rare in habitats without sand. Can locally occur to an elevation of 1500 m (e.g., in the Sinai).

**Habits** Active at dusk and at night. Shy, usually staying hidden. Sometimes found in early morning basking in the sun; more often seen at night – for example, when crossing a road – or found hidden under stones. Digs into sand with its shovel-shaped snout to bury itself. When it senses danger, it coils itself up and hides its head under its body.

**Diet** Mainly lizards and similar (especially nocturnal geckos); sometimes also large arthropods (crickets, grasshoppers or centipedes).

**Reproduction** Oviparous. Reduced breeding rate; 3–5 large elongated eggs in a clutch.

**Range** Desert areas from Mauritania and southern Morocco as far as western Iran, including Algeria, Tunisia, Libya, Egypt, Israel, Jordan, Syria, Iraq and the whole of the Arabian Peninsula; in the south as far as northwestern Senegal, Niger and probably Sudan.

**Geographic variation** The systematics of *Lytorhynchus diadema* is controversial. *L. diadema gaddi* Nikol'skij 1907, *L. d. arabicus* Haas, 1952 and *L. d. mesopotamicus* Haas, 1952 from the eastern part of the species' range are considered by several authors to be synonymous with *L. d. diadema*. *L. kennedyi* Schmidt, 1939 from the eastern Mediterranean deserts (Jordan, Syria and probably also Iraq and Saudi Arabia) is considered by most authors to be a rare morph of *L. diadema*, which has a back with widely spaced black transverse bars on an orange background. Some authors think that it is, in fact, a separate species.

**Typical snakes once placed in the genus *Elaphe***

As with the genus *Coluber*, the genus *Elaphe* has been completely dismantled following much phylogenetic work, all of which agrees in recognizing that the former classification was a mere grouping of medium-sized and large colubrid-type snakes with vague morphological similarities: often shiny scales, a calm character and, for some of the species, more or less pronounced tree-climbing habits. The old genus *Elaphe* included species from most of Eurasia and North America.

Genetic studies have shown that several genera separate the divergent lines of what was once *Elaphe*. Two genera occur in the Western Palaearctic:

▶ genus *Elaphe*, which still exists but includes only Asian species, three of which reach Europe, *Elaphe quatuorlineata*, *E. sauromates* and *E. dione*.

▶ genus *Zamenis*, which includes a few European and Asiatic species, including the Aesculapian Snake of western Europe, as well as the Ladder Snake, *Zamenis scalaris*, which had recently been placed in a genus dedicated to this species under the name *Rhinechis scalaris*.

# Western Four-lined Snake

*Elaphe quatuorlineata* (Bonnaterre, 1790)
**Family:** Colubridae
**Sub-family:** Colubrinae
**F.:** Couleuvre à quatre raies
**G.:** Westliche Vierstreifennatter

**Identification** Very large, quite thickset snake; average length 130–160 cm, occasionally exceeding 200 cm, with a diameter of 7 cm. Large head, very distinct from the body. Medium-sized eye; round pupils with a thin silver-grey rim in juveniles, golden in adults; black iris that leaves it indistinct from the pupil at first glance. Dorsal scales only slightly but markedly keeled, generally arranged in 25 rows (rarely 23 or 27) at mid-body, 1 high preocular scale above 1 (sometimes 2) small subocular, 8 supralabials (occasionally 9) on each side. 187–224 ventral plates in the male, 205–234 in the female, divided anal scale, 56–90 pairs of subcordals.

The adult has a yellowish- or reddish-brown back with 4 well-marked dark brown or black

*Western Four-lined Snake* Elaphe quatuorlineata, *adult. Croatia.* Cluchier.

longitudinal lines that progressively
fade towards the tail. Thick black
line from the eye to the corner of
the mouth. Labial plates are whitish,
yellowish or pale brown, without
markings. Juvenile's pattern is
very distinctive and quite different
from that of the adult: on a pale
background (white in very young
individuals) there are large, clearly
defined, black transversely oval
markings on the back; also black
markings on the flanks more or less
alternate with the dorsal ovals; a
black half-horseshoe-shaped band
goes from eye to eye on the front of
the top of the head; parietal scales
are black, labials white hemmed
with black and brown. It is only at
an age of 2 or 3 years that an adult
pattern is acquired; sub-adults have
a colouration intermediate between
that of juveniles and adults. The belly
is yellowish, often with dark speckles
or rows of dark markings.

**Venom** Harmless. Quite a calm snake,
disinclined to bite and often content to
hiss when it feels threatened.

**Habitat** Hillsides with patches of scree
and plenty of bushes, forest edges
and clearings, drystone walls, sides of
watercourses or ponds in marshy areas,
often near derelict buildings. Often
found in shade; prefers warm, wet
environments.

**Habits** Generally active through the
day and at dusk. Climbs and swims
well. Flees rather slowly, trusting in
its camouflage. Depending on region,
has a 3- to 6-month inactive period in
winter.

**Diet** Small mammals (rodents),
birds and their fledglings, lizards,
amphibians. May visit poultry sheds to

catch chicks. Kills prey by suffocating them in its strong coils.

**Reproduction** Oviparous. Mating between March and May. Lays clutch of 6–16 eggs in the ground, under stones or in a cavity. Incubation 50–60 days. Hatching occurs in September; the young then feed on young lizards and young rodents in the nest.

**Range** Central and southern Italy (not Sicily), the Balkan states near the coast from Slovenia to Albania, southern Macedonia, Greece and several islands in the Aegean Sea. Farther east it is replaced by the very similar *Elaphe sauromates*.

**Geographic variation** *Elaphe quatuorlineata* varies little. However, a few restricted, insular subspecies are recognised by several authors:

▶ *Elaphe quatuorlineata muenteri* (Bedriaga, 1882): the Cyclades Islands (Greece). Characterised by its small size (90–130 cm) and dorsal lines, which it acquires when 60 cm long, being particularly narrow.

▶ *Elaphe quatuorlineata scyrensis* Cattaneo, 1990: the Skyros Archipelago (Greece). Closely related to *E. q. muenteri*.

▶ *Elaphe quatuorlineata quatuorlineata* (Bonnaterre, 1790): the rest of the species' range.

The insular population on Paros (Greece) sometimes considered a subspecies *Elaphe quatuorlineata parensis* Cattaneo, 1990, is probably best considered as belonging to *E. q. quatuorlineata*. The population on Amorgos *E. q. rechingeri* F. Werner, 1932, characterised by many adults with a uniform colouration, without bands, belongs, however, to *E. q. muenteri*, based on molecular studies.

*Young individuals of the Western Four-lined Snake have a contrasting pattern composed of large dark markings on a clear background; this disappears slowly as the animal grows. It is only after two or three years that it obtains an adult pattern. Peloponnese (Greece). P. Geniez.*

# Eastern Four-lined Snake, Blotched Snake

*Elaphe sauromates* (Pallas, 1814)
**Family:** Colubridae
**Sub-family:** Colubrinae
**F.:** Couleuvre tachetée or Couleuvre à quatre raies orientale
**G.:** Östliche vierstreifennatter

**Identification** Morphologically and in its habits very similar to the Western Four-lined Snake *Elaphe quatuorlineata*. It is, however, smaller, usually 120–140 cm, rarely reaching 200 cm in length. It is distinctive especially in retaining the juvenile pattern when adult, instead of the 4 dark lines of the western species. It has a yellowish ground colour, often more pronounced than in *E. quatuorlineata*, sometimes even orange-brown. Sometimes difficult to recognise due to its often ill-defined, well-spread dark patterning. The back pattern consists of dark transverse oval or rhomboid markings, sometimes slightly intermingled; the flanks usually have large dark markings sometimes connected by

*The Blotched Snake* Elaphe sauromates, *when adult, retains a dorsal pattern of round or rhomboid spots reminiscent of the juvenile's pattern, but less contrasting. At all ages it has more yellowish tones in its colouring than* E. quatuorlineata. *Digor (northeastern Turkey).* P. Geniez and A. Teynié.

Elaphe sauromates. *Festovo (eastern shore of the Caspian Sea, Kazakstan). In this aged individual, the dark markings have almost disappeared. The slightly keeled dorsal scales, combined with a dark iris, are sufficient to identify this species in the local context.* Le Nevé.

small black lines. Older individuals have a diffuse pattern in which the marks are somewhat hidden by black longitudinal lines with yellow edges on each scale. A large black mark on the head is sometimes in the form of a V opened at the rear. However, in adults, the top of the head is almost completely black, hiding this pattern. A wide, dark temporal band from the eye to the rear angle of the mouth sometimes continues forwards to the nostril, separating the dark colour of the top of the head from the uniform yellow labial plates. The belly is yellow, pale brown or pearl-coloured, occasionally with dark faded markings. Black iris, leaving the pupil indistinct. The juveniles have a remarkably contrasting pattern, with large black-brown marks on the back and flanks on a yellow background. Large head, slightly distinct from the neck (a little smaller than in the Western four-lined Snake). The dorsal scales are less obviously keeled than in the other species, smooth and lightly keeled

on the flanks. More often than not there are 25 (rarely 23, 26 or 27) rows of dorsal scales at mid-body. 1 large preocular, occasionally divided, often 1 small subocular just underneath, 2 or 3 postoculars, 8 or 9 (rarely 6) upper labials. The number of ventral plates (197–228) and subcordals (61–72) is essentially the same number as in the Western Four-lined Snake; it has a divided anal plate.

**Venom** Harmless. A calm, slow-moving snake that, when threatened, hisses and feints strikes.

**Habitat** As varied as for the Western Four-lined Snake, but with a net preference for steppe habitats. It occurs from the edge of wetlands and riverbanks to deciduous forest, wooded and arid steppe and even desert areas. The habitats should, however be scattered, with stony outcrops or have drystone walls and stunted bushes. Prefers lower elevations but can in certain regions occur higher, to 2600 m in the Caucasus.

**Habits** Active during the day and early evening. Climbs and swims as well as the Western Four-lined Snake. Inactive for a few months in winter from early October until as late as mid-April.

**Diet** Small mammals such as mice, dormice and rats; birds and their eggs; occasionally lizards.

**Reproduction** Oviparous. Mating occurs after snakes emerge from the dormant period, in March or April, and after the first moult of the year. In June or July, the female lays up to 17 eggs; hatching occurs in September or early October, after about 60 days of incubation.

**Range** Northeastern Greece and the Greek Dodecanese islands, southwestern Bulgaria, Romania, Moldavia, southernern Ukraine and Russia, Transcaucasia, Armenia, Turkey, Syria, Lebanon around Mount Hermon, and farther east as far as western Iran, Kazakhstan and Turkmenistan.

**Geographic variation** *Elaphe sauromates* is a monotypic species.

## Dione Snake

*Elaphe dione* (Pallas, 1773)
**Family:** Colubridae
**Sub-family:** Colubrinae
**F.:** Couleuvre de Dioné
**G.:** Steppennatter, Dione-Natter

**Identification** To 100 cm in length (150 cm in central Asia). Slender snake. Relatively elongated, but with quite a thick head, hardly distinct from the body, with a wide, rounded snout. Eyes with round pupils, iris often pale grey, sometimes orange (but never red), leaving the pupil clearly visible. Smooth dorsal scales arranged in 23–25 (sometimes 26) rows at mid-body. 1 high preocular plate above a single small subocular, 8 supralabials (sometimes 7 or 9). 183–205 ventral plates.

Back pale grey almost white, beige, yellowish, pale brown or reddish-brown, with 3 clear and 4 dark bands that may be well-pronounced or, on the contrary, almost absent. These bands are covered with small, only slightly contrasting dark dorsal marks sometimes arranged in transverse bars. Some adults are completely without these marks (even on the head) and have just 4 longitudinal brown bands on a pale background. There are also small dark markings on the flanks. There is a U- or horseshoe-shaped mark on top of the head that is often paler in its centre. There is a brown or reddish temporal band with black edges on the side of the head that passes through the eye. 2 large parallel marks on the nape may be orange or reddish, touching or not

*Dione Snake* Elaphe dione. Kreiner.

*Dione Snake* Elaphe dione. *Near Srai, Mazandaran province (Iran). Note the five pale lines along the length of the back, interspersed with more or less apparent small transverse bars.* Heidari.

the two dark lateral, dorsal bands. Belly a little paler than the back, more often that not with dark speckling.

**Venom** Harmless.

**Habitat** A steppe species that is also present in open forest. Occurs on hillsides and mountainous areas,

scree with bushy or low vegetation, undisturbed edges of cultivated fields. In the western part of its range, hardly occurs above 500 m, but may occur to 2000 m in central Asia.

**Habits** Diurnal snake, discreet with hidden lifestyle. Readily seeks refuge under stones or in underground galleries of small mammals, in which it hunts them. Ground-living but can climb into high trees. When feeling threatened, the Dione Snake vibrates its tail with lateral movements, like a whip. Inactive during winter for several months.

**Diet** Small mammals, lizards and similar, snakes, frogs, fish, birds and their eggs. Suffocates its prey by constriction.

**Reproduction** The Dione Snake exhibits a first stage of ovoviviparity: when the eggs are laid, the young snakes are almost totally formed within the membranous-shelled egg, which they leave a few days later. Clutch of 3–24 eggs.

**Range** Vast range that is primarily Asiatic: extreme southern Ukraine, southern Russia, northern Caucasus region, Azerbaijan, extreme northwestern Turkey, and from there across central Asia from northern Iran and China as far as Mongolia, Korea and the Pacific coast of eastern Russia.

**Geographic variation** *Elaphe dione* is generally considered to be a monotypic species.

## Aesculapian Snake

*Zamenis longissimus* (Laurenti, 1768) (formerly *Elaphe longissima*)
**Family:** Colubridae
**Sub-family:** Colubrinae
**F.:** Couleuvre d'Esculape
**G.:** Äskulapnatter

**Identification** Large, slender snake that moves calmly and elegantly. Average length 130–150 cm, but it is said there have been records of 200 cm. Narrow head, only slightly distinct from the body with a wide, rounded snout, appearing slightly square when seen from above. Eyes quite large, pupils round, very obvious against the yellowish-grey to orange-brown (never red) iris. Dorsal scales are smooth and shiny, arranged in 23 (rarely 21) rows at mid-body, 1 large, high preocular plate, 2 postoculars, the upper larger than the lower, 8 (sometimes 9) supralabials on each side. Many ventral plates (211–250) with a slight keel on each side (the snake uses these to climb), a divided anal plate, subcordals divided and arranged in pairs.

Back is bright yellow-brown, olive, grey-brown or sometimes blackish-grey. A good number of the back and flank scales have a fine pure-white border, forming a pattern of interrupted small longitudinal lines, characteristic of the Aesculapian

*Aesculapian Snake* Zamenis longissimus, *adult female, Sierra de Montseny (northeastern Spain). Note the small white speckles, especially on the back (except on the spinal band), typical of this species.* P. Geniez.

Snake. There are sometimes 4 dark indistinct longitudinal lines on the back and flanks. In adults, the head is often yellower than the rest of the body, which has no markings, except for a wide temporal band, sometimes ill-defined, from the eye to the back of the temple. This dark band is crossed at the back by a yellow half-collar that becomes less and less distinct with age. The labials and belly are pale yellow, sometimes greenish-yellow. Juveniles are very differently coloured, with a more contrasting pattern: the head has a dark brown top with a dark ill-defined transverse band between the eyes, and a large black band from the eye to just

behind the temple that is unconnected with the black marking at the back of the mandibles. The labial plates are pale yellow, with a small black spot below the eye; a yellow half-collar forms 2 triangles that continue onto the nape without joining; this half-collar is at the origin of the confusion with the Grass Snake, *Natrix natrix* (see page 208), which is not, however, otherwise very similar. There is a large double black and mahogany marking on the nape in a U or V shape, open at the back; the back has 4 series of quite large brown markings that are close to each other and separated by numerous white or pale yellow streaks.

*Large male Aesculapian Snake. Near Nice (southeastern France). This yellow colouration is common in old males. Note also the four slightly darker bands that, in the past, gave rise to the wrong identification as a Western Four-lined Snake, which has been the case in France. Deso.*

**Venom** Harmless. Generally non-aggressive but can bite very strongly if it is captured in a brutal manner.

**Habitat** The most woodland-dwelling of European snakes. Occurs at low and mid elevations, abundant in certain forests – open chestnut woodland, for example. Readily frequents riverbanks and open riverside forest, forest clearings, wooded slopes with scree and dense scrub, ivy-covered drystone walls, ruins, unused quarries, the edges of farming areas (especially along hedgerows) and marshy areas. It needs a certain degree of humidity and shows a preference for a temperate climate without excessive variations in temperature. In mountainous areas, may locally occur to 1500 m, rarely more (1607 m in Switzerland).

**Habits** Mainly active on the ground or in low scrub, but can climb well in trees or on ruins using its keeled ventral scales; can even climb tree trunks with rough bark. Active during the day and at dusk, this snake often basks in the sun in the morning but retires to covered areas in very hot weather. Can be active during warm rain or at night, when it may be seen on roads and is unfortunately too often killed. Moves calmly and elegantly. Inactive during the winter for 4–6 months, depending on region and elevation.

**Diet** Small mammals, especially rodents; also birds, especially eggs and young from nests. Young Aesculapian Snakes prey on small lizards and young rodents in the nest. Prey is strongly enveloped and suffocated in the snake's coils before being ingested whole.

**Reproduction** Oviparous. Mating takes place during May, when there is a display that includes neck biting by the male. Often, during the mating period males perform ritual fighting; closely entwined, one tries to dominate the other. These fights always finish without injuries. During July, the female lays 2–18 soft-shelled, elongated eggs in damp soil or in rotting vegetation, under stones, in old tree stumps or in rock crevices. Hatching occurs in September.

**Range** From extreme northern Spain in the west to the Ukraine in the east, incorporating southern and central France, Switzerland, Austria, northern and central Italy, Germany (relict populations), the Czech Republic, Slovakia, southern Poland, the Balkans as far as Greece, Hungary, Romania and Bulgaria. Farther east and south, it occurs in northern Turkey (Black Sea coast and Mount Ararat region), in southwestern Russia (Krasnodar), western Georgia and western Iran. Past records from Sardinia are now considered doubtful. There are relict populations in Germany, among the most northern areas of the species' range, near Slangenbad in the Taunus massif, in southern parts of the Odenwald massif, on hill slopes overlooking the Danube near Passau, and along the lower reaches of the Inn and Salzach rivers.

*Juvenile Aesculapian Snake. Sainte-Croix-Vallée-Française (Lozère, France). Rufray.*

**Geographic variation** *Zamenis longissimus* is presently considered to be monotypic. In the extreme western part of its range (northeastern Turkey, for example) melanistic specimens are more common than elsewhere, blackish-grey with white stippling typical of the species but without any yellow tint in the general colour (the labial and dorsal plates are whitish-grey).

# Italian Aesculapian Snake

*Zamenis lineatus* (Camerano, 1891) (formerly *Elaphe longissima romana*)
**Family:** Colubridae
**Sub-family:** Colubrinae
**F.:** Couleuvre romaine
**G.:** Süditalienische Äskulapnatter

**Identification** Until recently considered a southern Italian subspecies of the Aesculapian Snake, with the name *Elaphe longissima romana*. General appearance and size (130–150 cm, rarely 170 cm) similar to that of *Zamenis longissimus*. Separated by having a red iris (like the Leopard Snake *Z. situla*) and by the 4 dark bands on the back and flanks that are generally better defined but thinner (narrower than the pale line that separates them). The small white speckles typical of the Aesculapian Snake are there, clear on the dark bands and nearly absent from the pale parts of the back and flanks. Some older individuals are nearly uniformly coloured, in which case they are paler than Aesculapian Snake, with the white markings more or less persisting. Belly yellowish-white with grey mottling as from mid-body; underside of tail uniform grey.

Juveniles have a much more contrasting colouration than adults with dark reddish-brown marks on the back and flanks. These markings are smaller than those of juvenile *Z. longissimus* and are already aligned in 4 longitudinal series, separate from the ground colour, which is a paler grey than in the other species; the white speckles are poorly defined (very visible and numerous in juvenile *Z. longissimus*). Finally, the head pattern is very distinct: a very narrow black bar between the eyes contrasts with the uniform pale grey or yellowish ground colour, and there is a well-marked black longitudinal line along the joint between the 2 parietal plates (in *Z. longissimus* this line is blurred and indistinct); the black marking on the temple is narrower and generally connected to the black marking at the back of the mandibles (usually separated in *Z. longissimus*). As the snake grows, the back markings become progressively more attached and eventually form longitudinal bars that become less distinct with age. Back scales are smooth and glossy, arranged in 23 rows at mid-body, 225–238 ventral plates.

**Venom** Harmless. Non-aggressive; can bite if caught and not handled with care.

**Habitat** Usually found on plains and at low elevation, but as high as 1600 m in places. Readily occupies riverbanks and other places near water, but often in habitats drier than those occupied by the Aesculapian Snake. Prefers clearings and edges of open woodland. Habitats should include stone walls, rock outcrops and bushes to hide in. Also found in ruins and overgrown gardens.

**Habits** Active during the day and at dusk. Very wary especially on the ground but can climb, like the Aesculapian Snake, into bushes or low trees. Basks in the sun mainly in the morning; more active from late afternoon until nightfall. Inactive in winter for several months starting in October.

**Diet** Mice, rats, young rabbits, birds and their eggs, lizards and geckos.

**Reproduction** Oviparous. During mating, in April and May, the male grasps the female's neck with its mouth. Sometimes males engage in ritual fighting, entwined together. Females lay up to 8 eggs; after an incubation period of about 2 months, they hatch in the very early autumn.

**Range** South of Italy, to the south of a line between Naples and Monte Gargano, and Sicily. Replaced farther

*Italian Aesculapian Snake* Zamenis lineatus, *adult. To distinguish from the Aesculapian Snake, note the red iris, narrow temporal band, four quite well-defined dorsal bands, and white speckling visible only on the dark bands.* Kreiner.

north by the Aesculapian Snake. Presumed hybrids between the two species occur in their contact zone.

**Geographic variation** *Zamenis lineatus* is considered to be a monotypic species.

# Persian Ratsnake

*Zamenis persicus* (F. Werner, 1913) (formerly *Elaphe persica*)
**Family:** Colubridae
**Sub-family:** Colubrinae
**F.:** Couleuvre de Perse
**G.:** Persische Kletternatter

**Identification** Closely related to the Aesculapian Snake, with the same elegance and general appearance but a more easterly distribution. Also more variable in colour, even within the same population, with three morphs: brown, grey and black (melanistic). The brown morph is quite similar to the Italian Aesculapian Snake but lacks the dark longitudinal bands

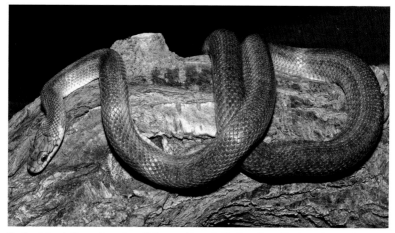

*Persian Ratsnake* Zamenis persicus. *Note resemblance to the Aesculapian Snake, and especially the Italian Aesculapian Snake (red iris, narrow temporal band). Z. persicus has ill-defined more or less transverse diffuse markings on the back, and the scales at the back of the body are slightly keeled.* Orlov.

on the back and flanks. General colouring pale brown to pinkish-grey or sometimes brick-red, with indistinct transverse markings alternating with similar markings on the flanks and small white speckles that are hardly visible (less than in *Zamenis longissimus*) between the markings.

Large individuals are almost uniform pale brown. Juveniles resemble those of the Italian Aesculapian Snake *Z. lineatus*: pale, quite uniform head with very narrow black transverse bar at the front of the frontal plate that does not touch the eyes; a narrow, dark, bridle-like mark extends from in

*Persian Ratsnake* Zamensis persicus. *Near Lenkoren (southeastern Azerbaijan). As seen in these photos, the Persian Ratsnake is much more variable than the Aesculapian Snake.* (Above left and right) Sculz, (below) Orlov.

front of the eye to as far as behind the temples, separating the brown of the head from the white labial plates and the throat: there is an indistinct, very narrow black bar under the eye; no pale half-collar but a pale mark on each side at the very front of the neck; red eye as in *Z. lineatus* and *Z. situla*; the dorsal makings are quite well pronounced in very young individuals but quickly fade with age. The grey morph is identical to the brown form except that the overall colour tone is greyish and blackish. The melanistic morph is blackish in juveniles, with the patterning more or less apparent, whereas the adults are entirely black, including the underparts and the eye. Certain black individuals have a pure white or pinkish-white head, base of flanks and belly, without pigment, which is called the piebald morph. There are also intermediates between all of these morphs.

The maximum size is less than in the Aesculapian Snake (mean length 70–90 cm, maximum 120 cm). The preocular plate is large and high, there are generally 2 postoculars, 8 supralabials (rarely 7), on each side. The dorsal scales are smooth and quite shiny, very slightly keeled towards the tail and generally arranged in 23 (occasionally 21 or 25) rows at mid-body, 207–233 ventral plates, divided anal plate.

**Venom** Harmless.

**Habitat** Low- and middle-elevation wooded, mountain habitats, relatively humid deciduous forests, densely bushed slopes, clearings and meadows; also around villages, in gardens. Locally present on the Caspian Sea coast, at low elevation. In Iran occurs to 1600 m in Hamadan province and recorded to an elevation of 4000 m on Mount Elbourz, records that are very much doubted by some authors.

**Habits** Quite similar to that of the Aesculapian Snake. Climbs well; as in the other species, in doing so it uses the angular external edges of its ventral plates to ascend branches or walls.

**Diet** Mainly small rodents. It is possible that it takes birds on the nest when the possibility presents itself.

**Reproduction** Oviparous. In late May or June, lays 2–8 eggs that hatch 50 days later.

**Range** Endemic to the Elbourz range, high mountains on the southern side of the Caspian Sea, in northern Iran, to the east as far as Gorgan, in Golestan province; to the west it reaches the Western Palaearctic, southeastern Azerbaijan, in the Talysh mountains (Lenkoran region).

**Geographic variation** *Zamenis persicus* is considered to be monotypic.

# Leopard Snake

*Zamenis situla* (Linnaeus, 1758) (formerly *Elaphe situla*)
**Family:** Colubridae
**Sub-family:** Colubrinae
**F.:** Couleuvre léopard (Couleuvre léopardine)
**G.:** Leopardnatter

**Identification** One of the most beautiful European snakes, with a variegated colouration and bright pattern of red or orange markings with black or dark grey borders. There is a series of large orange or red oval marks on the back that may be divided into 2 rows on a pale background of anything from ivory white to pearl grey, yellowish or beige. Some individuals are beautifully patterned; their dorsal marks are joined into 2 continuous or partially fragmented orange bands. There is a row of smaller black markings along the flanks, sometimes with orange centres, that alternates with the back spots. In certain large individuals, the back and flank markings are subdued by 4 more or less distinct dark longitudinal bands, with the initial markings showing through. The head colour is similar to that of *Zamenis lineatus*: pale above, same colour as the back, with a small black mark at the top of the rostral plate, a black transverse mask, curved forwards between the eyes, and an oblique black band on each side from the edge of the frontal plate along the temple as far as the corner of the mouth. There is a large red and black fork-shaped mark on the nape,

*Leopard Snake* Zamenis situla.
*Croatia.* Alexandre Cluchier.

175

*Leopard Snake with markings subdued by obscure longitudinal bands. Near Vrisses, Crete (Greece).*
*P. Geniez.*

orientated backwards, that protrudes in a point onto the head as far as the frontal scale. A vertical black bar on the 2nd supralabial; another under the eye, directly below a black mark on the 3rd supralabial. The belly is yellowish-white towards the front, becoming progressively marbled and eventually densely spotted with dark markings at the rear. Juveniles are similarly coloured to adults but with more gleaming, contrasting colours.

Quite a large, slender snake (normal length about 70–90 cm, occasionally as long as 116 cm). The thin head is hardly distinct from the body, except when in a defensive attitude. Medium-large eyes with a red iris (as

in *Z. lineatus*). Smooth dorsal scales, usually arranged in 27 (rarely 25 or 29) rows at mid-body. 8 (more rarely 7 or 9) supralabials on each side, 1 large preocular that is almost twice as high as large. 215–260 ventral plates and a divided anal plate.

**Venom** A calm, quiet snake that tends not to try to escape; it may bite if irritated, but the bite is not venomous. If it feels threatened, it sometimes vibrates its tail.

**Habitat** Plains and hillsides, especially in Mediterranean garigue. Rarely above 500–700 m, but locally as high as 1600 m in the southern Balkans. Occurs in a variety of sunny habitats but with scrub, grasses and saplings, rocks and drystone walls that offer a multitude of possibilities to shelter or find refuge. Also field edges, scree, riverbanks, around water tanks or springs, road embankments, areas with ruins or gardens.

**Habits** Diurnal, ground-dwelling snake that can climb with ease in rocks and brush. Has a well-defined daily rhythm: basks in the sun in the morning, rests in shade around midday, followed by the most active period, from late afternoon until night. During hot periods, this species can be found on roads in the early nighttime hours. Not very quick; when in danger it stays immobile for a while before fleeing. Inactive for several months in winter, disappearing quite early, sometimes in late September or early October. Much prized for terrariums due to its bright colours and relative rarity, it is locally menaced with extinction despite being a totally protected species.

**Diet** Especially small mammals, occasionally birds or lizards; juveniles eat mainly lizards.

**Reproduction** Oviparous. Mating occurs after considerable displaying that may last for several hours. Small clutches of 2–8 eggs are laid in moist soil under rocks or in burrows. Hatching occurs in August.

**Range** In separated areas: Sicily, southeastern Italy (Apulia), Malta, Adriatic coast from Croatia (north to southern Istria and the Krk and Cres islands) as far as Albania, Greece, Crete, several Ionian, Aegean and Dodecanese islands, Rhodes, southern Bulgaria, western and southwestern Turkey. A very isolated population in the southern Crimean Peninsula. Found once in Cyprus.

**Geographic variation** *Zamenis situla* is considered to be a monotypic species despite its vast fragmented distribution.

*Leopard Snake, adult. Crete, near Varvara*. P. Geniez.

# Caucasian Snake

*Zamenis hohenackeri* (Strauch, 1873) (formerly *Elaphe hohenackeri*)
**Family:** Colubridae
**Sub-family:** Colubrinae
**F.:** Couleuvre de Hohenacker
**G.:** Transkaukasische Kletternatter

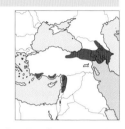

**Identification** Quite small (generally 60–70 cm, to 85 cm), slender snake, the smallest member of the genus *Zamenis*. Head little distinct from body. Medium-large eyes with round pupils that are obvious on the pale orange (not red) iris. Smooth and quite shiny dorsal scales on the front part of the body, slightly keeled on the hind part, arranged in 23 (rarely 25) rows at mid-body. 8 supralabial plates on each side (rarely 7 or 9), 1 large preocular, 2 postoculars. 195–226 ventral plates. Very variable colouration; from afar, the general aspect is reminiscent of that of the Smooth Snake *Coronella austriaca*. The back is between grey and beige with 2 rows of small reddish,

Zamenis hohenackeri hohenackeri. *Kajaran region (southern Armenia).* Cheylan.

Zamenis hohenackeri. *This individual is probably of the special form found in the Levant countries.* Schweiger.

brown or blackish marks, merged into transverse bars on the neck. A thin paler vertebral line often occurs between the 2 lines of dorsal marks that continue to the tail tip (which often occurs in *C. austriaca*, with which it often cohabits locally). Flanks have dark bars and spots, smaller than the dorsal markings, that are sometimes merged into dark longitudinal bands. There is a dark, quite thin temporal band from the eye to the corner of the mouth and a small black mark below the eye. On the nape, there is an orange-red backwards-facing forked mark that is often studded black. The underside of the chin is often spotted with black and is sometimes completely black. The belly ranges from grey-brown to yellow-brown, sometimes blackish, marbled with dark red or orange (colouration similar to that of the Smooth Snake). The subspecies *Z. hohenackeri tauricus*, from southern Turkey, has a completely different dorsal pattern: the dorsal markings are much larger, clearly defined, orange or reddish-brown studded with black on a very pale silver-grey background, recalling the colours of the Leopard Snake *Z. situla*. The chin and undersides of the head are black. Individuals from Mount Lebanon have a colour intermediate between that of these two subspecies.

**Venom** Harmless.

**Habitat** Dry mountain hillsides with scrub and isolated trees. Open forest, narrow river valleys, gorges. Also mid- to high-elevation steppe plateaus. Can locally occur to almost 3000 m.

**Habits** Active during the day, but stays hidden. Keeps to dense scrub or under rocks. Probably active at night when it

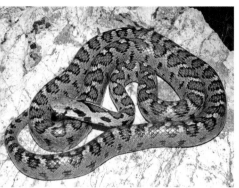

Zamenis hohenackeri tauricus. *Mugla province (southern Turkey). Note the bright colours, recalling those of the Leopard Snake. It is easily distinguished by having a black chin; the head is slightly more angular.* Mebert.

is warm or humid. Very shy, not at all aggressive.

**Diet** Small mammals, especially rodents. The young also feed on young lizards.

**Reproduction** Oviparous. Lays 3–8 eggs that normally hatch in August.

**Range** Central and eastern Turkey, the Caucasus area, Transcaucasia as far as extreme northern Iraq and northwestern Iran. Also present, in an isolated population, on Mount Lebanon (northern Lebanon) and Mount Hermon (eastern Lebanon, western Syria and northern Israel).

**Geographic variation** *Zamenis hohenackeri* is a quite variable species, with several populations with distinct colourations:

▶ *Zamenis hohenackeri hohenackeri* (Strauch, 1873): northeastern Turkey,

Transcaucasia and from there as far as northern Iraq and northwestern Iran.

▶ *Zamenis hohenackeri tauricus* (F. Werner, 1898): mountainous areas of southern Turkey. Morphologically and genetically quite different from *Z. h. hohenackeri*. The populations on Mounts Lebanon and Hermon have a colouration that is more or less intermediate between that of *Z. h. hohenackeri* and *Z. h. tauricus*; even though their taxonomic status is not yet established, we now know that they form a genetically distinct assemblage that is, however, closer to *Z. h. tauricus* than to *Z. h. hohenackeri*. However, several authors consider that there are no subspecies within the Caucasian Snake, despite the obvious differences within the three forms highlighted by genetic studies.

# Ladder Snake

*Zamenis scalaris* (Schinz, 1822) (formerly *Elaphe scalaris*,
*Rhinechis scalaris*)
**Family:** Colubridae
**Sub-family:** Colubrinae
**F.:** Couleuvre à échelons
**G.:** Treppennatter

**Identification** Quite a thickset snake
when adult, with 2 noticeable dark
longitudinal lines along the sides
of the back on a pale, yellowish or
reddish-brown background. The
juveniles have a very contrasting
black on straw-yellow or very pale
beige pattern, the black in the form of
successive Hs joined together to form
a ladder. The ladder bars progressively
disappear as the snake ages, eventually
leaving just 2 dark lines. A series of
loose irregularly spaced small, vertical,
black marks on the flanks tend to
disappear with age, sometimes leaving

just an indistinct longitudinal band
on either side, making 4 longitudinal
bands in all (hence, probably, its local
name of Four-lined Snake in southern
France). There are a few black marks
on the top of the head, at the edges of
the head scales, that tend to disappear
in the adult. On the side of the head,
there are a few incomplete black bars
on the labials, a black mark below
the eye and a wide black oblique
temporal band that ends at the
corner of the mouth. These also fade
in the adult and disappear in certain
aged individuals, especially males.

*Left: Ladder Snake* Zamenis scalaris, *adult male. Fos-sur-Mer (southern France). Grégory Deso.
Right: Ladder Snake, female. Montfragüe (Spain). Évrard.*

*Ladder Snake, juvenile. Causse d'Aumelas, near Montpellier (southern France).* P. Geniez.

There exists, in effect, a slight sexual dimorphism; males have yellowish colouring on the sides of the head and neck, whereas females partially retain their juvenile colouring (but not transverse bars on the back). Juveniles have large, dark more or less checkered markings on the belly that disappear when adult, the belly then becoming a uniform pale yellow.

Length often around 100 cm, rarely 150–165 cm. The head is little distinct from the body, with a quite pointed snout. Eyes have black irises; the pupil is hardly distinguishable. Smooth and shiny back scales, generally arranged in 27 (sometimes 25 or 29) rows at mid-body. The rostral plate is remarkably pointed and predominant, protruding farther than the lower mandible, especially in adults, and is inserted in a point between the internasals; 1 large preocular, 2 postoculars, 7–9 supralabials on each side, the 4th and 5th in contact with the lower edge of the eye. 198–228 ventral plates, a divided anal plate.

**Venom** Quite a heavy, lethargic snake. It can become aggressive if handled without caution, giving strong but non-venomous bites.

**Habitat** Exclusively Mediterranean: garigue, the edge of maquis, dry slopes with scree and scrub, open forest, vineyards and wild gardens, field edges, drystone walls, areas with ruins, road embankments, undisturbed areas between crops. It is a species of plains or low elevation, but has been found in France at an elevation of 1056 m in the eastern Pyrenees and 2246 m in the Baetic Sierras in southern Spain.

**Habits** Quite lethargic terrestrial snake, active in the day and at dusk and, in hot periods, at night. In summer and autumn, the Ladder Snake may be seen moving in the rain. Quite a good climber. Has a 4- or 5-month inactive period in winter.

**Diet** Especially small mammals, including small rabbits that it routs out in their burrows and nests; also birds and their nestlings. The young feed on small lizards, crickets and grasshoppers. Kills larger prey by strongly constricting them in its coils until dead.

**Reproduction** Oviparous. Long mating sessions (sometimes more than an hour) occur in April and May; lays a clutch of 4–24 eggs in July or August; eggs are often stuck together and hatch 7–11 weeks later.

**Range** Iberian Peninsula and southern France, only in Mediterranean areas. Also present in the Baleares: Menorca, where it was introduced several centuries ago, as well as Mallorca, Ibiza and Formentera, to which it was introduced more recently. Only 1 record from Italy, from Ventimiglia, very close to the French border.

**Geographic variation** *Zamenis scalaris* is considered to be monotypic.

**Genus *Coronella***

A genus including only three species, *Coronella austriaca* and *C. girondica*, well known snakes in the Western Palaearctic, and a third species, *C. brachyura* of western India, whose place in the genus *Coronella* has not yet been proved by genetic analysis.

The smooth snakes are small, discreet, not very colourful species that are often hidden. The genus *Coronella* is the sister-group of *Zamenis*, in a larger clade, along with many other species, within the *Elaphe* group.

# Smooth Snake

*Coronella austriaca* Laurenti, 1768
**Family:** Colubridae
**Sub-family:** Colubrinae
**F.:** Coronelle lisse or Couleuvre lisse
**G.:** Schlingnatter oder Glattnatter

**Identification** Small snake (40–70 cm, rarely to 80 cm total length), quite slow-moving. Relatively small, oval head, wider at the rear, which may appear triangular when the snake is in a defensive attitude; not very

*Left: Smooth snake* Coronella austriaca austriaca. *The black band crossing the eye, which continues to the snout as well as the eye, touching the third and four upper labials, allows it to be differentiated from the Southern Smooth Snake* C. girondica. *Reinhard.*
*Above: Newly hatched Smooth Snake* Coronella austriaca austriaca *swallowing its first meal, an adult Wall Lizard* Podarcis muralis. *Le Frau, near Fraisse-sur-Agout (southern France).* P. Geniez.

distinct from the body. Small eyes with a round pupil that contrasts with the iris, which is orange coloured at the top. Smooth dorsal scales arranged in 19 rows at mid-body. The rostral plate is quite prominent, higher than wide, and inserts in a point between the internasals: it is thus visible seen from above. 7 (rarely 8) supralabial scales, the 3rd and 4th in contact with the lower edge of the eye. 150–200 ventral plates, the anal plate divided, the subcordals divided and occurring in pairs. The back is grey, brown, yellow-brown or rusty brown, sometimes brick-red, with dark paired or staggered markings that may resemble the shape of a bow tie, and 2 indistinct longitudinal lines on the rest of the back. There is often a vague, thin, pale longitudinal band along the spine. There is a more or less triangular large dark (black in juveniles) mark on the top of the head, square-cut at the rear or in the form of an open V at the back of the nape. A dark (often black) band on either side of the head extends from the nostril to the neck, passing through the eye. The flanks are often marked with small dark points or spots that are sometimes hardly visible. The belly is uniformly dark, from black to brick-red passing through grey, brown, orange or yellowish, often with small dark and pale speckling that can, in some cases, give a marbled effect to the underparts. Juveniles much resemble adults, but with more contrasting colours (the markings are black, not brown or grey). At birth they measure 12–21 cm.

**Venom** Harmless. A quiet, unafraid snake that may bite if captured. The bite is, however, insignificant considering the size of the snake.

**Habitat** A mid-European snake that avoids Mediterranean areas. It occurs in open, sunny sites affording cover. Heathland, vegetated coastal dunes, riverside forest, woodland clearings, bushy forest edges, scrub-covered hillsides, railway embankments, scree, ditches with brush, vineyards, ruins,

overgrown gardens, any spaces with piles of stones and abundant ground vegetation. Often occurs near human habitations. More at low and mid elevations in the northern part of its range. In the south, occurs much more in mountains, over 2000 m in the Alps and Pyrenees. To 2790 m in the Sierra Nevada (southern Spain), where it is considered a relic of a boreal past.

**Habits** Essentially diurnal but discreet and often hidden under vegetation, in a burrow, in rocks or under a pile of stones. Generally active in mild and cloudy conditions or after rain. It may also hunt at dusk during hot or wet weather. Relatively slow; flees only at the last moment as it obviously has confidence in its camouflage. Usually on the ground but may climb into low brush. Very loyal to its home range. Hibernates for several months.

**Diet** Mainly lizards, including slowworms, and small snakes;

*Left*: Coronella austriaca acutirostris, *the western Iberian subspecies of the Smooth Snake. Note the very protruding rostral plate that gives the snout a pointed appearance, and the two dark bands that join the occipital spot to the first dorsal mark. Serra da Estrela (Portugal). P. Geniez.
*Right*: A particularly red juvenile Smooth Snake Coronella austriaca cf. austriaca. Near Hopa (northern Turkey). P. Geniez and A. Teynié.

occasionally young rodents and young birds in the nest. Seeks out prey where they're hiding (e.g., lizards in rock crevices). Kills large prey by constriction, swallows small prey whole and alive.

**Reproduction** Ovoviviparous. Mating occurs after winter rest, in April or May. Young born after 4–5 months' gestation. The 2–15 (rarely as many as 19) young are born in August or September within a transparent membrane from which they free themselves by wriggling. In cooler climates the gestation period may be prolonged by a year, the young being born the following year. In the south there may be a second mating period, in the summer, the females giving birth at the end of the following winter.

**Range** From southern England (only isolated populations) and Scandinavia (north as far as 62°N), through central and eastern Europe (including Russia to the western part of the Urals) as far as mountainous parts of Turkey and as far west as the higher mountains of Iberia (as far south as the Sierra Nevada), Italy, Sicily and the Balkans, including the Peloponnese. Its vast distribution extends southeast as far as the Caucasus, to Azerbaijan, Kazakhstan and northern Iran.

**Geographic variation** *Coronella austriaca* is very variable geographically. It is probable that studies on all the populations, including those farthest east, will show that significant morphological and genetic differences exist. At the present two subspecies are recognised:

▶ *Coronella austriaca austriaca* (Laurenti, 1768): most of its range.
▶ *Coronella austriaca acutirostris* Malkmus, 1995: northwestern Iberian Peninsula.

Genetic studies have shown a strong structuring across the Iberian Peninsula (three lineages from west to east, of which *C. a. acutirostris* is the most western); the eastern Iberian lineage reaches France as far as the French Pyrenean foothills, including the Corbières range. These three Iberian lineages are very similar in colouration. For example, the wide dark mark on the top of the head is often connected to the first dorsal mark, with the rostral plaque more prominent and more pointed; in France, excepting the Pyrenees, the nape mark is always separated from the first dorsal marks.
▶ *Coronella austriaca fitzingeri* Bonaparte, 1840: southern Italy and Sicily; is not now recognised by most authors, but this name could apply to an Italian lineage, very different from *C. a. austriaca*.

# Southern Smooth Snake

*Coronella girondica* (Daudin, 1803)
**Family:** Colubridae
**Sub-family:** Colubrinae
**F.:** Coronelle girondine
**G.:** Girondische Schlingnatter or Girondische
Glattnatter

**Identification** Small, slim snake with an average length of 45–65 cm, rarely to 80 cm (but on Oléron, on the Atlantic coast of France, at the northern limit of the species range, the species is larger, to 95 cm). Quite similar to Smooth Snake *Coronella austriaca* but often a little slimmer, with a rounder snout and the head more distinct from the body. The dark markings on the top of the head are different; they do not form a large dark mark on the nape. The dark band from the eye to the neck does not pass the eye towards the snout: instead it continues onto the forehead from eye to eye. More often than not, there is a small vertical dark mark (a "tear") under the eye (absent in *C. austriaca*). The back has dark marks positioned transversely, especially on the neck, which could recall an ill-defined ladder pattern (based on which there is confusion with Ladder Snake *Zamenis scalaris*, which is otherwise quite different). The general colour is usually a beautiful

*Left: Southern Smooth Snake* Coronella girondica, *adult. The Bec de l'Estéron (southeastern France).*
Deso.
*Right: Close-up of Southern Smooth Snake from the Ifrane region (Morocco). On this photo, we can clearly see the black bar on the side of the head that reaches the eye but does not go beyond, the dark bar from eye to eye, the "tear" mark under the eye and its position in contact with the fourth and fifth supralabial plates – all criteria that lead to a sure separation from the closely related Smooth Snake.* P. Geniez.

187

*Southern Smooth Snake, Ile d'Oléron, at the northern limit of its range (western France). Note the particular colouration, as well as the scales between the eye and the supralabials, apparently distinctive characters of the Oleron population, which probably represents an until now undescribed subspecies. Évrard.*

mouse grey, often with pink tints. Certain individuals are more yellowish, others more brown. Bicoloured ventral surface; quadrangular black spots in a checkered pattern or aligned on the sides of the belly contrast strongly with the ground colour, which varies from pale yellow to pink or orange (in *C. austriaca*, the ventral surface appears uniformly dark).

Smooth dorsal scales, nearly always arranged in 21 rows at mid-body (19 in Smooth Snake); 8 supralabial plates on each side, the 4th and 5th touching (rarely not) the lower edge of the eye (3rd and 4th in Smooth Snake). Rostral plate wider than it is high, not inserted between the internasals and thus only slightly visible from above. 167–198 ventral plates, a divided anal plate, divided subcordals arranged in pairs.

**Venom** Harmless. Calm, unobtrusive snake that rarely bites when captured.

**Habitat** Especially in lowland areas, but occurs in mountains in south of range to 1600 m in the Pyrenees, 2470 in the Spanish sierras and 2900 m in the Moroccan High Atlases. Prefers dry areas with scree and scrub that also have ground vegetation. Forest clearings, verges of tracks with bushes, rocky hillsides, railway embankments, disused quarries, scrub between cultivated areas, wild gardens, edges of wetlands. One of the snakes most tolerant of man: in southern France and the northern Iberian Peninsula, this snake is not rare in most towns, whereas most other species of snake do not occur in extensive urban areas. Certain individuals occupy the roofs of houses, finding shelter, warmth and food (e.g., geckos) amongst the tiles.

More often than not, the two species of smooth snakes occupy distinct areas, the southern species occurring in the south and on exposed south-facing hillsides and in dryer, more Mediterranean habitats. It is, however, possible to find the two species together in the same area when the habitat is intermediate between the preferences of the two species. In North Africa, a similar situation arises with Southern Smooth Snake and false smooth snakes *Macroprotodon* spp.; there it is the *Macroprotodon* species that occur in the most arid and hot areas, *Coronella girondica* occupying the more mountainous, cooler and more humid areas.

**Habits** Active mainly at dusk and at night. After rain, or in the evening; when it is raining, it may come out during the day.

**Diet** Especially lizards, even large ones, and geckos; also other snakes and sometimes arthropods; more rarely, young rodents and young birds in the nest.

**Reproduction** Oviparous (does not produce live young like Smooth Snake). Mates from April to early June; lays 1–10 eggs, rarely as many as 16, in July; hatching occurs in late August or September, after 6–9 weeks incubation.

**Range** Iberian Peninsula, southern and southwestern France, central and northern Italy; North Africa, where it occurs mainly in the mountains of Morocco, Algeria and Tunisia. Northern limit reached on the Atlantic coast of France, as far as the Ile d'Oléron (like the Ocellated Lizard *Timon lepidus*, with which it shares more or less the same range in France and the Iberian Peninsula).

**Geographic variation** The Southern Smooth Snake is considered to be monotypic, even though the species can be quite variable, especially considering the results of molecular studies. The populations in the Baetic Sierras (southern Spain) are generally the most distinct; among other traits, they are the only populations in which a large percentage of individuals have 19 rows of dorsal plates instead of 21: they probably merit subspecies status.

**Genus *Lampropeltis*, kingsnakes**

Kingsnakes occur naturally in North America, Central America and northern South America. Of the 21 existing species, one, *Lampropeltis californiae*, has been introduced to the Canary Islands (Gran Canaria). The genus *Lampropeltis* is closely related to the American genera *Pantherophis* and *Pituophis*, once of the genus *Elaphe* and part of a large clade that includes Old World snakes of the ex-*Elaphe* genus and the present-day genus *Elaphe*.

# California Kingsnake

*Lampropeltis californiae* (Blainville, 1835)
**Family:** Colubridae
**Sub-family:** Colubrinae
**F.:** Serpent-roi de Californie
**G.:** Kalifornische Kettennatter

**Identification** Of North American origin; cylindrical, head hardly distinct from neck, generally with dark ring markings on a white background. Total length: 75–122 cm. The rostral plate is quite high and triangular, slightly wedged within the internasals, 1 large and high preocular, 2 postoculars, 7 supralabials on each side, the 3rd and 4th in contact with the eye. Eyes not very large; round pupils with a narrow white border, not very distinct from the black iris. Smooth, shiny dorsal scales arranged in 23 (rarely 25) rows at mid-body. Undivided anal plate. The back is pure or creamy white, with 20–32 wide, beautiful chocolate-maroon to nearly black rings. Some individuals may have, in place of the rings, a pale vertebral line on a maroon background. Others, of captive origin, are albino, with a suggestion of the pattern in pale yellow on a pinkish-white background.

**Venom** Harmless.

**Habitat** In its native range, it occupies a wide variety of habitats, from open conifer forests to almost desert areas and all types of biotope between these with plenty of vegetation, including agricultural or damp coastal areas. In Gran Canaria, where it is introduced, it occurs mainly in wet and hot sites: for example, orchard terraces separated by drystone walls and bottoms of wooded ravines.

**Habits** Mainly diurnal, it is also active at dusk and at night in summer. Mainly terrestrial, it may climb into bushes.

**Diet** Has a reputation for feeding on other snakes. In its native range, it

*California Kingsnake* Lampropeltis californiae. *Huelva province (southwestern Spain). A specimen found "in the wild."* González de la Vega.

has even been known to devour the dangerous rattlesnake, as large as itself; it is, in fact, immune to its bites. It also eats lizards, rodents, birds and their eggs. On Gran Canaria, where there are no indigenous snakes, it will take anything it can find, especially the endemic lizards, the Gran Canaria Giant Lizard *Gallotia stehlini*, the Gran Canaria Skink *Chalcides sexlineatus* and Boettger's Wall Gecko *Tarentola boettgeri*, species with an unsure future. In areas occupied by the California Kingsnake, medium or large Gran Canaria Giant Lizards have already disappeared. Eradication campaigns have been organised. In addition to these three endemic reptiles, it also catches the House Mouse.

**Reproduction** Oviparous; lays 3–21 eggs in Gran Canaria. On the island, mating occurs from March to May; eggs are laid at the start of summer and hatch at the end of summer after 9–10 weeks of incubation.

**Range** It originates from the eastern United States and northeastern Mexico. Very much prized in the pet trade for its beautiful colouration, its calm character and the fact that it is easily reared in captivity, it has been unintentionally released in the Canaries, on Gran Canaria, since 1998 at least. It now occurs over an area of 50 square km, from an elevation of 23–620 m. During the eradication campaigns conducted over the last few years, several hundred kingsnakes have been captured. Some 30% of females found in the wild were carrying eggs. At the same time as this introduction in the Canaries, three specimens of *Lampropeltis californiae* were found in the wild in southern Spain, one in 2003 at Huelva, one in 2005 at El Portil, 10 km to the southwest of Huelva, and another in 2011, again at Huelva (personal communication with Juan Pablo González de la Vega).

**Geographic variation** *Lampropeltis californiae* is a monotypic species. Until recently it was considered to be a subspecies of *L. getula* (Linnaeus, 1766).

### Genus *Telescopus*, cat snakes

Cat snakes are small to medium-sized colubrid snakes (rarely more than 100 cm total length), with remarkable, quite globular eyes with vertical pupils, as in vipers. Active mainly at dusk and at night, they are quite slow-moving; they have an opisthoglyphous dentition (grooved venomous fangs at the rear of the upper maxillaries). They have a wide head that is well distinct from the neck. Their scales are smooth, shiny and iridescent. Most species have beautiful, contrasting colours.

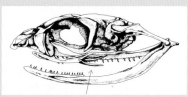

*A cat snake skull, genus Telescopus, showing the venomous teeth located at the rear of the mouth on a slightly mobile backwards-pointing bone.*

The European Cat Snake is the only European colubrid snake with vertical pupils. Within the area covered here, cat snakes are represented by two groups: the *Telescopus fallax* group, among other features characterised by the loreal plate being in contact with the eye (three species: *T. fallax*, *T. nigriceps* and *T. hoogstraali*); and the *T. dhara* group, characterised by the loreal and eye being separated by a preocular plate (three species: *T. dhara*, *T. obtusus* and *T. tripolitanus*). The genus *Telescopus* is part of quite a large group within Colubrinae that includes the genera *Boiga*, *Crotaphopeltis*, *Dipsadoboia*, *Toxicodryas* and *Dasypeltis*.

# European Cat Snake

*Telescopus fallax* (Fleischmann, 1831)
**Family:** Colubridae
**Sub-family:** Colubrinae
**F.:** Couleuvre-chat d'Europe
**G.:** Europäische Katzennatter

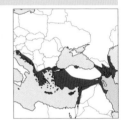

**Identification** A gentle snake, smooth and shiny, of short to medium length (generally 50–60 cm, sometimes to 106 cm). The body is slightly rectangular in section, higher than it is wide. The ovoid, flat head, is obviously distinct from the body. Eyes quite large, pupils with a vertical slit, resembling cat's eyes, the iris generally mustard yellow or orange, contrasting with the general grey colour of the head. Smooth dorsal scales arranged in 19 (sometimes 21) rows at mid-body. A remarkable feature: the elongated loreal scale touches the front edge of the eye (in most colubrid snakes, the loreal scale is separated from the eye by 1 or more preoculars). 1 preocular above the loreal; 2 postoculars; 8 (rarely 7 or 9) supralabials on each side; 200–230 ventral plates (a few less on Cyprus);

the anal plate normally divided; divided subcordals arranged in pairs.

Back usually true grey or mouse grey, sometimes also yellow or brownish-grey, generally paler than the flank colour and ornamented with a series of bright, quadrilateral dark marks located transversely. There are vertical, well spaced, irregular markings on the flanks that alternate with the back markings. The top of the head is almost a uniform pale grey, sometimes with small black speckles. On the side of the head, there is a vague, oblique dark band behind the eye. A wide dark mark on the nape resembles a wide collar; it continues in a point to just behind the parietal scales.

The belly is yellowish-white, pale pink or cream with dark marking,

*European Cat Snake* Telescopus fallax fallax, *a very large adult (about 100 cm long). Southern Turkey.* Teynié.

generally in a checkered pattern and darkening the underside more and more towards the rear and under the tail.

**Venom** A slightly venomous opisthoglyphous snake; the fangs, implanted at the back of the maxillaries, have a furrow that is used to transport the venom. Usually quite a slow-moving species; however, certain individuals can bite when captured when they try to swallow a finger in order to inoculate the venom. Even though the venom appears to be relatively harmless to man, it is better not to be bitten.

**Habitat** Dry sunny slopes, from lowlands to mountains, with bushes and scattered rocks and rocky outcrops, often garigue-type Mediterranean habitats. Also in river valleys and open oak woods. In the eastern part of its range, also present on steppe plateaus without trees but with plenty of stones and grassy vegetation. In mountainous areas, to 2000 m in Armenia.

Telescopus fallax fallax. *Skyros Island (Greece). Note the vertical pupil and the very elongated loreal plate in contact with the eye, a feature characteristic of the* Telescopus fallax *group.* P. Geniez.

Telescopus fallax syriacus *individual ready to moult. Lebanon. Note one of the characteristics of this southeastern subspecies: there are few and widely separated black bars.* Cluchier.

**Habits** A terrestrial snake that can, however, climb up almost vertical rocks or buildings. Mainly active at dusk and at night, but may lie out in the day's first sunshine. Hides under stones or narrow rock crevices or cracks in walls during the day. Calm, indisposed to fleeing when disturbed. Active from mid-March to mid-October. There is a 3- to 6-month inactive period during winter, depending on region and elevation.

**Diet** Especially skinks, lizards and geckos, but also small snakes and, more rarely, rodents or birds. Slowly stalks its prey, "moving like a cat", suddenly throwing itself at the victim and holding it until the venom from the rear-mouth fangs takes effect.

**Reproduction** Oviparous. Lays 5–9 eggs in June or July that hatch in late August or September. As soon as the first moult, a few days after hatching, young cat snakes start hunting, searching for small lizards.

**Range** Along the Adriatic coast from very southwestern Italy through Slovenia, Croatia, Albania, southern Macedonia as far as Greece and extreme southern Bulgaria. Also present on Crete, many of the Ionian and Aegean islands, Rhodes and Cyprus. Farther east in Turkey, the Caucasus area as far as the Caspian Sea and northern Iran, northwestern Syria, Lebanon and northern Israel (now considered extinct in Israel; unknown from Jordan). Also present on Malta, where it is considered as having been introduced. It reaches the extreme northwestern part of its vast range in Italy, in the Rosandra Valley, a few kilometres from the Slovenian border.

**Geographic variation** *Telescopus fallax* is a quite variable species that occupies a vast range; several subspecies are recognised, whereas others, such as *T. f. nigriceps* and *T. f. hoogstraali*, are now considered to be separate species:

▶ *Telescopes fallax fallax* (Fleischmann, 1831): Balkans, Malta, many of the

Ionian and Aegean Islands, southern Bulgaria, western and central Turkey. Characteristics, among others, include a large number of markings on the back (45–64, rarely fewer in southern Turkey, minimum of 36), and with much space between each one (on average 2 or 3 pale scales).

▶ *Telescopus fallax pallidus* Štěpánek 1944: Crete and its satellite islands, Gavdos and Elasa, Christiana Island near Santorini. Characterised by its relatively small and numerous (more than 50) dorsal markings that are, however, well separated (on average by 4 scales) and especially by the generally pale colour, often yellowish; the dorsal marks are pale brown, not black.

▶ *Telescopus fallax cyprianus* Barbour and Amaral, 1927: endemic to Cyprus. Characterised by a small number of ventral plates (194–212), a general brown colour with a large number (39–>50) of more or less contrasting reddish-brown dorsal bars.

▶ *Telescopus fallax iberus* (Eichwald, 1831): eastern Turkey, the Caucasus (Armenia, southern Georgia, extreme southeastern Dagestan, Azerbaijan), northern Iran. Characterised, among other features, by the small number of dorsal markings (32–48) that are close to each other, separated on average by only 1 or 2 scales.

▶ *Telescopus fallax syriacus* (Boettger, 1880): southwestern Turkey, Syria, Lebanon, northern Israel. Characterised, among other features, by very few dorsal markings (24–46) that are well spaced (4 or 5 scales between each mark).

The validity of subspecies *Telescopus fallax intermedius* Gruber, 1974 from the Greek island of Antikythira, *T. f. multisquamatus* Wettstein, 1952 from the Greek island of Kufonisi (opposite Crete) and *T. f. rhodicus* Wettstein, 1952 from Rhodes has yet to be clearly established.

Telescopus fallax iberus, *juvenile. North of Dogubayazit (northeastern Turkey). Note the large quadrilateral markings on the back, quite close to each other, typical of this northeastern subspecies.* Teynié.

# Black-headed Cat Snake

*Telescopus nigriceps* Ahl, 1924
**Family:** Colubridae
**Sub-family:** Colubrinae
**F.:** Couleuvre-chat à tête noire
**G.:** Schwarzkopf Katzennatter

**Identification** Closely related to *Telescopus fallax* and considered a subspecies of that taxon until recently. Characterised by being smaller on average (maximum recorded length 67 cm), having a smaller number of ventral plates (172–196 compared to more than 200) and a distinct colouration. The head is black, blue-grey or beige-grey with numerous small black speckles. A wide black collar (hardly distinguishable in individuals with a pure black head) separates the grey head from the beige body. The back is beige-grey, sandy beige or pinkish-grey with a small number of dorsal markings (15–34 and 5–8 on the tail), most connected with an oblique black bar on the flanks, thus forming well marked and widely spaced (6–10 rows of pale scales between each ring) thin black rings. The iris is mustard yellow, contrasting with the grey head. The belly is black, without a checkered pattern.

**Venom** An opisthoglyphous, slightly venomous snake with fangs located at the rear of the maxillaries. When irritated, Black-headed Cat Snake coils up, raising up to a third of its body length upright, its neck folded into an S, showing its black underside.

**Habitat** Arid steppe with sparse vegetation and scattered rocks; dry Mediterranean habitat with stands of pine.

*Two Black-headed Cat Snakes* Telescopus nigriceps. *Shawbak (Jordan). Note that the head is not always black: it can be dark grey with numerous thin black speckles. The well-defined, straight-sided black rings are a good identification feature.* Cluchier.

**Habits** A nocturnal species usually found whilst hiding under stones or rocks or in rodent burrows.

**Diet** Lizards, young of small rodents, bird eggs. Probably also arthropods.

**Reproduction** Oviparous. The eggs hatch in late summer.

**Range** Arid areas, but not desert, in northeastern Lebanon, western Jordan, Syria, northwestern Iraq (near Ar Ruthbah) and southeastern Turkey.

**Geographic variation** *Telescops nigriceps* is considered to be monotypic. It is quite variable in colour. In low and arid areas it has a black head, whereas in the Mediterranean zone the head is normally grey. In certain sites, however, black-headed and grey-headed individuals have been found together, in total cohabitation (e.g., near Shawbak in Jordan and in Turkey).

## Sinai Cat Snake, Gray's Cat Snake

*Telescopus hoogstraali* Schmidt and Marx, 1956
**Family:** Colubridae
**Sub-family:** Colubrinae
**F.:** Couleuvre-chat du Sinaï
**G.:** Hoogstraals Katzennatter

**Identification** General appearance as in *Telescopus nigriceps* but differentiated by having many more ventral plates: 242–254 compared to 172–196. The head is a little thicker and wider. More often than not it has 8 supralabials on each side (usually 9 in *T. nigriceps*). Head dark grey with numerous fine black flecks that can cover almost the whole of the top of the head, which then appears black. A wide black collar continues diffusely on the back of the head, without a clear demarcation between the colour of the head and the collar; the back is beige-grey to pale caramel-beige, with 28–56 dark grey rings (not black, at least in adults) sometimes with a slight pale border, narrow and less regular than in *T. nigriceps*. The belly is marbled dark grey and white, and 1 ventral plate in every 3 to 5 has dark markings on the sides (the belly is completely black in *T. nigriceps*).

An elongated loreal plate touches the eye, 1 preocular, 2 postoculars, 19 rows of scales at mid-body, divided anal plate. Maximum length: 102 cm.

**Venom** A slightly venomous opisthoglyphous snake with venom fangs located at the rear of the maxillaries.

**Habitat** Arid or desert slopes with stones, totally sandy or with sparse vegetation.

**Habits** A rarely seen nocturnal species. If threatened, the Sinai Cat Snake adopts the same stance as the Black-headed Cat Snake, hissing and raising the front part of the body, the neck bent in an S.

**Diet** Lizards and occasionally small birds or small mammals.

**Reproduction** Oviparous.

**Range** Endemic to a small part of the eastern Mediterranean area: Sinai, southern Israel and southwestern Jordan.

*Right: Close-up of a Sinai Cat Snake* Telescopus hoogstraali. *Negev desert (Israel).* Martínez del Mármol Marín.

**Geographic variation** *Telescopus hoogstraali* is considered to be monotypic.

**Remarks** Another species of cat snake has been reported from eastern Iraq: *Telescopus tessellatus* (Wall, 1908), represented in Iraq by the subspecies *T. t. martini* (K. Schmidt, 1939). This taxon, part of the *T. fallax* group, is characterised by the constant presence of 21 rows of dorsal scales at mid-body (usually 19 in *T. fallax*), a large number of ventral plates (226–242 compared to 200–230 in *T. fallax*) and by the anal plate being undivided in a certain number of individuals. Several authors consider this taxon to be close to or synonymous with *T. f. iberus*.

Sinai Cat Snake Telescopus hoogstraali. *Negev desert (Israel). Note that the dorsal bars are less regular then in* T. nigriceps. *The other criteria, not visible in the photo, is the larger number of ventral plates.* Martínez del Mármol Marín.

# Israeli Cat Snake

*Telescopus dhara* (Forskål, 1775)
**Family:** Colubridae
**Sub-family:** Colubrinae
**F.:** Couleuvre-chat d'Israël
**G.:** Israelische Katzennatter

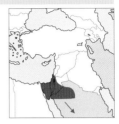

**Identification** 80–97 cm in length. Flat, elongated head, clearly distinct from body, with a square-shaped snout when seen from above. The nasal scales are slightly prominent. Slim, very elongated body. Smooth dorsal scales, generally arranged in 21 rows at mid-body, rarely 19. 1 loreal scale much longer than wide, separated from the eye by the preocular; 1 preocular, 2 postoculars, 9 (sometimes 10) supralabials, of which 3 (rarely 2) touch the lower edge of the eye: the 3rd, 4th and 5th. A large number of ventral plates: 228–274. An undivided anal plate in most cases; the subcordals are divided and arranged in pairs. The back is pale reddish-yellow to brick-red, sometimes sandy beige, with often indistinct large transverse bands, close to each other and separated by a small pale band; normally only slightly contrasting colours, sometimes uniform. The head is often redder than the body; there are no black bands on the side of the head. The iris is often pinkish, orange or grey. The belly is uniformly pale.

**Venom** A slightly venomous opisthoglyphous snake with fangs at the rear of the maxillae. There are recorded attacks on humans.

*Above: Close-up of* Telescopus dhara. *Eilat (Israel). Note the absence of dark bands on the cheeks (this differentiates it from* T. obtusus) *and the long loreal plate separated from the eye by a preocular scale (which differentiates it from species of the* T. fallax *group). Crochet.*
*Right: Israeli Cat Snake* Telescopus dhara. *Israel.* Werner.

**Habitat** More or less sandy desert areas, rocky slopes, rocky wadis, sometimes near agricultural areas. Penetrates slightly into the Mediterranean zone in certain wadis.

**Habits** Terrestrial. Active at dusk and at night. A quiet, slow-moving snake; however, males can be more aggressive than females.

**Diet** Mainly small birds and their eggs, lizards and, more rarely, small mammals. Kills its prey with its venom. It maintains its hold until its swallowing movements place the prey where the fangs at the back of the upper mandibles can act.

**Reproduction** Oviparous. Lays 6–20 large, elongated eggs.

**Range** Asiatic. Sinai (Egypt), Israel, western Jordan, mountainous areas of the Arabian Peninsula.

**Geographic variation** *Telescopus dhara* is considered to be a monotypic species. *T. d. obtusus* is now considered to be a separate species: *T. obtusus* (see next species).

# Egyptian Cat Snake

*Telescopus obtusus* (Reuss, 1834)
**Family:** Colubridae
**Sub-family:** Colubrinae
**F.:** Couleuvre-chat d'Egypte
**G.:** Ägyptische Katzennatter

**Identification** Closely related to *Telescopus dhara*, of which it is often considered a subspecies despite distinct and constant morphological features. A little larger than *T. dhara* (maximum length 110 cm). The head is obviously less elongated. More often than not, 23 rows of scales at mid-body, sometimes 21. There are 9–11 supralabials on each side, normally 3 touching the eye: the 4th, 5th and 6th. There are 230–278 ventral plates; the anal plate is normally divided. Normally 50 large, square transverse marks on the back, often well marked, very close one to the other and separated by a pale bar. 2 rows of alternately placed marks on the flanks. A short dark bar from the snout to the temple passes through the eye, clearly separating the dark colour of the head from the pale underparts (probably the best character for separating *T. obtusus* from *T. dhara*). There are individuals showing little contrast and even uniformly pale individuals, but the bar on the side of the head remains clearly visible. The general colour is decidedly less brick-red than in most *T. dhara* specimens. The iris is normally mustard yellow.

**Venom** A slightly venomous opisthoglyphous snake with fangs at the rear of the maxillae.

**Habitat** *Telescopus obtusus* is a desert-dwelling snake found mainly near cultivated areas, in oases and gardens, along vegetated riverbanks and on the edge of urban areas. It hides in ruins, old houses, piles of stones and old walls.

**Habits** Terrestrial; active at dusk and at night. A calm, slow-moving snake.

**Diet** Probably the same as *Telescopus dhara*.

**Reproduction** Oviparous.

**Range** Eastern and northeastern Africa: Egypt, where it is restricted to the Nile Delta and Valley and Djebel Elba, Sudan, Somalia, Tanzania and Kenya. From there, it occurs farther west along ranges to the southern Sahara: in Chad and southern Algeria, where it is little known.

**Geographic variation** There are two subspecies of *Telescopus obtusus*:

▶ *Telescopus obtusus obtusus* (Reuss, 1834): over the major part of its range.
▶ *Telescopus obtusus somalicus* (Parker, 1949): actually known due to a few examples from Kenya and southern Somalia, and considered by some authors as a separate species. Head and body uniform lead grey or brown with much iridescence: no dark band on the sides of the head, fewer ventral plates (200–219) and dorsal scales (20 or 21 rows at mid-body); general aspect more like that of aged examples of *T. tripolitanus* but clearly different in numbers of scales.

*Egyptian Cat Snake* Telescopus obtusus. *Note the dark band on the side of the head, which differentiates it from* T. dhara, *which has no such mark.* Modrý.

# Tripoli Cat Snake

*Telescopus tripolitanus* (F. Werner, 1990)
**Family:** Colubridae
**Sub-family:** Colubrinae
**F.:** Couleuvre-chat de Tripolitaine
**G.:** Tripolitanische Katzennatter

**Identification** A small to medium-sized snake (normally 40–60 cm, maximum 97 cm total length). Characterised by its entirely bright black head with much paler body. Back beige, yellowish-brown or red-brown with a series of dark ill-defined markings, often indistinct, sometimes quite noticeable. The flanks are uniform and pale more often than not, but well-marked individuals have irregular dark spots. In large specimens the black on the head and the pale body turn progressively more uniform dark grey, without any

*Tripoli Cat Snake* Telescopus tripolitanus, *two sub-adults from southwestern Morocco, one from Tiniffift (right), the other Assa.* García-Cardenete *(left)* and Martínez del Mármol Marín *(right).*

contrast, highlighting the iridescence of the scales. In rare cases, the top of the head and the upper temples are not black; the black is limited to the sides of the head and the nape. The belly is uniform white under the neck, changing progressively to pale brown or reddish towards the rear and under the tail. Under the throat there can be an indistinct, horseshoe-shaped black collar.

More often than not there are 21 rows of dorsal scales at mid-body, sometimes 22 or 23. Generally 9 supralabials on each side, the 4th and 5th touching the eye. 207–248 ventral plates. Divided anal plate.

**Venom** A slightly venomous opisthoglyphous snake with fangs at the rear of the maxillae.

**Habitat** A snake of desert areas usually found not far from human habitations with gardens. Also present in oases and wadis with vegetation.

**Habits** Terrestrial. Principally crepuscular and nocturnal; rarely seen to be active during the day. Sometimes found under stones.

**Diet** Unknown, but at least lizards and similar.

**Reproduction** Oviparous.

**Range** The Sahara's western and southern edges, where it has a fragmented distribution: northwestern Libya, Tunisia, southern and southwestern Morocco (including the Atlantic Sahara), the Tindouf region in extreme western Algeria (unknown from elsewhere in that country), Mauritania, Senegal, Mali, northern Burkina Faso, Niger, southern Chad and 1 record from Sudan.

**Geographic variation** *Telescopus tripolitanus* is considered to be a monotypic species.

**Genus *Dasypeltis*, egg-eating snakes**
This genus of egg-eating snakes contains at least 13 different species, two of which occur in the Western Palaearctic: *Dasypeltis sahelensis* in Morocco and *D. bazi* in Egypt. All are nocturnal snakes, slow-moving, not at all aggressive, slim with a small head, pupils with vertical slits and strongly keeled dorsal scales. They are specialised in eating birds' eggs. The genus occurs throughout much of Africa as well as the extreme southwestern part of the Arabian Peninsula.

The genus *Dasypeltis* is placed in the same group as the genus *Telescopus*, along with *Boiga*, *Crotaphopeltis*, *Dipsadoboa* and *Toxicodryas*.

# Sahel Egg-eater

*Dasypeltis sahelensis* Trape and Mané, 2006
**Family:** Colubridae
**Sub-family:** Colubrinae
**F.:** Serpent mangeur d'oeufs du Sahel
**D.:** Gewöhnliche Sahelische Eierschlange

Distribution of *Dasypeltis sahelensis* (in red) and *Dasypeltis bazi* (blue triangle)

**Identification** Calm and shy, this is a small snake of distinct general appearance with a small round head, eyes with vertical pupils and very strongly keeled dorsal scales. Total length normally 40–55 cm, to 62 cm. Small, quite short head, not very distinct from the body, and a very rounded snout. Medium to large eyes with a vertically slit pupil. Long, quite slim body appears quadrangular; the back is narrower than the top of the flanks. No loreal plate, which is rare in the *Colubridae*, and the very large nasal plate directly touches the prefrontal. Normally a single, high and well-developed preocular: rarely, there is a 2nd very small preocular located directly below the larger one. 2 postoculars, 7 supralabials on each side, the 3rd and 4th touching the eye. The supraocular plates are smaller than in other members of the *Colubridae*, leaving space for the very large frontal plate, almost as wide as it is long. Dorsal scales quite matte and very keeled, arranged in 21–23 rows at

Sahel Egg-eater Dasypeltis sahelensis. *Near Sidi Ifni (southwestern Morocco). Compare the resemblance of this harmless snake to that of the redoubtable White-bellied Saw-scaled Viper* Echis pyramidum leucogaster *(see page 352).* P. Geniez.

*Close-up of* Dasypeltis sahelensis *from Sidi Ifni surroundings. Note the globular head with a short rounded snout, the absence of a loreal plate, the nasal plate touching the preocular, the large eye with vertical pupil (as in cat snakes) and the strongly keeled body scales.* P. Geniez.

mid-body. Remarkably, the 4th, 5th and 6th rows of scales counting from the bottom are oblique and have serrated keels; these rows of keeled scales are separated from the ventral scales by 2 or 3 rows of serrated but non-keeled scales arranged horizontally. When frightened, the egg-eater coils itself up and rubs its flanks together where they touch, thus rustling its keeled scales, which produce a rattling sound. 207–237 ventral plates, an undivided anal plate, divided subcordals arranged in pairs.

There are 2 brown forwards-pointing chevron marks on the head that continue onto the temple and front of the neck. The beige, almost white labial plates are covered in quite large brown markings. There is a row of about 60 large, more or less rectangular, long chocolate-brown markings on the back, separated from one another by white bars. Each

white bar connects to irregular dark vertical bars arranged regularly along the flanks. The flanks have a more or less orange ground colour between the vertical bars and beige towards the bottom of the body. The belly is brilliant white with dark mottling on the outer edges of certain ventral plates.

**Venom** Non-venomous.

**Habitat** The Sahel zone.

**Habits** Nocturnal. When dusk arrives, actively searches for bird's nests and the eggs that make up its diet entirely.

**Diet** As all members of the genus *Dasypeltis*, the Sahel Egg-eater eats only eggs.

**Reproduction** Oviparous.

**Range** The Sahel belt of West Africa.

**Geographic variation** *Dasypeltis sahelensis* is a monotypic species. Until recently, it was included in *D. scabra*, which is now split into several distinct species.

# Egyptian Egg-eater

*Dasypeltis bazi* Saleh and Sarhan, 2016
**Family:** Colubridae
**Sub-family:** Colubrinae
**F.:** Serpent mangeur d'oeufs d'Egypte
**D.:** Gewöhnliche Ägyptische Eierschlange

At the end of the 19th century, there were several citations of the presence of egg-eating snakes in Egypt, in the Faiyum oasis. One of the specimens, ascribed to the Common Egg-eater, *Dasypeltis scabra* (Linnaeus, 1758), is held in the British Natural History Museum in London and illustrated by a drawing in J. Anderson (1898). Since the rediscovery of the Faiyum population by Mostafa Saleh, in 1997, egg-eaters had never been found since in Egypt and were considered extinct in that country and thus the Western Palaearctic. More recently, Mostafa Saleh and Moustafa Sarhan, examining 25 specimens, conclude that they do not refer to *D. scabra* but to a new species they named *D. bazi*.

**Identification** General appearance close to *D. sahelensis* but with special morphological features. Compared with the other species, it has fewer dorsal rhombs (38–49 against more than 50), a fewer number of dorsal scale rows (20–22 against 21–23 in *D. sahelensis*), a more rounded snout, giving the head a somewhat rectangular shape, a frontal shield clearly longer than wide (as long as wide in *D. sahelensis*), and on the flanks wide small blotches (narrow more or less vertical stripes in *D. sahelensis*). Contrary to *D. sahelensis* which exhibits distinct V marking on the head, *D. bazi* has more or less unpatterned large dark blotches. Dorsal blotches are larger than in the other species of the genus *Dasypeltis*, less regular and darker, sometimes entirely blackish. Finally, ground colour of the body is grey or greyish while it is brown, brownish or sandy in *D. sahelensis*. Total length is on average lower than *D. sahelensis*, around 50 cm but reaching at least 64.7cm.

**Venom** A snake with a non-venomous dentition; harmless, very calm and indolent, which lets itself be handled.

**Habitat** All specimens have been found in an ancient natural drainage system with patches of agricultural lands, in a desert depression 52 m below sea level, in a hot and dry climate.

**Habits** As *D. sahelensis*, a nocturnal snake that, once dusk arrives, becomes active searching for bird's nests.

**Diet** As all the members of the genus *Dasypeltis*, the Egyptian Egg-eater eats birds eggs. However, one *Dasypeltis bazi* captured and placed in a bag with a skink *Chalcides sepsoides* swallowed it. But it is hard to tell if this is a normal feeding behavior or not for this snake (M. Saleh personal communication).

**Reproduction** Oviparous.

*Egyptian Egg-eater* Dasypeltis bazi *from the vicinity of Ibshwai, in the depression of Faiyum (Egypt). Note the dark unpatterned top of the head, the large dark blotches on the dorsum and the small blotches on the flanks, not stretched vertically, and on a greyish ground color.* Saleh.

**Range** Only known in Egypt, from a small area of about 300 km², in the vicinity of the village of Ibshwai in the Faiyum depression, 100 km southwest of Cairo.
**Geographic variation** *Dasypeltis bazi* is a monotypic species. From a phylogenetic point of view, it represents a very distinct clade in the genus *Dasypeltis*, as well as *D. sahelensis* which also is very distinct from other species of egg-eater snakes.

### Genus *Natrix*, semi-aquatic snakes

The semi-aquatic snakes belong to a special sub-family, the Natricinae, within the family Colubridae. The Natricinae are placed on the edge of the Colubridae and are so divergent from the other sub-families that certain authors advocate their being treated as a separate family, the Natricidae. The Natricinae contains about 36 genera dispersed over all continents. There are just four species in the genus *Natrix*, all present in the Western Palaearctic: *N. natrix*, *N. astreptophora*, *N. tessellata* and *N. maura*. They are small to medium-sized, characterised by strongly keeled dorsal scales (like those of vipers) and very little aggressive behaviour, which is compensated for by their ability to extrude a very offensive-smelling substance from their anal glands and by their semi-aquatic lives.

# Grass Snake

*Natrix natrix* (Linnaeus, 1758)
**Family:** Colubridae
**Sub-family:** Natricinae
**F.:** Couleuvre à collier; **G.:** Ringelnatter

**Identification** A medium-sized to large semi-aquatic snake; attractive, generally greyish with a white, yellow or sometimes orangish collar composed of 2 more or less attached wide transverse marks and bordered at the rear by 2 black triangular marks, sometimes joined on the nape. Males are smaller than the females, 70–90 cm on average; females may be much larger and especially more thickset, measuring on average 80–120 cm but to 140 cm for a weight of 1 kg; there are doubtful records of specimens of 180–200 cm. The head is clearly distinct from the body, and is quite oval in young animals, becoming thicker, wider and more triangular in large individuals, especially females. The medium-sized eyes have a round pupil and a grey, yellowish, orangish or red iris, depending on subspecies. The dorsal scales are clearly keeled, arranged in 19 rows at mid-body.

The Grass Snake is the most common snake in central Europe. Easily recognised by the white or yellowish half-moon markings, edged with dark, on each side of the back of the head. Natrix natrix persa, *a female photographed near Çubuk, northeast of Ankara (Turkey). This individual lacks the usual pale dorso-lateral lines that normally characterise the subspecies persa.* P. Geniez and A. Teynié.

*Grass Snake* Natrix natrix natrix. *The two yellow half-moon marks with black border are typical of the species; the small mottling, which does not form clear markings, is characteristic of the type subspecies.* Franzen.

1 preocular is higher than it is wide, 3–4 postoculars (sometimes 2), generally 7 supralabial plates on each side, 150–200 ventral plates, subcordals divided and arranged in pairs.

Quite variable in colour, more often than not slate or mouse-grey, but also greenish, yellowish-brown, orange or brick-red. There are normally well-separated black spots or bars on the back and flanks that occur in a regular, alternating manner. There are black vertical bars on the sides of the labial plates, characteristic of this species. Some populations have 2 pale lines along the length of the body (e.g., within subspecies *persa*; see "Geographic variation"). Melanistic individuals occur rarely in some populations, quite commonly in others (e.g., *Natrix natrix persa* in Turkey). The belly is whitish-grey or yellowish with black marks normally in a checkered pattern, sometimes with a large black band with pale edges, sometimes entirely black.

**Venom** Non-venomous. Harmless and non-aggressive, although certain large individuals adopt a menacing posture and hit the aggressor with a closed mouth.

**Habitat** Associated with water, but can occur far from wetlands: banks of lakes and fishponds, ponds, rivers, marshy areas with reedbeds, wet woodlands and alluvial forests. Also found some distance from standing water: gardens, meadows, quarries, clearings, woodland edge, in areas with ruins and even on islands at sea. It is resistant to cold and is thus found at elevations to 2006 m in the French Alps, 2300 m in the Italian Alps, 3200 m in the Sierra Nevada (southern Spain).

**Habits** Mainly diurnal. Swims and dives well, but less strictly associated with water than the Viperine Snake *Natrix maura* (see page 220) or the Dice Snake *N. tessellata* (see page 217). Lies for long periods in the sun in the early morning before starting

*With the Grass Snake, cases of melanism (black pigment replacing the usual pigmentation) are common, especially in the eastern part of the species' range. Here a female* Natrix natrix persa *feigns death. Boludagi Geçidi (northwestern Turkey). P. Geniez and A. Teynié.*

Natrix natrix scutata, *European part of Kazakhstan. It looks very similar to N. n. natrix but is often darker and genetically quite distinct. Crochet.*

to hunt in aquatic habitats. Very shy and secretive. When threatened, flees into water or another hideaway. When captured, it does not bite but, as a defensive reaction, produces an extremely foul-smelling, yellowish-white fluid from the anal gland; the smell can last for a long while on skin or clothes. Often, it reverts to a surprising reflex, which is quite typical of the species: it feigns death, becoming totally limp, whilst lying on its back with belly uppermost, head hanging, mouth open and the tongue sticking out of the mouth. It even goes as far as letting its pupils slide downwards, accentuating the appearance of a long-dead snake. This is an annoying behaviour for photographers who want a natural-looking shot. If the snake is left alone for a few minutes, it suddenly "wakes up" from apparent death and tries to escape rapidly.

Active from March to October in the south, from May to September in the north. There is an inactive period in winter for a few months, when it hides away from frosts, in the ground, in compost heaps, under tree stumps, in burrows or in deep rock crevices. Like all snakes, the Grass Snake has many predators: birds of prey, storks, herons, owls, crows, polecats, martins, badgers, otters, foxes, wild and domestic cats, hedgehogs, other snakes, pike and perch. The young are sometimes eaten by large beetles, frogs, predatory fish and even birds (chickens, sparrows, and Blue Rock Thrush have been recorded).

**Diet** Frogs, toads, newts, fish, more rarely lizards, small rodents and birds. The young eat worms, insect larvae, slugs, tadpoles, small fish and newt larvae. The Grass Snake catches its prey, which it detects by their movements, and swallows it whole in one piece, even the largest items (e.g., an adult Common Toad that may be 10 times the volume of the snake's head). A Grass Snake surprised whilst swallowing a large prey may regurgitate it, still alive, in order to escape more easily.

**Reproduction** Oviparous. Emerges from hibernation in April or May and mates after the first moult. At that time, many Grass Snakes may come together (rarely up to 50–60 individual) at mating sites, where males, which far outnumber the females, try to mate with the largest females. Paired animals stay attached for a long time. If threatened, the female flees and drags the much smaller male along with her.

Eggs are laid in July and August, in a compost heap, stack of rotten wood, pile of dead leaves or decomposing reeds, under rotting tree stumps, sites of peat extraction, etc. A female lays 6–50 white eggs, each with a soft, leathery shell; eggs are often stuck together. Particularly favourable nest sites may contain the eggs of several females, sometimes several hundred eggs together (there are quotes of up to 3000 eggs; they often lay communally in a favoured place so large "clutches" are often reported). After 4–8 weeks incubation, the young hatch from the eggs, tearing apart the shell with a few movements aided by their small horny growth called the "egg tooth". Hatching may last several hours.

**Range** A vast range: most of Europe except for Scotland, Ireland, the Balearic Islands, and the far North (but occurs as far north as the Arctic Circle in Finland). Present on many of the Mediterranean Islands, including Corsica, Sardinia and Cyprus. To the west, its range continues across Turkey, northern Iraq and Iran, central Asia and Siberia as far as Lake Baikal, to Mongolia and northern China. In the south, it is present in Mediterranean countries in the Levant (northwestern Syria, Lebanon, Palestine and northern Israel, where it is very rare)

Natrix natrix persa, *from southeastern Europe and Asia Minor (here, in Montenegro), easily recognised by the two pale lines running along the length of its back.* Cluchier.

Natrix natrix schweizeri, *a subspecies endemic to the Milos Archepeligo (off Greece), is closely related to* N. n. persa, *from which it is easily distinguished by the large black markings on the back and flanks and by the partial or complete absence of the two pale lines on the back. Gruber.*

and in North Africa (northwestern Morocco, extreme northern Algeria and northern Tunisia), where it is not common.

**Geographic variation** The Grass Snake is an extremely variable species; many subspecies have been described. Genetic studies have shown the existence of three main groups: one from Iberia and North Africa *Natrix (natrix) astreptophora*, the most divergent of the three, which was recently considered to be a separate species; a western European group (*N. n. helvetica*, as well as *N. n. cetti* of Sardinia and *N. n. corsa* of Corsica); and an oriental group (*N. n. natrix*, which contains many subspecies, including *N. n. persa*). The following list of subspecies is considered to be provisional. It is based essentially on differences in morphology and colour, but genetic sudies show that several of the subspecies are of very divergent lineages and, as well, are paraphyletic (that is, they contain lineages that can be found in other subspecies described only on morphological grounds). The Large-headed Water Snake *N. megalocephala* Orlov and Tuniyev, 1986, once thought to be a distinct species, is now considered to be a melanistic variant of *N. n. scutata* (or of *N. n. persa*, according to some authors). It was described from the Caucasus region from particularly robust females with large heads that exist throughout the species' range.

### *Natrix natrix natrix* group
▶ *Natrix natrix natrix* (Linnaeus, 1758): central and western Europe, to the east as far as western Russia and the Crimea, to the west as far as northern Holland and western Germany. Characterised by a quite slim aspect, a marked collar and an almost uniform grey colouration or with often indistinct small black markings, as well as small, pale yellowish speckles; the black bars on the supralabials are often joined in pairs in the upper half. Juveniles have a large, wide, black bar on the temples; this pattern shows up very well against the pale bar in front of the eye, another behind the eye, and a 3rd character, the pale mark of the collar. Melanistic individuals exist and in some populations are abundant. Individuals with 2 pale lines along the back are rare and sporadic.
▶ *Natrix natrix scutata* (Pallas, 1771): Russia to eastern Dniepr, Kazakhstan, Transcaucasia and extreme northern Turkey, where some individuals have

*Left:* Natrix natrix helvetica, *a large female, 86.5 cm in total length. Note its very robust stature and the vertical black bars on the flanks.* Nicolas.
*Right:* Natrix natrix helvetica, *juvenile on the edge of a pond in the Causse Méjean (southern France).* P. Geniez.

2 longitudinal yellow lines that are typical of *N. n. persa*, suggesting that there is an integration zone of the two subspecies. Morphologically and genetically very similar to *N. n. natrix*, but often darker. Melanistic individuals are often common, even the majority in certain populations. Specimens of *N. megalocephala* are, in fact, melanistic individuals of *N. n. scutata*.

► *Natrix natrix persa* (Pallas, 1814): the Balkans as far as northern Croatia, Serbia and Bulgaria. Asia Minor (Turkey and the Levant countries), Iran, and from there to south-central Asia. It is very similar to *N. n. natrix* except that most individuals have the 2 wide yellowish-white or pale yellow lines along their back. *N. natrix syrica* (Hecht, 1930), a subspecies from the Levant, is now considered synonymous with *N. n. persa* by most authors. However, examples of *N. n. persa* from Iran and Azerbaijan are genetically quite different. The same applies to *N. n. persa* from the western Balkans, which is very different from its counterpart in Asia Minor. It is therefore probable that the

Natrix natrix lanzai, *subadult. Elba island, Marina di Campo (Italy).* P. Geniez.

subspecies *N. n. persa* concerns only Grass Snakes from Iran and western Azerbaijan.

► *Natrix natrix cypriaca* (Hecht, 1930): endemic to Cyprus, where it is very rare and almost certainly endangered. It is genetically very similar to *N. n. persa* from western Turkey, Bulgaria and Roumania. It does, however, have a very unusual colouration, recalling certain individuals of *N. n. cetti* and *N. n. corsa* of Sardinia and Corsica. It does not

Natrix natrix sicula. *Caltagirone (Sicily). Note the orange snout, the absence of a collar and that the dorsal bars more or less form rings around the body.* P. Geniez.

Natrix natrix cetti, *a rare subspecies endemic to Sardinia, photographed in the Monte dei Sette Fratelli. Note the absence of a collar and that the dorsal bars more or less form rings.* P. Geniez.

have yellow longitudinal dorsal lines. It often has alternating large black marks that may be separate or form irregular rings; head scales may all have irregularly black borders, as is the case in Corsica and Sardinia.

▶ *Natrix natrix schweizeri* Müller, 1932: a very restricted subspecies confined to the Milos archipelago (Milos, Kimolos and Poliagos) off Greece. Genetically very close to the lineage from the western Balkans. No yellow lines on the back; large black, alternating, oval or quadrilateral markings; a collar that tends to disappear in adults. Partially or entirely melanistic individuals exist.

### Natrix natrix helvetica group

▶ *Natrix natrix helvetica* (Lacépède, 1789): Britain, continental France, Belgium, extreme western Germany, Switzerland, northern Italy, western Slovenia and Istria (northwestern Croatia). A large subspecies; the females are very robust, generally slate-grey (sometimes yellowish or

reddish), with a well marked collar and, on the back and especially the flanks, alternating transverse and vertical black bars. The iris is grey or yellowish in the north and east, bright orange in the southwest, suggesting there is integration with *N. n. astreptophora*. Melanistic individuals are rare and widely scattered.

▶ *Natrix natrix lanzai* Kramer 1970: central Italy and the island of Elba. Genetically and morphologically close to *N. n. helvetica*. General colour more often yellowish-brown than grey; the tip of the snout suffused with orange; pale orange iris; vertical black bars on the flanks; pale collar tends to fade in older adults.

▶ *Natrix natrix sicula* (Cuvier, 1829): Sicily and southern Calabria. Quite a massive appearance; snout heavily diffused with orange; pale orange iris; dorsal markings form transverse bands that are connected, or not, to the bars on the flanks and can form more or less well-defined rings; the pale collar fades and eventually

Natrix natrix corsa, *endemic to Corsica, photographed at Ospedale. Other than the absence of a pale collar and bars on the back, which form rings, the scales on top of the head are bordered with or are completely black.* Rufray.

disappears with age. Commonly melanistic, but the snout of black individuals is always suffused with orange.

▶ *Natrix natrix cetti* Gené, 1838: endemic to Sardinia, where it is localised and rare. Colouration quite different from most other types of Grass Snake: absence of pale collar (at all ages); the back markings of the collar are fragmented, not well defined or limited to a dark mark on the side of the neck; head scales have black borders; dorsal markings form bars often connected with those on the flanks, thus forming black rings; body background colour generally slate-grey; bluish-grey iris. Reputed to be more nocturnal than the other subspecies.

▶ *Natrix natrix corsa* (Hecht, 1930): endemic to Corsica, where it is not rare. Almost identical to *N. n. cetti*, including genetically. Can be distinguished by a few differences in scale numbers and in the black bars being on average thinner and more numerous. It may well be

synonymous with that subspecies, in which case *N. n. cetti* would be the name of the subspecies on Sardinia and Corsica.

### *Natrix (natrix) astreptophora* group

▶ *Natrix (natrix astreptophora)* (Seoane, 1884): the subspecies that is the most distinct from the others genetically; recent genetic data indicate that it is in fact a separate and full species, *Natrix astreptophora*. Within *N. astreptophora* there are two very distinct lineages that could represent two subspecies: one from the Iberian Peninsula and the other from Tunisia (there is no genetic data from Morocco or Algeria). Thus, it is possible that Grass Snakes in Algeria and Tunisia belong to a separate subspecies: *Natrix astreptophora algericus* (Hecht, 1930).

Occurs across North Africa (Morocco, Algeria and Tunisia) and the whole of the Iberian Peninsula (Spain and Portugal); from there it has penetrated into southernmost

215

Natrix (natrix) astreptophora, *adult in the region of Zamora (Spain). Note the red iris, the thin, well-spaced black bars on the labials, the absence of the white part of the collar and body markings.* Évrard.

France. Characterised when adult by the near total disappearance of the collar. The iris is bright red at all ages, and dorsal markings are reduced to spots (not bars or, if present, very small) disposed regularly and alternating on the back and flanks. Adults are often quite uniform, their markings all but vanished; the head is then often bluish-grey, whereas the body is more greenish-grey or pale almond-grey, but there are also individuals that are orange or brick-red.

In juveniles, the back half of the top and sides of the head, in front of the pale collar, is uniform black or very dark. In southwestern France (Aude departement), there is an area of overlap of *N. n. astreptophora* and *N. n. helvetica* with scarce morphologically intermediate individuals or presumed hybrids. The coexistence of these two lineages, very distinct genetically, and the scarcity of hybrids indicate that they are two distinct species (personal observation).

Two distinct subspecies are not recognised by all authors, and genetic studies show that they are hardly distinct: *N. n. gotlandica*

*A supposed example of the Large-headed Water Snake* Natrix megalocephala. *It is relatively easy to see in the photo that this is simply a nearly totally melanistic Grass Snake (intermediate between* Natrix natrix scutata *and* N. n. persa). Kreiner.

Nilson and Andrén, 1981, endemic to Götland island, in Sweden, of which a large proportion of individuals are melanistic, is clearly part of *N. n. natrix*; and *N. n. fusca* Cattaneo, 1990, endemic to the Greek island of Kea, is part of the western Balkan group within *N. n. persa*. Thus, this name could apply to the *N. n. persa* clade of the western Balkans.

## Dice Snake

*Natrix tessellata* (Laurenti, 1768)
**Family:** Colubridae
**Sub-family:** Natricinae
**F.:** Couleuvre tessellée; **G.:** Würfelnatter

**Identification** A medium-sized (60–90 cm in total length, very old females can exceed 130 cm) aquatic snake. A slim snake very similar to the Viperine Snake (*Natrix maura*, see page 220). A narrower and more elongated head, appearing triangular. The nostrils and the eyes are slightly orientated upwards, allowing it to breathe and see better when swimming on the surface of the water. The dorsal scales are well-keeled, arranged in 19 (rarely 21) rows at mid-body. There are generally 2 preoculars (sometimes 3), 3 postoculars (sometimes 4 or 5, in which case they almost form a half-crown around the eye). 8 supralabials on each side (rarely 7 or 9), the 4th and 5th touching the eye, sometimes only the 4th (the 3rd and 4th in Viperine Snake),

*Dice Snake* Natrix tessellata. *Male photographed in Lebanon.* Cluchier.

Natrix tessellata, *large female, 97 cm long. Talysh mountains (southwestern Azerbaijan). Despite its resemblance to the Viperine Snake (which has a different range), it is easily differentiated by the position of the eye, situated above the fourth and fifth supralabials (third and fourth in the Viperine Snake) and by the larger number of postoculars (two in the Viperine, always more in the Dice Snake).* P. Geniez.

160–187 ventral plates (rarely to 198, or many less, 130–140 in certain Swiss populations); divided anal plate.

Quite variable colouration, bluish-grey, greenish-grey, yellowish-olive or reddish-brown. An alternating checkered pattern on the back may be arranged longitudinally or in rough transverse bands. There are dark marks on the flanks, often aligned vertically. The back and flank markings may be reduced to mottling or, on the contrary, may be large and separated by narrow spotted yellow lines, as in the Viperine Snake. However, distinct from that species, the top of the head is often uniform, without markings; sometimes it has a backwards-pointing V-shaped mark on the nape. The belly is whitish, yellowish or pinkish-grey, quite often even bright red, with dark quadrilateral markings that become more and more pronounced towards the rear of the belly, forming a checkered pattern or a wide, dark central band. Totally melanistic individuals are quite rare (rarer than in Grass Snake, but commoner than in Viperine Snake).

**Venom** Harmless. Not venomous. As in the other two species of the semi-aquatic genus *Natrix*, it is not aggressive; it simply hisses, forms its head into a triangular shape and, if it is held, extrudes a fowl-smelling substance from its cloacal gland. Some individuals, when captured, feign death, but in a less spectacular way than Grass Snake.

**Habitat** Very much associated with water, in the same habitats as the Viperine Snake, with which it shares the same overall appearance of a drab aquatic snake. Banks of slow-flowing rivers and lakes with abundant riparian vegetation. Calm, warm waters with a moderate current are preferred to deeper, cooler waters with a rapid current, but this species is sometimes found on the sides of small streams at some elevation. Around the Black Sea and on certain islands of the Adriatic coast, will go into seawater. A species of low and mid elevations, but may locally reach as high as 2200 m.

**Habits** A real aquatic snake: swims and dives for long periods and can remain submerged for several hours, even more than the Viperine Snake. Readily lies in the sun on branches overhanging water. When it feels threatened, it freely falls into flowing water silently slipping across the surface or moving away by diving.

*The Dice Snake can exude blood when feigning death, a strange auto-bleeding behaviour seen here. Iran.* Martínez del Mármol Marín.

Inactive for several months during winter.

**Diet** Fish, frogs, newts and tadpoles.

**Reproduction** Oviparous. Mating occurs in April; eggs are laid in June, a little away from water. The female lays 5–25 eggs (sometimes to 37) under rotting tree stumps, in damp soil or in rotting vegetation. Hatching occurs in August or September. Newly hatched individuals greatly resemble adults in colouration and markings.

**Range** More eastern than the Viperine Snake: Italy (except for Liguria, Calabria and Sicily), Alpine countries (including Switzerland, where it reaches the westernmost point of its range, on the shores of Lake Geneva), southern Poland, the Czech Republic, Slovakia, Hungary, Romania, Bulgaria, southern Ukraine, the Balkans including Greece, certain Mediterranean islands, in particular Crete and Cyprus, Turkey, the Levant countries (from Syria to Israel). Reaches the African continent in Egypt, where it occurs only in the Nile Delta. Farther east: southern Russia, the Caucasus region, in Asia as far as extreme northwestern India and northwestern China. In Germany, at the extreme northwestern part of its range, it occurs in relict fragmented populations isolated from the rest of the species' range, along the Rhine, Moselle, Nahe and Ahr rivers. These populations are considered in danger of extinction.

**Geographic variation** The Dice Snake is considered by most authors to be a monotypic species despite its vast range. Studies of its genetic variability have shown a strong "structuration" (a scientific term that designates the existence of divergent lineages). On Serpilor island (also known as Snake Island), in the Black Sea, 35 km off the Danube Delta, there is a distinctive population described as a subspecies *Natrix tessellata heinrothi* (Hecht, 1930). It is characterised by having 21 rows of dorsal scales at mid-body (instead of 19), often 6 postocular scales (instead of 3 or 4), and by a high proportion of melanistic individuals.

# Viperine Snake

*Natrix maura* (Linnaeus, 1758)
**Family:** Colubridae
**Sub-family:** Natricinae
**F.:** Couleuvre vipérine; **G.:** Vipernatter

**Identification** Small to medium-sized aquatic snake generally with a yellowish-brown colouration. Normally quite slim, but large females can be quite thickset. Males 45–60 cm long, females 50–80 cm, occasionally to 93 cm. Quite an elongated, somewhat pointed head, distinct from the body; in females, it is massive and triangular, an appearance that is reinforced when it takes up an aggressive attitude. Medium-sized eyes slightly orientated upwards; the round pupil is black with a fine orange border. The body scales are highly keeled, arranged in 21 (rarely 19 or 22) rows at mid-body. 2 preocular scales, 2 (rarely 3) postoculars, 7 supralabials on each side, the 3rd and 4th touching the eye. 142–163 ventral plates, divided anal scale.

Ground colour variable but generally with a yellowish shade: greenish-grey, olive, yellowish-grey, brownish, yellowish-orange, sometimes reddish-brown. There is a series of dark angular markings along the back that may join to form a zigzag band. Along the vertebrae there is often a band of small yellow spots that alternate with the back markings. A series of large, dark oval marking on the flanks alternate with the back markings. The flank ovals can be uniformly dark or may have a pale centre, giving them an eye-like appearance that exists in neither Dice nor Grass Snakes. 2 dark V-shaped

*Viperine Snake*
Natrix maura, *adult.*
*Monfragüe (Spain).*
Évrard.

bars, open towards the back of the head, continue onto the temple. The labial scales possess wide dark bars rimmed with black (never entirely black as in Grass Snake). The belly is whitish, yellow, orange or reddish, decorated with large, dark quadrangular markings that may form a checkered pattern. Some individuals have 2 remarkable longitudinal lines along the back and tail; this particular form, found in the same populations as non-lined individuals, is sometimes called the "*bilineata* morph".

**Venom** Harmless. A non-venomous snake that shows no aggressive behaviour, never even biting a finger if handled. Its defence consists of hissing, of flattening its body and enlarging the back of its head, which thus appears triangular like that of a viper (which is the reason for it often being confused for a viper and thus killed by humans). Its best defence is the vile-smelling substance it produces from a cloacal

gland via the anus, mixed with faecal matter.

**Habitat** A species strictly reliant on wetland habitats. Occurs in all sorts of water bodies, from which it rarely wanders and then in order to reach a different aquatic area. They are sometimes numerous on the banks of stagnant water or slow rivers with an abundance of riperine vegetation; also on the sides of very stagnant water and ponds, even far from any other water body, as well as at mid-elevation rivers, water meadows, marshes and reedbeds. Any wet habitat from coastal plains to lower mountains, including hilly areas. Occurs to 1700 m in the French Pyrenees, 2035 m in the Baetic sierras in southern Spain and 2600 m in the High Atlases in Morocco.

**Habits** Mainly diurnal, but can be found hunting underwater during hot summer evenings or when moving from one site to another. Spends

221

*Viperine Snake, subadult. Covelàes (northern Portugal).* P. Geniez.

much time in the sun on exposed banks. When threatened, swims and dives silently and very well. An inactive winter period of 3–5 months depending on elevation.

**Diet** Essentially aquatic animals that it hunts by stealth, lurking on the bottom: fish, frogs, toads, newts, fish fry and tadpoles. Also soft invertebrates such as worms or slugs.

**Reproduction** Oviparous. The female lays 4–15 (sometimes as many as 20) eggs in July in slightly humid soft soil.

*Close-up of the Viperine Snake. Les Bains d'Avène (southern France).* Nicolas.

The eggs hatch after 40–45 days. As in other *Natrix* species, the males, smaller and more numerous, group together around large females in the hope of mating.

**Range** Extreme western Italy (Liguria), southwestern Switzerland, the southern two-thirds of France, the whole of the Iberian Peninsula (Spain and Portugal) and the whole of non-Saharan North Africa (Morocco, Algeria, Tunisia and extreme northwestern Libya). Also present on Sardinia and the Balearic Islands, where it is considered to have been introduced.

**Geographic variation** *Natrix maura* is considered to be a monotypic species despite its vast range and the large genetic differences that exist between different populations. The most divergent populations are those of Tunisia, Algeria and northeastern Morocco, and more especially those of the rest of Morocco. Indeed, genetic

*Young Viperine Snake blowing up its neck in order to look menacing. Strangely, this behaviour is nearly always seen in Morocco, more rarely in Spain, and apparently never in France. Boumia (Morocco). P. Geniez.*

studies have shown that Viperine Snakes from Sardinia are identical to those of Tunisia, which suggests an introduction from northeastern North Africa.

The Viperine Snake and the Dice Snake are often given as a very good example of vicariance – that is, two very similar species, both morphologically and phylogenetically, that have similar ecological habits, but have a different and often complementary distribution. However, genetic analyses have shown that the Dice Snake is more similar to the Grass Snake than to the Viperine Snake, the last being the more divergent of the three.

*Viperine Snake of the "bilineata" morph, with the two yellow dorsal bands. Spain. Cluchier.*

# Family Lamprophiidae

Of all the snake families, this is the most heterogeneous from a morphological point of view. It is a group closely related to the Elapidae; together the Lamprophiidae and Elapidae are closely related to the Colubridae (see also the overview of grass snakes, page 207). Of the seven sub-families in the Lamprophiidae family, four are found in the Western Palaearctic: the Lamprophiinae, the Psammophiinae, the Atractaspidinae and probably the Aparallactinae if it is accepted that *Micrelaps muelleri* is actually a member of this sub-family.

## Genus *Boaedon*

This small genus is represented by at least six species distributed across the major part of sub-Saharan Africa, with two relict populations – one in Morocco, the other in the Arabian Peninsula (in western Yemen). All are snakes of tropical or subtropical climates, nocturnal, of medium size (around 100 cm in length), with a vertical slit to the pupil and smooth, shiny scales. The genus *Boaedon* appears to be closely related to the genera *Lamprophis* and *Lycodonomorphus*, within the sub-family Lamprophiinae and the family Lamprophiidae.

# Brown House Snake

*Boaedon fuliginosus* (Boie, 1827) (formerly *Lamprophis fuliginosus*)
**Family:** Lamprophiidae
**Sub-family:** Lamprophiinae
**F.:** Couleuvre commune d'Afrique or Lamprophis des maisons
**G.:** Afrikanische braune Hausschlange

**Identification** A medium-sized (70–120 cm), quite slim snake, with an entirely iridescent bright black body when young, turning to a milky coffee colour when adult. The belly surface is very dark in juveniles but quickly turns uniform whitish with age. The head is quite long, a little flattened, slightly distinct from the neck. Medium-sized eyes, with vertical pupils and a dark iris in juveniles, brown in adults. 1 loreal plate, 1 preocular, 2 postoculars, 8 (rarely 9) supralabials on each side. The scales are smooth and shiny, very

*Brown House Snake* Boaedon fuliginosus, *adult. Between Sidi Ifni and Foum Assaka (southwestern Morocco).* García-Cardenete.

225

*Close-up of* Boaedon fuliginosus, *showing clearly the vertical pupil, which is quite similar to that of the cat snakes (genus* Telescopus). García-Cardenete.

Boaedon fuliginosus, *juvenile near Tantan-Plage (southwestern Morocco). Note the very dark overall colour, the highly iridescent blue scales and the vertical pupils, so narrowed that they are hardly visible.* M. Geniez.

numerous, arranged in 29–33 rows at mid-body. Very many ventral plates, 201–243; an undivided anal plate. The subcordals are divided and arranged in pairs.

**Venom** Non-venomous. Harmless; calm and rarely aggressive, it is easily manipulated.

**Habitat** In the Sahel, it occurs in savanna and forest; also in villages. In Morocco, where it is apparently rare, it occurs in steppe along the Atlantic coast, and in very arid areas but with frequent sea spray influence and a mild climate with little variation in temperature. During the day, it hides in a rodent burrow or sometimes under a large stone.

**Habits** A crepuscular and nocturnal species that readily enters houses searching for rodents, which make up a major part of its diet.

**Diet** Lizards and geckos, small rodents, shrews and snakes. Hunts on the ground by stealth; kills its prey by constriction.

**Reproduction** Oviparous. The female can lay up to 10 eggs in a clutch several times a year.

**Range** Present in much of Africa, but its distribution is poorly known due to confusion with similar species. In the area covered here, it is only known from southwestern Morocco, relict of a tropical past.

**Geographic variation** *Boaedon fuliginosus* is considered to be a polytypic species, but many of the subspecies are now considered to be separate species. In Morocco, it is the nominate subspecies *B. f. fuliginosus* that occurs.

## Sub-family Psammophiinae

Psammophiinae represents a group of snakes that are unusual in more ways than one; they share so many unique characters that it has been proposed by several authors that this sub-family should in fact be considered a family in its own right. The origin of these snakes and their centre of diversification is Africa. From there a few genera have extended their range: *Malpolon* into southern Europe and southwestern Asia, *Rhagerhis* into southwestern Asia and *Psammophis* into desert and steppe areas of central Asia. The genus *Mimophis* is found only in Madagascar.

Psammophiinae is represented by seven or eight genera. Characteristics that differentiate them from other Western Palaearctic Colubridae include spermatogenesis in sping, when for most other European snakes this occurs in summer. And especially the morphology of the nasal plates, with their valvular nostrils, a feature unique amongst reptiles, and also that of the hemipenis, which is small, thin and without projections; on the contrary, in the Colubridae the hemipenis is highly developed, fleshy, quite globular, and has many spines, resulting in a firmer hold during copulation. The Psammophiinae appear to be closely related to the Prosymninae; these two sub-families are placed between all the other Lamprophiidae and the Elapidae.

### Genus *Malpolon*

This is the only genus of the sub-family Psammophiinae to occur in Europe and Asia Minor. It is represented by just two closely related species, the Western Montpellier Snake *Malpolon monspessulanus* and the Eastern Montpellier Snake *M. insignitus*. Their range covers most of the areas surrounding the Mediterranean except for the "boot" and northeastern Italy. Among the Psammophiinae, the genus *Malpolon* is the closest to the genus *Rhagerhis*, the Moïla Snake, within a small clade that also includes the genus *Rhamphiophis*, as opposed to all other members of the Psammophiinae.

# Western Montpellier Snake

*Malpolon monspessulanus* (Hermann, 1804)
**Family:** Lamprophiidae
**Sub-family:** Psammophiinae
**F.:** Couleuvre de Montpellier occidentale
**G.:** Westliche Eidechsennatter

**Identification** Large, robust, impressive snake that appears quite slim, with males quite often reaching 150–170 cm, females 100–130 cm; rarely, some males can exceed 200 cm. (verified record: 223 cm). The narrow head is quite high and not very distinct from the body. The underside of the head is slightly concave. The eyes are very large, with round pupils encircled with whitish, golden, yellowish or orange rim; this pale colour extends on to the top of the iris; the rest is black. The large supraocular scales extend beyond the top of the eyes. Very elongated frontal plate, said to have the form of a clove, is more often than not in contact with the large preoculars. 1 nasal scale, often half divided, has valvular nostrils (a characteristic unique to the genus *Malpolon* within the sub-family Psammophiinae ), 2 loreals, 2 or 3 postoculars. It has 1 very large and high preocular, concave in front of the eye; its upper part not only curves and thus touches the frontal, but also passes the upper edge of the eye, continuing in a point farther back. Even more than the large supraocular scales it's the unusual form of this big preocular which passes above the eye like a prominent eyebrow, that gives the snake a "severe", "piercing" or "eagle-eyed" expression. 8 (rarely 9) supralabial plates on each side, the 4th and 5th touching the eye. The back scales are smooth, matte, and grooved (a unique feature among Western Palaearctic snakes, but more visible in adults than in juveniles), and arranged in 19 rows at mid-body. 159–195 ventral plates, divided anal scale, divided subcordals arranged in pairs. In the Western Montpellier Snake, there is very distinct sexual dimorphism as

*Western Montpellier Snake* Malpolon monspessulanus monspessulanus, *a large, 160-cm-long male. Saint-Jean-de-Védas (southern France). The large black mark on the back of the neck is typical of adult males of this species. It is also present in females, but less prominently.* P. Geniez.

*Adult female Western Montpellier Snake* Malpolon monspessulanus monspessulanus. *Near Saint-Maurice-de-Navacelles (southern France). As seen in this photo, the barred markings are distinct from those of the male.* Nicolas.

regards colour and size. Adult males are much larger and more thickset than the females. The head and front of the neck are greenish-brown, olive-green or almond-green; the back of the neck and the front of the body are black or blackish (this large black mark is sometimes called the "saddle"). The rest of the body is uniform brown or grey-brown, sometimes greenish, but never as bright as on the head and front of neck; the large flank scales are blue-grey with wide black borders; the belly is yellowish-white or yellow, with more and more black the older the individual becomes. The females have a beautiful colouration termed "meshed", marbled with black, whitish and mahogany brown on a reddish

background; the forehead has large pale markings with white spotting on the edges of the head scales, a white mark on the preocular and a white mark with black spotted edges on each labial; the dark saddle appears progressively in older females; the throat and belly are white or pale yellow with small orange marks, each with a black border; the whole is quite beautiful. Young males are pale grey with the same mottling as in the female, but less contrasting; young females are coloured even more brightly than adults.

**Venom** A venomous opisthoglyphous snake; the fangs have a groove for dispersing the venom, located at the back of the maxillae, which means

*Juvenile male Western Montpellier Snake Malpolon monspessulanus monspessulanus. Near Aspiran (southern France). Note the pale grey colouration enhanced with a few markings that are far less contrasting than in females. Note also the presence of two loreal scales, the very elongated frontal scale, and the fact that the bottom of the head is slightly concave, characters typical of the genus Malpolon. P. Geniez.*

that it is hardly dangerous to man. Cases of poisoning are quite rare and nearly always concern people who have handled a snake. Symptoms are limited to a reddening and swelling of the bitten limb. The Western Montpellier Snake is very timid, retiring into its hiding place at the slightest disturbance or fleeing at amazing speed. If cornered, it hisses loudly and for a long time as a warning; but, if captured, it can bite ferociously, even causing bleeding, and it aims for the face of its attacker.

**Habitat** The Western Montpellier Snake is the ultimate Mediterranean snake, even more so than its close relative the Eastern Montpellier Snake *Malpolon insignitus*. It occupies open, sunny sites in depressions or on hillsides, as well as dry slopes with abundant bushes with some stones and ground vegetation, valleys, open forest, maquis, waste ground in cultivated areas, and neglected gardens. In open Holm oak woodland, it occurs to 1400 m in France (Provence), 1600 m in the Spanish Pyrenees, 2406 m in the Baetic mountains in southern Spain and 2100 m in the Moroccan High Atlas.

**Habits** Very timid and discreet, fleeing at the slightest sign of danger, particularly shy of man; when lying in wait it can detect danger tens of metres away thanks to its large eyes. Lives mainly on the ground but will climb into bushes or small trees without difficulty.

The two species of Montpellier snake display an original and complex territorial behaviour. In spring, males travel long distances looking for females or in defence of their territory; it is at this time that many are killed on roads. Once a partner has been found, the male traces a territory around the hiding place of their choice by doing several rounds of inspection (sometime followed by the female), during which time a thin, colourless film of a dry secretion of pheromones, previously applied to the ventral scales, is disposited on the ground as the belly is rubbed against the ground. In order to mark their territory, the males, and to a lesser extent the females, when in their hiding place, apply a colourless, watery substance containing pheromones to the ventral scales and tail from a valve in the nostril. This substance is produced by a special nasal gland (one on each side of the snout, at the level of the loreal scales). As it presses this bulbous gland against its belly, the snake pushes the substance through a tiny hole at the exterior of the gland as soon as it has closed its nostril and the snake starts an act termed "auto-rubbing". This behaviour consists of two series of brief movements of the head (up to 100 in 90 seconds) during which the snake applies the substance to its belly scales. The particles serve as a chemical signal, a reminder, that the male and congeners detect using their mobile tongue, which is connected to the Jacobson's organ. For a territorial male, it serves to make sure, despite the thinness of his hemipenis (a characteristic of the

Psammophiinae), long uninterrupted mating bouts. These pheromone substances help to avoid continued fighting with other males, as the battle has already been decided.

During numerous matings, the pair stays immobile but vigilant, often with their heads raised, their eyes immediately spotting anything unusual. The hemipenis, which is slim and smooth at all occasions, allows for an immediate separation of partners in case of danger or if potential prey arrives within sight of the pair. During 20–40 days, in May and June, the male leaves all prey for the female, which, helped and escorted by the male, hunts throughout egg development until she moults in late June; this precedes egg laying by some 10 days.

**Diet** Lizards, snakes, small mammals (rodents, sometimes young rabbits) and birds. Juveniles, and occasionally adults, eat insects (e.g., large grasshoppers, butterflies

*Below left: Western Montpellier Snake* Malpolon monspessulanus monspessulanus, *juvenile female. Lignan-sur-Orb (southern France). In the young female, the colouration is almost the same as in the adult female.* Nicolas.

*Right: Close-up of Western Montpellier Snake* Malpolon monspessulanus monspessulanus, *juvenile female, showing the colour of the underparts, which are just as colourful as the upperparts. Coustouges (southern France).* P. Geniez.

*Western Montpellier Snake* Malpolon monspessulanus saharatlanticus, *the Atlantic Sahara subspecies, southwestern Morocco. This is an adult male of typical colouration: black spangled with white spots. Between Guelmin and Tan-Tan (southwestern Morocco). P. Geniez.*

and hymenoptera). The Western Montpellier Snake hunts by sight, head raised like a periscope, following its prey, at which it throws itself with incredible speed. It kills larger prey by holding it firmly in its mouth whilst suffocating it with powerful constriction movements until the fangs at the rear of the mouth deliver its strong venom. Small prey is swallowed directly.

**Reproduction** Oviparous. A clutch of 3–15 eggs, depending on the size of the female, laid under large stones, trunks or rubble or in rodent burrows. Eggs hatch in late August or early September. Newly hatched young are already quite large compared to those of other European snake species: 25–35 cm total length.

**Range** Western Mediterranean, from extreme western Italy (Liguria) to northwestern Algeria, passing, via Mediterranean France, into the whole of the Iberian Peninsula (Spain and Portugal) and the whole of Morocco, to the south along the edge of the Atlantic Sahara as far as Dakhla.

**Geographic variation** The Eastern Montpellier Snake *Malpolon insignitus* of the eastern Mediterranean is now considered to be a separate species, *Malpolon insignitus* (see following species), on the basis of significant morphological differences and genetic separation, this in spite of a zone (northeastern Morocco and northwestern Algeria) where morphologically intermediate specimens occur (adult males have a hardly visible dark saddle on the neck, slightly bluish scales at the bottom of the flanks but not contrasting black and near-white).

The Western Montpellier Snake is, however, considered to be polytypic, with two subspecies:

▶ *Malpolon monspessulanus monspessulanus* (Hermann, 1804): most of the species' range, south as far as the Moroccan High Atlas.
▶ *Malpolon monspessulanus saharatlanticus* Geniez, Cluchier and De Haan, 2006: along the oceanic coast of the Moroccan Atlantic Sahara. Adult males are characterised by the black saddle extending along almost the whole body, except for the head, from foreneck to tail. Each black scale is marked with a white or pale spot, giving the individual a remarkable spangled appearance. Females resemble those of the nominate subspecies but are darker and more spangled with white. Morphologically intermediate individuals are found in the Souss Valley (Agadir region in Morocco) and in the Jbel Siroua. Its status as a subspecies has been confirmed genetically.

*Magnificent Western Montpellier Snake, intermediate between* Malpolon monspessulanus monspessulanus *and* M. m. saharatlanticus. *Note the overall colour, which is quite similar to that of* M. monspessulanus *(greenish-grey with a large black saddle) and numerous white spots, typical of* M. m. saharatlanticus. *Askaoun (Jbel Siroua, southern Morocco).* García-Cardenete.

# Eastern Montpellier Snake

*Malpolon insignitus* (Geoffroy Saint-Hilaire, 1827)
**Family:** Lamprophiidae
**Sub-family:** Psammophiinae
**F.:** Couleuvre de Montpellier orientale
**G.:** Östliche eidechsennatter

**Identification** Similar to Western Montpellier Snake. Distinguished at all ages by absence of black saddle on the neck. Head appears a little less elongated, and the snout less arched. Nearly always 8 supralabials, 17 or 19 rows of dorsal scales at mid-body, depending on distribution and subspecies. Usually smaller, rarely longer than 150 cm (but 182 cm has been recorded, in Egypt). The males are generally less olive-green, more pure grey, blue-grey, yellowish-beige or sandy. The throat and neck are decorated with reddish flecking arranged in longitudinal lines. The scales at the bottom of the flanks have regular pale borders on each side, giving the snake a lined appearance; 2 or 3 longitudinal dark lines run along each flank as far as the tail (more marked in females than in adult males). The females are marbled and "meshed" as in the Western Montpellier Snake, but often integrating more orange tones. The underside is often reddish, more or less marbled, even in some males.

**Venom** A venomous opisthoglyphous snake; the grooved fangs are placed at

*Eastern Montpellier Snake* Malpolon insignitus insignitus, *young adult male. Note the complete absence of a dark area at the front of the body.* P. Geniez and C. de Haan.

*Eastern Montpellier Snake* Malpolon insignitus insignitus, *large adult male. To the north of Beer Sheva (Israel). Note the reddish colour, a colouration that does not occur in male Western Montpellier Snakes* M. monspessulanus. *Crochet.*

the rear of the maxillae, as in the other Montpellier snake species.

**Habitat** Mediterranean garigue, steppe with bushes or herbaceous tufts, rocky slopes, agricultural areas with hedgerows, damp depressions with scrub. In the south, occurs to the edge of desert areas and to the east into arid steppe.

**Habits** Very similar to those of the Western Montpellier Snake.

**Range** Eastern Mediterranean, from the Adriatic coast (from the northern to the extreme southern part of the Istria Peninsula) as far as the high plateaus of Algeria and eastern Morocco, including the whole of the Balkans, western and southern Turkey, southern Armenia, Transcaucasia, western Kazakhstan, northern and western Iran, northern Iraq, the Levant

countries (Syria, Lebanon, Jordan, northern Israel) and northeastern Africa. *Malpolon insignitus* is absent between southeastern Jordan and the northeastern Sinai and from the Nile Delta in Egypt. It is present on Lampedusa island (Italy), Cyprus and on several Ionian and Aegean Sea islands (Greece).

**Geographic variation** *Malpolon insignitus* is represented by two subspecies:

► *Malpolon insignitus insignitus* (Geoffroy Saint-Hilaire, 1827): range to the southern Mediterannean area, as far north as Israel, Jordan and Syria. Characterised by having 19 rows of dorsal scales at the mid-body. Nevertheless, individuals exist with 17 rows of dorsal scales, especially on the

235

*Eastern Montpellier Snake Malpolon insignitus fuscus, adult male. Montenegro. Note the absence of a black saddle on the back and of dark and pale scales on the flanks. Cluchier.*

Algerian and Moroccan high plateaus and on Lampedusa island (Italy), mixed with individuals with 19 rows.

▶ *Malpolon insignitus fuscus* (Fleischmann, 1831): the species' range to areas around the northern part of the Mediterranean, to the northeast as far as Kazakhstan, as far as Iran to the east, south to Lebanon and to the northwest as far as Istria and several Adriatic islands. Characterised by the presence of 17 rows of dorsal scales at mid-body.

**Genus *Rhagerhis***
The genus is represented by just one species, the Moïla Snake *Rhagerhis moilensis*. It is placed alongside the genus *Malpolon*, of which it was considered to be part not long ago.

## Moïla Snake

*Rhagerhis moilensis* (Reuss, 1834)
**Family:** Lamprophiidae
**Sub-family:** Psammophiinae
**F.:** Couleuvre de Moïla
**G.:** Moilanatter

**Identification** A medium-sized opisthoglyphous snake, with a quite pointed snout and 2 more or less visible dark oblique bars on each side of the back of the head. Length 70–95 cm, occasionally to 150 cm. A relatively slim snake, large individuals may appear quite thickset. Triangular head not very distinct from body; the pointed snout projects well beyond the lower mandible. A large flattened neck like that of the cobras; it is enlarged and flattened as soon as the snake feels threatened. Large eyes with round pupils and bright orange iris. The supraoculars slightly pass the eyes, but in a less pronounced manner than in the Montpellier snakes (*Malpolon monspessulanus* and *M. insignitus*). The dorsal scales are smooth and matte, sometimes slightly concave, arranged in 17 rows at mid-body. The snout scale, the rostral, is very developed, pointed, quite visible from above, and is inserted in a point between the internasals. The nasal, often divided, has a nostril valve (as in all other members of the Psammophiinae). 1 loreal scale, 1 very high and grooved preocular, of which the top part continues in a point above the front edge of the eye, 2 or 3 postoculars, 7 or 8 supralabials on each side. It has 151–176 ventral scales, and a divided anal scale; the divided subcordals are arranged in pairs. General colour reddish-yellow to sandy grey; the bottom of the flanks is very pale, nearly white. A latticed pattern on the back is more or less marked with brown-, reddish- or violet-tinted spots, often quite contrasting in juveniles. On the flanks are 2 series of alternating marks, the bottom ones smaller. There are 2 oblique dark bars on each side of the back of the

*Moïla Snake* Rhagerhis moilensis, *adult. Near Assa (southwestern Morocco). Note the neck enlarged into a shroud, somewhat like that of a cobra, and the double dark bar at the back of the head.* García-Cardenete.

*Close-up of Moïla Snake Rhagerhis moilensis, Wafrah Farms (Kuwait).* P. Geniez.

head, separated by a yellowish-white bar, the hind one is larger and better marked. This characteristic pattern of the Moïla Snake may fade, but the 2 wide bars are always visible, sometimes reduced to blurred marks. There is a dark mark on the 4th and 5th supralabials, under the eye. The belly is uniform creamy-white, the central part more colourful, between pale yellow and orange, sometimes with reddish speckling.

**Venom** A venomous opisthoglyphous snake; the grooved fangs are located at the rear of the maxillae. There is no known case of poisoning in humans. A relatively calm snake, it is, for example, less aggressive than the Montpellier snakes.

**Habitat** Stony deserts with sparse, bushy vegetation, sometimes close to human habitations. Very arid steppe, usually almost void of vegetation. Also at the foot of rocky slopes with scree and in dry wadi beds. Locally may occur to 1500 m.

**Habits** Diurnal, shy and quite rapid; however, not as quick as the Montpellier snakes. If threatened, it adopts a very particular form of intimidation, lifting the front third of its body such that it maintains at an angle (not in an S, as in cobras), flattening its neck in a shawl-like shape. It is only when captured that it tries to bite, with the mouth wide open. If it feels the threat has passed, it flees quite slowly, dropping its neck whilst keeping it inflated. It also shows auto-rubbing behaviour as in the Montpellier snakes. Ritual fighting between males has been recorded. As in many Saharan species, this one does not have a dormant period in winter.

**Diet** Lizards, snakes, small mammals (rodents) and birds. Juveniles also eat large insects.

**Reproduction** Oviparous. Mating occurs between April and June. The female lays 4–18 eggs in July or August, deep in a soft substrate. Hatching occurs in late September or October.

**Range** The whole Sahara and the Arabian Peninsula. To the north as far as the foothills of northern Algerian mountains, Israel, Jordan, central Syria, Iraq and southwestern Iran; to the south, as far as the Sahel belt.

**Geographic variation** *Rhagerhis moilensis* is considered to be a monotypic species despite its vast range.

**Genus *Psammophis***

This genus regroups some very slim, very rapid snakes, with an elongated head and very large eyes. Composed of no less than 34 species, it is distributed and has diversified, especially in Africa. A few species occur as far north as the Arabian Peninsula, central Asia and even Southeast Asia. Snakes in this genus also practice auto-rubbing, somewhat like the genus *Malpolon* but in a way that they can keep their balance even in trees. There, on branches, ritual fighting between males has been observed, as well as the females that, in order to drive off rival females without any violence, mark their male on the back with a viscous saliva that comes from a temporary outlet made in their 4th or 5th lower labial scales, on each side of the jaw (personal communication with Cornelius de Haan). The genus *Psammophylax* is in a clade that includes, and is closely related to, the genera *Psammophylax*, *Hemirhagerrhis*, *Disina* and *Mimophis*. This large clade is separated from a less diversified clade containing the genera *Malpolon*, *Rhagerhis* and *Rhamphiophis*.

# Forskål's Sand Snake

*Psammophis schokari* (Forskål, 1775)
**Family:** Lamprophiidae; **Sub-family:** Psammophiinae
**F.:** Psammophis de Forskål; **G.:** Forskåls Sandrennatter, Schlanke Sandrennatter

**Identification** 100–120 cm in length, to 148 cm. A very slender desert snake, thin and very elegant, with a very long, tapered tail, which moves extremely quickly. Head long and thin, distinct from body. Very large eyes with round pupils rimmed with white and orange-brown iris. Smooth and quite matte dorsal scales, arranged in 17 rows at mid-body. Frontal scale thin and elongated. The nasal scale, often divided, has a valved nostril. 1 well-developed and concave preocular that continues on from the long loreal, 2 (rarely 3) postoculars, 9 (rarely 8 or 10) supralabials. There are 160–180 ventral scales, a divided anal scale, and divided subcordals arranged in pairs. Quite variable colouration. One part of the population, found especially in Mediterranean areas, is strongly striped with 4 dark and 5 pale longitudinal lines. In certain individuals, these stripes are almost black and white, giving the snake a very elegant appearance. In adults, the dark stripes are chocolate-brown, the pale ones cream and brick-red. The head and nape also have a bright pattern of large, dark, elongated marks on a pale background. In steppe and especially desert regions, most individuals are uniform sandy beige or pinkish-grey, with dorsal lines. There are also populations that have very dull bands, reduced in some cases to a series of lines of small black spots and in others of small pale spots. In certain medium-arid areas, uniform individuals occur alongside lined, bicoloured individuals. All individuals, whether uniform or with contrasting colours, have a dark narrow bar that begins at the nostril, passes through the eye and continues to the front of the neck, separating the relatively dark colouring of the top of the head from the paler underparts. White labials, some tinted with yellow or orange and often with dark speckling. The undersides are generally very pale, with a yellow or dark marbled centre, more or less contrasting with the white sides. In other individuals, the belly is lead-grey, bordered by a yellow or white line and a black line.

**Venom** A venomous opisthoglyphous snake; the grooved fangs, located at the rear of the maxillae, can be implanted in the victim if the snake has the time to champ on the bitten part until the fangs come into contact with the flesh. In humans, there are reported cases of the bite causing intense pain, vomiting and swelling, but it is not fatal. Even though Forskål's Sand Snake is shyer and less aggressive than the Montpellier snakes, it should only be handled with caution.

*Forskål's Sand Snake* Psammophis schokari, *a magnificent lined specimen. Wadi Rum (Jordan).* M. Geniez.

*Forskål's Sand Snake* Psammophis schokari, *uniformly coloured, swallowing a Moïla Snake* Rhagerhis moilensis. *Note the white throat with fine black spotting. These two characters, when present, are enough to distinguish it from the similar but long-ignored P.* aegyptius. *Near Figuig (southeastern Morocco). Aymerich.*

**Habitat** At home in Mediterranean habitats, steppe or true desert habitats: open garigue, edges of orchards, open steppe with scattered bushes, wadis, oases, sandy desert. In the Sahel region, south of the Sahara, it also occupies open dry savanna.

**Habits** Diurnal; active during the day, but also at dusk. Very mobile and extremely rapid. Often occupies rodent burrows; sometimes found under small shrubs or large stones.

**Diet** Mainly lizards and similar; also other snakes, small birds and sometimes small rodents. Maintains its bite on its prey until it can put its rear-located fangs into action.

**Reproduction** Oviparous. Lays 5 or 6 eggs in June or July.

**Range** A very extensive range, described as representative of the Saharo-Sindian region: almost all of North Africa (except for extreme northwestern Morocco), as far south as the Sahel (Senegal, Mauritania, Mali, Niger); in Egypt only in the north and in the Sinai, then to the east along the Red Sea coast as far as Somali; farther east, Israel, Jordan, Lebanon, Syria, the whole of the Arabian Peninsula, Iraq, Iran, Turkmenistan (Koper Dag), southern Afghanistan, Pakistan and extreme northwestern India.

**Geographic variation** *Psammophis schokari* is a monotypic species (there are no subspecies) despite its vast range and immense colour variation. This conforms to genetic data that show low variability, suggesting that the species very rapidly colonised all the desert regions and the Mediterranean margins in contact with desert where it occurs.

## Egyptian Sand Snake

*Psammophis aegyptius* Marks, 1958
**Family:** Lamprophiidae
**Sub-family:** Psammophiinae
**F.:** Psammophis d'Égypte; **G.:** Ägyptische Sandrennatter

**Identification** Very similar to Forskål's Sand Snake *Psammophis schokari* in its general appearance and quite matte dorsal scales. Distinguished by its loreals being even more elongated with a more pronounced longitudinal groove, and by having more ventral scales: 183–203 instead of 160–180 in *P. schokari*. 17 or 19 rows of dorsal scales at mid-body (it seems there are always 17 in *P. schokari*). Quite a different colour: back a uniform yellowish beige, pinkish-grey or reddish-brown, with traces of dark spots arranged in 2 lines on the side of the back, sometimes invisible. The underparts are mostly uniform bright brick-red, without a line of markings. It can reach a total length of 150 cm.
**Venom** A venomous opisthoglyphous snake; the grooved fangs are located at the rear of the maxillae.

**Habitat** Truly a desert species. Sandy or stony deserts without vegetation. Also in open acacia forest on the sides of oases. To 1500 m in the southern Sinai.
**Habits** Diurnal snake, active mainly during the day but also at dusk. Just as quick, if not more so, as *P. schokari*, it is capable of crossing open desert between two bushes at an incredible speed, making it difficult to catch. Climbs in trees very well, in search of small birds resting whilst on migration.
**Diet** Clearly specialised in preying on migrant birds that pass in thousands through the eastern Sahara during their migration.
**Reproduction** Oviparous.
**Range** Eastern part of the Sahara: especially known from Egypt, where it is present almost throughout with the exception of the far north, where only *Psammophis schokari* is found. Present

The Egyptian Sand Snake Psammophis aegyptius *is closely related to the Forskål's Sand Snake* P. schokari. *The underside of the head and neck are brick-red and without dark markings, typical of this species. Note also the grooved and very elongated loreal.* P. Geniez.

also in the southern Sinai and extreme southern Israel. Could be present farther west in the southern Sahara and the northern Sahel: southern Libya and southern Algeria (Hoggar and Tassili n'Ajjer), Niger, Chad and Sudan. Until recently confused with *P. schokari*; the exact distribution of *P. aegyptius* is still poorly known.

**Geographic variation** *Psammophis aegyptius* is considered to be monotypic.

The Egyptian Sand Snake Psammophis aegyptius *specialises in hunting small migrant birds that it is capable of capturing very high in trees. Here, in Egypt, in a Spiny Acacia, an adult swallows a Common Whitethroat* Sylvia communis. *Crochet.*

# Striped Sand Snake

*Psammophis sibilans* (Linnaeus, 1758)
**Family:** Lamprophiidae
**Sub-family:** Psammophiinae
**F.:** Psammophis rayé; **G.:** Gestreifte Sandrennatter

**Identification** Quite a long, slender snake, but less so than *Psammophis schokari* and *P. aegyptius*. Normally 100–120 cm in length, may reach 144 cm in Egypt and 175 cm in the Sahel. Longish head, not particularly distinct from body. Large eyes with round pupils on an orange background. Smooth, shiny dorsal scales (duller in *P. schokari* and *P. aegyptius*), arranged in 17 rows at mid-body. Head scaling as in the two other species, 8 upper labials on each side. 158–172 ventral scales, on average fewer than in *P. schokari* (160–180) and far less than in *P. aegyptius* (183–203).

The back is olive, yellow-brown or dark brown. In the middle of the back is a thin line of dark and pale dashes contrasting with the dark grey of the back; there is another pale line at the bottom of the flanks. On the side of the neck is a series of more or less distinct large markings

*A young Striped Sand Snake* Psammophis sibilans. *Note the bright scaling and small pale vertical bars at the back of the head and on the neck, which allow this Afro-tropical species to be distinguished with certainty from* P. schokari. *Trapp.*

*Left: close-up of a young Striped Sand Snake* Psammophis sibilans *with a quite contrasting pattern, Djikoye (Senegal). Right: adult, of uniform yellow-brown colouration. Diogo (Senegal). Trape.*

that fade to disappear in the front third of the body. The head pattern is markedly distinct from that of *P. schokari* and *P. aegyptius*: it has transverse pale lines that divide the large dark temporal patch into oval marks with dark edges, especially on the sides of the head. This pattern fades considerably in adults. The belly is yellow, greenish-yellow or porcelain-coloured, sometimes with a striped pattern. Young Striped Sand Snakes have a beautiful contrasting pattern that fades with age. Most adults are almost uniform, including the head. To separate them from uniform examples of *P. schokari*, the scaling is more shiny in the latter species, and the dark-dotted vertebral line more or less persists in the adult.

**Venom** A venomous opisthoglyphous snake; the grooved fangs are located at the rear of the maxillae.

**Habitat** A Sahel species that favours well-vegetated savanna, particularly on the edges of wetlands or watercourses. In the area covered by this book, present in Egypt, along the Nile Valley, where it is found in all sorts of relatively damp habitats: cultivated ground, meadows, sides of canals and fish farms, edges of reedbeds.

**Habits** A diurnal terrestrial snake that may climb into low bushes. Shy and rarely agile. If captured, it bites immediately.

**Diet** Small mammals (rodents), lizards and similar, frogs and birds. Pursues its prey, head raised to increase its field of view. Bites its prey and with movements of the jaw holds it in its mouth until the venom has been inoculated.

**Reproduction** Oviparous. Very prolific, with clutches of up to 30 eggs.

**Range** A species of Afro-tropical origin present in much of Africa. Occurs northwards along the Nile Valley as far as the Mediterranean coast, but spreading no further; it thus occurs in the Western Palaearctic and the region covered by this book. Found several times on the northern edge of the Sinai, where it has probably been released by man. Recently discovered on the Banc d'Arguin (northwestern Mauritania) on Tidra island, some 600 km north of its previously known range in southern Mauritania. The islands of the Banc d'Arguin are traditionally associated with the Western Paleartic and thus are part of the area covered by this book.

**Geographic variation** *Psammophis sibilans* is considered to be polytypic (it has several subspecies), but its systematics are still controversial, as several subspecies probably represent full species. In Egypt, this is true of the nominative subspecies, *P. s. sibilans*.

**Remarks** To be comprehensive on the genus *Psammophis* in the Western Paleacrtic, it should be noted that *P. lineolatus* (Brandt, 1838) has been reported once from Azerbaijan, in the Nakhïchevan region. This species is widespread in central Asia, from the eastern shores of the Caspian Sea as far as Iraq, northwestern China and Mongolia. This record is thus the only one from west of the Caspian Sea: it is an isolated record, more than 500 km from the rest of the species' range and thus needs confirming.

## Sub-family Atractaspidinae

A group of very strange burrowing snakes, with a cylindrical body almost triangular in cross section, with bright scales and without any pattern on the back or the head. The general colour is always shiny black or dark brown (except for the two species in the genus *Homoroselaps*); some species have a slightly paler head. The head is small and indistinct from the body. The very small eyes have round pupils. Species of the genus *Atractaspis* are venomous and have the particularity of having 2 very long, slightly articulated fangs, located at the front of the maxillae. The snake can project its fangs from the mouth, even when it is closed, and thus easily inoculate its venom – for example, on anyone who handles the species without the necessary precaution.

The Atractaspidinae snakes have undergone a multitude of changes in classification within families, which has proved very difficult when taking into account their anatomy and morphology. Genetic studies have shown that they are a part of a large clade that also includes the Colubridae, the Lamprophiidae and the Elapidae. They are closely related to the Apallactinae within the family Lamprophiidae. Some authors consider that they belong to a family in their own right, the Atractaspididae. There are two genera within the Atractaspidinae, *Atractaspis* (21 species) and *Homoroselaps* (two species in South Africa). Their total range covers most of sub-Saharan Africa, and two species have reached the Arabian Peninsula, one of which, *A. engaddensis*, occurs in the Western Palaearctic.

# Palestine Burrowing Asp

*Atractaspis engaddensis* Haas, 1950
**Family:** Lamprophiidae
**Sub-family:** Atractaspidinae
**F.:** Atractaspide d'Ein Gedi
**G.:** Israelische Erdviper

**Identification** Peculiar, almost completely shiny black snake with a small conical head and flattened snout; there is no demarcation between the head and the rest of the body. Total length 50–70 cm, to at least 75 cm for the largest individuals. The rostral scale is very developed, conical and prominent (adapted to the burrowing behaviour of the species); a very wide trapezoid frontal scale. The supraoculars are small and narrow, the parietals smaller than the frontal scale, a small and trapezoid preocular, a single postocular, 6 supralabials on each side, the 3rd and 4th are highest, in contact with the eye. Small eyes with a black iris, making the round pupil hardly visible. The body is slightly triangular in cross section, flattened on the belly. The dorsal scales are smooth and shiny, arranged in 23–29 rows at mid-body, 260–285 ventral scales, an undivided anal scale, 31–39 undivided subcordals, which is the contrary to most snakes, which generally have them arranged in pairs. The normal colour is shiny black, sometimes dark brown, especially around the head. The ventral scales are iridescent black.

**Venom** Calm, venomous snake, that may get angry if captured; then it will try to bite the aggressor's skin with the very long fangs that stick out of the sides of the mouth. The backwards-pointing fangs are located at the front of the mouth; they are articulated and slightly mobile. This snake is capable of inoculating its venom by turning one of its fangs outside the mouth whilst keeping the mouth closed, and making a lateral movement of the head – more of a sting than a real bite. The venom, similar to that of vipers, is highly toxic and can be very dangerous for humans. Rare cases of death from poisoning are known for other species *Atractaspis* snakes from sub-Saharan Africa. A bite from *A. engaddensis* produces painful swelling of the bitten limb, which may become necrotic and gangrenous, possibly leading to amputation (at least of a finger). The victim also suffers from side effects: nausea, heavy sweating, abdominal pains, loss of consciousness, hypertension, haemorrhaging, heart trouble and others. A case of death has been recorded near Riyadh, in Saudi Arabia: a two-year-old girl, after having been bitten by a closely related species, *A. andersonii*. No antivenin exists against the venom of *Atractaspis* species.

*Palestine Burrowing Asp* Atractaspis engaddensis. *Israel. Note the pointed snout, very small eyes, very small and narrow supraocular scales, thick neck and bright black scaling.* Haimovitch.

**Habitat** Arid and desert steppe with some vegetation, or with small shrubs or bushes, sometimes near oases and farms; also found in wide wadi beds. Prefers soft, even sandy soils.

**Habits** The *Atractaspis* species are burrowing snakes, rarely seen in the open. They are sometimes found moving on the surface at night after rain.

**Diet** Prefers other snakes, sometimes of equal size, especially members of the Leptotyphlopidae, but also others such as *Eirenis coronelloides*.

**Reproduction** Oviparous. Mates in June; lays 2 or 3 eggs in September, very late in the season; hatching occurs 2–3 months later.

**Range** This species, of Afro-tropical origin, has a relict distribution limited to western Jordan, Israel, the Sinai (Egypt) and extreme northwestern Saudi Arabia.

**Geographic variation** *Atractaspis engaddensis* is a monotypic species. Some authors consider that it is very closely related to *A. andersonii*, of the central and southwestern Arabian Peninsula, to such an extent that the two are of the same species.

**Genus *Micrelaps***

The *Micrelaps* are small, opisthoglyphous, burrowing, elongated, cylindrical snakes with shiny scales and a very small, flattened, round head, indistinct from the neck; they have very small black eyes with hardly visible dark or variegated round pupils.

Three species occur in sub-Saharan Africa; the fourth, *Micrelaps muelleri*, is endemic to a small area in the Levant countries. The genus Micrelapsis is placed in the sub-family Aparallactinae , within the family Lamprophiidae. However, the position of one of the species, *Micrelaps bicoloratus*, at the base of the evolutionary tree that regroups the Lamprophiidae and the Elaphidae, would suggest that this genus could constitute a separate family within its own right, that of the Micrelapsidae.

## Müller's Ground Viper

*Micrelaps muelleri* (Boettger, 1880)
**Family:** Lamprophiidae
**Sub-family:** Aparallactinae?
**F.:** Micrélaps de Müller
**G.:** Müllers "Erdviper"

**Identification** Small, slim and cylindrical burrowing snake with general appearance of a colubrid snake; shiny with black and pale rings. Back with creamy or pinkish-white or pale yellow ground colour with about 40 black or very dark brown rings that are often wider than the ground colour that separates them. These rings may be regular, sometimes with scalloped edges, or sometimes irregular and incomplete. In juveniles, the spacing between rings is narrower (2 or 3 rows of scales) than in adults (5 or 6 rows of scales). In a few individuals, there are 7 or 8 rings on the neck; the rest of the back is yellow without rings, and the flanks are entirely black (see photo). The rings touch the belly, which is shiny black. The head is entirely black or a violet black-brown, separated from the first back ring by a yellow collar. Length 30–37 cm, rarely to 52 cm; body cylindrical, but flatter on belly. Tail quite short and thick, only slightly tapered, finishing in a stump, not in a thin point, as in most snakes. Small, slightly flattened head follows the silhouette of the body, without a marked neck. The eyes are very small and round or slightly oval, with a black iris. The dorsal scales are smooth and shiny, arranged in 15 rows at mid-body. The snout scale (rostral), large and convex, rises in an obtuse angle between the 2 internasals. The frontal scale is smaller than in most other colubrid snakes, is hexagonal and is inserted at an acute angle at the rear of the parietals. No loreal or preocular scales, only 1 postocular (very small), 7 supralabials on each side, the 3rd and 4th in contact with the lower edge of the eye. The 3rd, 4th and 5th supralabials are very high, the 5th sometimes nearly touching the parietal. 242–279 ventral scales, a divided anal scute, 20–32 pairs of subcordals.

**Venom** A venomous opisthoglyphous snake with only 3 teeth of increasing size on each maxillae, the 3rd (the fang) being the largest. The venom is probably very virulent. Even though it is a very small and rather calm species, caution is recommended.

**Habitat** Steppe with spiny bushes and semi-desert, often in stony or rocky areas. In mountainous areas to 2000 m.

*Left: Müller's Ground Viper* Micrelaps muelleri. *Israel*. David Modr. *Above: close-up of* M. muelleri, *Israel*. García-Cardenete.

*Müller's Ground Viper* Micrelaps muelleri, *an atypically marked individual, Israel*. García-Cardenete.

*Tchernov's Ground Viper* Micrelaps muelleri tchernovi. *Israel.* Haimovitch.

**Habits** Hardly known. Nocturnal, terrestrial snake. Readily burrows in the ground or installs itself in small mammal burrows.

**Diet** Lizards and snakes, including those its own size (e.g., the Striped Dwarf Snake *Eirenis lineomaculatus*).

**Reproduction** Oviparous.

**Range** A little-known species, present in Israel, Lebanon, extreme northwestern Jordan and western Syria. It is considered to have a relict distribution of Ethiopian origin.

**Geographic variation** Another species of the genus *Micrelaps* has recently been described that is present in Israel and the Mediterranean steppe zones at low elevation in northwestern Jordan: *M. tchernovi* Werner, Babocsay, Carmely and Thuna, 2006. In this taxon the black transverse bars do not reach the ventral scales, forming saddles rather than rings. They are more numerous: about 58 compared to 34–45. The description of this species is simply based on colour rather than on genetic differences; the validity of *M. tchernovi* as a species is contested by several authors. In the meantime, without further information, we consider this taxon as a subspecies: *M. muelleri tchernovi*.

# Family Elapidae (cobras and sea snakes)

The Elapidae resemble the harmless colubrid snakes. However, they have 2 fangs located at the front of the maxillae (proteroglyphous dentition) and have highly toxic, quick-acting venom that acts principally on the nervous system (neurotoxic). Certain species in this family are amongst the most venomous and dangerous snakes in the world (e.g., the Common Death Viper and Inland Taipan of Australia, the King Cobra of Southeast Asia and the mambas of Africa).

Located at the front of the upper maxillae, the fangs are less developed than those of vipers. They are fixed (not articulated like those of vipers), and the venom flows through an almost completely closed duct during a bite. Another characteristic of the Elapidae is they have no loreal scale (unlike most other snakes). A few species, such as the spitting cobras, can spit their venom several metres and attain a target as precise as human eyes.

Their range extends over Central and South America, Africa, Asia, Indonesian archipelagos such as New Guinea, the Australsian archipelagos and Australia. Four species of Elapidae are known to occur in the area covered by this book: the Egyptian Cobra *Naja haje*, present in North Africa; the Nubian Spitting Cobra *N. nubiae*, along the Nile Valley in Egypt; the Desert Cobra *Walterinnesia aegyptia*, found in northeastern Egypt and the Levant countries; and Morgan's Black Desert Snake *W. morgani* of more eastern distribution.

**The sea snakes**. All the sea snakes are members of the family Elapidae. In the past, they were placed in their own family, the Hydrophiidae, then in two sub-families of the Elapidae, the Hydrophiinae and the Laticaudinae. Thanks to the advent of molecular (genetic) studies, we now know that sea snakes are not monophyletic (that is, they are not all directly related) but are spread among various groups that, for some, include terrestrial members of the Elapidae.

In the Western Palaearctic (the area covered by this book), several species occur in the Persian Gulf (between the Arabian Peninsula and Iran) having come from coastal India. Four species have been recorded to date, all well within the Persian Gulf, on the coasts of Kuwait and extreme southeastern Iraq: *Hydrophis cyanocinctus*, *H. gracilis*, *H. ornatus* and *H. platurus* (= *Pelamis platura*). Five other species reported from farther southwest in the Persian Gulf are likely to be recorded in Iraq, Kuwait or extreme northeastern Saudi Arabia in the future: *H. schistosus* (= *Enhydrina schistosa*), *H. lapemoides*, *H. spiralis*, *Lapemis curtus* and

*Praescutata viperina*. In those countries, sea snakes are sometimes seen in very shallow waters, during a falling tide. Although normally quite calm, they do present a danger for humans and should not be handled. Their venom is one of the most dangerous, and bites can be fatal.

Sea snakes are not treated in this book, although they are included in the country lists (page 370 and 371).

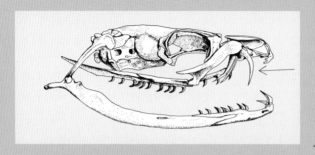

*Skull of an elapid showing the tubular venom fangs at the front of the maxillae.*

# Egyptian Cobra

*Naja haje* (Linnaeus, 1758)
**Family:** Elapidae
**F.:** Cobra d'Égypte
**G.:** Uräusschlange

**Identification** One of the most elegant of Western Palaearctic snakes. Large and quite slender, shiny, general appearance of a colubrid snake. Total length: 150–180 cm in the Western Palaearctic, to 194 cm in Egypt, 181 cm in Morocco (but probably over 200 cm, according to reliable sources). Can reach up to 260 cm in the African Sahel. Quite a large head, quite pointed at the front and only slightly distinct from the body when the animal is in a non-threatening posture. Medium-sized eyes appear small in adults, with round pupils and a very dark iris so that the pupil is hardly visible. It has a thick neck the same width as the head. As soon as the snake is alarmed, it enlarges its neck in the form of a flattened "cape". A well-developed, pointed rostral plate, going beyond the front of the mouth; no loreal, as in all Elapidae; 1 long preocular that replaces the loreal, 1–3 small scales under the eye that generally separate it from

*Egyptian Cobra. Tahoua (Niger). Note the pointed, prominent snout and the pale ground colour with irregular dark markings, typical of* Naja haje haje. *Trape.*

*Moroccan Cobra* Naja haje legionis, *an almost entirely black subspecies found nearly exclusively in Morocco. The Tan-Tan region, not far from Drâa Wadi (Morocco). Note the extreme elegance of the snake.* Aymerich.

the supralabials, 2 postoculars, 7 supralabials on each side, the 7th surmounted by a large, oblique temporal scale, which prolongs the line of the supralabials upwards. Very large and shiny scales under the head, similar to the colubrid snakes: 2 internasals, 2 prefrontals, 2 supraoculars, 1 frontal and 2 parietals. The back scales are smooth, quite shiny and very elongated, giving the snake a distinct aspect compared with colubrid snakes. They are arranged in 21 rows at mid-body, 197–216 ventral scales in Egypt, 201–212 in Morocco, 199–226 in West Africa; an undivided anal scute; the divided subcordals are arranged in pairs; rarely, the 2nd and 3rd rows may be fused into a single subcordal plate. Colouration is quite variable, depending on age and geographic origin. In Egypt and the Sahel, they vary from straw-yellow to pale brown to quite dark brown.

*Moroccan Cobra in a threatening position. Tan-Tan (southwestern Morocco).* Cayuela.

*Two newly hatched Moroccan Cobras Naja haje legionis. In juveniles, the head and the whole of the hood are black, and the rest of the body is pale. Near Sbouya (left) and near Labyar (southwestern Morocco). García-Cardenete.*

This colour may be uniform or have irregular small black markings that may cover a large part of the back, giving the snake a dirty appearance. The juveniles and subadults have more contrasting colours than adults; the head and neck are black or very dark. The belly is a dirty beige, with a large blackish band of variable extent under the neck towards the back of the hood. There is a black or dark mark under the eye on the 3rd and 4th upper labials. In Morocco, individuals become very dark with age (see "Geographic variation", below).

**Venom** A venomous snake with fixed 8–12 mm-long fangs located at the front of the upper jaw. The neurotoxic venom is quite virulent, acting primarily on the musculature of the respiratory organs. It also contains haemolytic components that provoke severe necrosis that may result in the necessity of amputation of the affected limb (even a leg) if the victim has survived the neurotoxic symptoms. Without the use of a serum and hospitalisation, most bites are usually eventually fatal.

**Habitat** Varied, although always dry habitats with scattered rocks and bush. Steppe and savanna, the edges of deserts, mountain slopes with rocky scree and bushes, also littoral steppe influenced by sea spray, oases, wadis with sparse vegetation, edges of cultivated fields, roadside ditches and brownfield sites with ruins. Primarily present in regions with mild winters and more even climate.

**Habits** A terrestrial snake mainly active at dusk and at night. In the morning, it may be seen in the sun close to its burrow, into which it quickly retreats at the slightest sign of danger. Very sedentary, it may occupy the same burrow or live in the same termite mound or under the same pile of rocks for years. It can be very rapid and mobile, and can have very sharp reactions. It will flee at the slightest alert. However, if cornered, it will adopt the typical menacing attitude of cobras, with the hood stretched, but it will bite only if provocation continues. It has a short, inactive spell in winter, but is active during warm and humid evenings in February in Morocco. The Egyptian Cobra is very much sought after by snake charmers and animal dealers due to its elegance and spectacular behaviour. Where it occurs in the Western Palaearctic, it is of relict distribution; populations are low, and it is menaced with extinction within the more or less short term.

**Diet** Small mammals, birds and their eggs, lizards and similar, snakes, frogs and toads. An Egyptian Cobra in Morocco, on being handled, regurgitated a young Puff Adder *Bitis arietans* (Gilles Trochard, personal communication). Kills its victims with venomous bites.

**Reproduction** Oviparous. Lays 8–20 eggs in the stump of a hollow tree, in a hole in the ground, in a rodent burrow or under boulders. Young Egyptian Cobras measure 23–34 cm at hatching.

**Range** A widely distributed species in sub-Saharan Africa. In the Western Palaearctic, it appears to be isolated to the northern Sahara except for its presence in the Nile Valley, which constitutes an important corridor allowing tropical species access to areas farther north. Present in Morocco, to the southern part of the Anti-Atlas chain and along the Saharan Atlantic coast, as far as Boujdour. Unknown in most of Algeria, except for the Aurès hills, in the northeastern part of the country. From there it is present in Tunisia and along the Mediterranean coast in Lebanon and northern Egypt; present in the Nile Delta and along the Nile Valley as far as its frontier with Sudan.

**Geographic variation** *Naja haje* is often considered to be monotypic, especially due to very little genetic difference between populations. However, new species have been recognised due to populations showing very distinct morphological and genetic characteristics. We accept the existence of two subspecies:

▶ *Naja haje haje* (Linnaeus, 1758): Egypt and Libya and further south to Sahelian Africa. Generally pale, a sandy or brown colour.

▶ *Naja haje legionis* Valverde, 1989: Morocco and probably northeastern Algeria. A less developed rostral scale, less pointed than in individuals in Egypt or Sahelian Africa. Generally bright black or very dark mahogany-brown, sometimes iridescent or with a blue sheen. The throat and underside of the head are paler, pale yellow or whitish. The belly is violet-black. Juveniles have a bright black head and neck (the hood); the rest of the body is abruptly pale beige or yellowish, marbled with numerous thin dark markings. Some individuals have a wide yellow band under the middle part of the hood. Generally the snout and throat are yellow. In Tunisia, Egyptian Cobras appear to have an intermediate colouration, quite dark, with much of the body covered with dark scales on a pale background. Many authors do not recognise *N. h. legionis* as a valid subspecies due to the small genetic difference between *legionis* and other populations, and this despite constant and unique colouration characters, which are within the framework

of recognition of a population as a subspecies. Furthermore, it is possible that the distribution of *N. h. legionis* isolates and separates it from other Egyptian Cobra populations. This is very important in terms of conservation, as the Moroccan Cobra, as it is often called, is in imminent threat of extinction, faced with being ruthlessly and incessantly captured for members of the Aïssaoua fraternity and snake charmers.

## Nubian Spitting Cobra

*Naja nubiae* Wüster and Broadley, 2003
**Family:** Elapidae
**F.:** Cobra cracheur de Nubie
**G.:** Nubische Speikobra

**Identification** A medium-sized spitting cobra, belonging to the sub-genus *Afronaja* (*Naja haje* belongs to the sub-genus *Naja*). Total length less than that of the Egyptian Cobra: to 151 cm in Egypt (Djebel Elba), 148 cm in Sahelian Africa, but normally to 120 cm. A large, quite short, round head, slightly distinct from the body (more so than in *N. haje*). Round pupils and quite a dark iris. The round rostral plate is non-protruding and far less developed than in *N. haje*; 1 long, triangular preocular replaces the loreal; sometimes a second small preocular; generally 3 postoculars; 7 supralabials on each side, sometimes 6 or 8, the 4th touching

the eye. The dorsal scales are smooth, not as long as in *N. haje*, and more numerous, arranged in 25 (sometimes 23 or 27) rows at mid-body. 214–228 ventral plates, an undivided anal scute, the subcordals divided and arranged in pairs.

Uniform pale head, between sandy beige and straw-coloured, with a narrow oblique black band under the eye (*N. haje* has a large, more or less square dark mark). There is a black collar on the nape, of variable width depending on the individual, in front of a large grey transverse band that is followed by another wide dark band. The rest of the body is of a uniform

*Nubian Spitting Cobra* Naja nubiae, *juvenile. Sudan. Note the uniform colouration of the back, in contrast to young Egyptian and Moroccan Cobras, which have numerous dark markings on the body.* Mazuch.

shiny mahogany-brown in adults, more sandy-beige in juveniles. The belly is a uniform yellowish-white or brownish-yellow. On the underside of the lower neck (under the hood) there are 1 or 2 wide black or dark bands.

**Venom** As in all cobras, it possesses a strong neurotoxic venom. What's more, it is capable of projecting a jet of venom at the face of the aggressor from a distance. The venom is a strong irritant if it gets in the eyes, which must be washed immediately; if not, the resulting damage may lead to blindness.

**Habitat** Dry savannas and semi-desert, arid mountainous regions in which, however, it occupies the more humid,

well-vegetated areas. In Egypt, on the edges of cultivated areas of the upper Nile Valley, the rocky shores of Lake Nasser and desert areas around it.

**Habits** Mainly active during dusk and at night, rapid and very wary. When worried, it raises its head and the front third of its body, widening its neck into a hood, showing off the 2 dark bars under its neck, before spitting its venom and then biting if provoked further or captured.

**Diet** Probably quite similar to that of its close relative, the Black-necked Spitting Cobra *Naja nigricollis*: amphibians, reptiles, rodents and probably birds.

**Reproduction** Oviparous.

*Nubian Spitting Cobra Naja nubiae, an adult confiscated by customs officials at London; origin unknown). Compared to the colubrid snakes, note the distinct scale pattern on the head: no loreal, the preocular directly in touch with the nasal scale, and the very high third supralabial that touches the preocular. Wüster.*

**Range** A species of dry savanna of East Africa, from Eritrea as far as the Aïr Mountains via Somalia, Kenya, Tanzania, Sudan and Chad. It extends northwards along the Nile Valley in Egypt as far north as Durunkah, just to the south of Assiout, some 540 km to the southern Mediterranean.

**Geographic variation** *Naja nubiae* is a monotypic species. Confused with *N. nigricollis* in the past and then with *N. pallida*, two species of sub-Saharan spitting cobras, it was identified as a separate species as recently as 2003.

## Black Desert Snake, Desert Cobra

*Walterinnesia aegyptia* Lataste, 1887
**Family:** Elapidae
**F.:** Cobra noir du désert
**G.:** Schwarze Wüstenschlange

**Identification** A strange desert snake, almost entirely shiny black with the general appearance of a thick colubrid snake. Average length 90–110 cm, to 132 cm in Saudi Arabia. Large, thick head that is only slightly distinct from the body, with a large, flat snout. Small eyes with round pupils and dark iris. No loreal plate, 1 very long preocular that touches the rear nasal scale, 2 postoculars above a small subocular, 7 very high supralabials on each side, the 3rd and 4th touching the eye.

*Black Desert Snake, or Desert Cobra* Walterinnesia aegyptia. Heckes.

The shiny dorsal scales are numerous and arranged in 24–29 (rarely 23) rows around the nape, 23 at mid-body; they are smooth on the neck, becoming progressively more keeled towards the rear of the body. 178–210 ventral plates, a divided anal scute (in contrast with cobras and atracaspides); most of the subcordals are divided and arranged in pairs, but a few at the start of the tail are undivided (4–22 in males, 1–5 in females).

The body is a uniform very dark brown or shiny black. The snout and sides of the head and neck are often paler, more a yellowish-brown, but in some individuals the whole head is the same colour as the body. The belly is brownish or greyish-yellow. Juveniles are a uniform almost black, dark reddish-brown.

**Venom** A venomous elapid; the fangs are located at the front of the maxillae. It has a very toxic neurotoxic venom.

Cases of humans being bitten are rare, as it is quite calm, discreet, burrows and is nocturnal. In most known cases, the resulting symptoms appear not to have been too serious. A bite has been known to provoke enormous haemorrhaging around the bitten area. A fatal case is known from Israel: a small girl inadvertently sat on a snake and was bitten; she died 20 hours later; however, she did not receive medical care.

**Habitat** Desert and arid areas, stony or sandy, with sparse bushy vegetation; wadis, oases and abandoned gardens. In Jordan, it penetrates slightly into the Mediterranean zone. It spends part of its time under boulders, in cracks in the soil or in small mammal burrows. It is especially fond of the banks of streams and water bodies occupied by the Green Toad, a favourite prey species. It can even swim.

**Habits** Nocturnal, much of its time is spent underground. It has poor eyesight and detects its prey using its acute sense of small. Quite slow and calm during the day, it can be aggressive at night. When surprised, it flattens its body on the ground, hiding its head under its body coils. When threatened, it hisses loudly and strikes with its head, generally with a closed mouth.

**Diet** Mainly Green Toads *Bufotes viridis*; also lizards and similar, snakes smaller than itself, occasionally young rodents in the nest or small birds. Kills its prey by holding them in its jaws and poisoning them.

**Reproduction** Little known. Oviparous. Newly hatched young snakes have been observed in September.

**Range** Desert and semi-arid areas in northeastern Egypt, in the Sinai, Israel, western Jordan and northwestern Saudi Arabia.

**Geographic variation** *Walterinnesia aegyptia* is a monotypic species. Eastern populations have recently been ascribed to another species, *W. morgani* (see next species).

## Morgan's Black Desert Snake

*Walterinnesia morgani* (Mocquard, 1905)
**Family:** Elapidae
**F.:** Cobra noir de Morgan
**G.:** Morgans Schwarze Wüstenschlange

**Identification** Very closely related to *Walterinnesia aegyptia*, from which it was separated as recently as 2007. It has fewer rows of scales around the nape (21–23 compared to 24–29, rarely 23), which also means it has a slightly slimmer neck. Often 23 rows of scale at mid-body, sometimes 21. The subcordals are less numerous in the females (39–44 compared to 44–46). And note that the juveniles have 25–33 transverse reddish-white bars across the back. The adults appear identical to those of *W. aegyptia* at first sight.

**Venom** A proteroglyphous elapid with a neurotoxic venom that probably has the same properties as that of *W. aegyptia*.

**Habitat** Steppe with mugwort, wheat and pea fields as well as rocky, arid steppe. To 2000 m on the southern face of the Elbourz chain (northern Iran).

**Habits** Probably identical to *W. aegyptia*.

**Diet** Probably identical to *W. aegyptia*.
**Reproduction** Little known. Oviparous.
**Range** Desert and arid areas in the
eastern half of the Arabian Peninsula,
north to Kuwait; also in western and
northwestern Iran, Syria and extreme
southeastern Turkey (Sanluirfa region),
which is at the northern limit of the
range of the genus *Walterinnesia*.
**Geographic variation** *Walterinnesia
morgani* is a monotypic species.

*Morgan's Black Desert Snake* Walterinnesia morgani, *adult (southeastern Turkey). Note that in the
adult the entirely black colouring is the same as that in* W. aegyptia. Göçmen.

# Family Viperidae (vipers)

This family contains three sub-families: the true vipers (sub-family Viperinae), the pit vipers (sub-family Crotalinae) and the Fea's vipers (sub-family Azemiopinae), this last sub-family, the most primitive, contains only two species, both in a single genus: *Azemiops feae* and *A. kharini*; their range covers part of Southeast Asia.

One of the more spectacular features of the vipers is their venom apparatus, the most advanced in the reptile world. They have 2 – sometimes more – large, tubular venom fangs at the front of the upper jaw that act like a syringe (solenoglyphous fangs, see page 24) and are positioned on a mobile bone. When not in use, they are folded backwards in a groove in the mucous membrane. When the snake is biting, the fang points forwards, penetrates the victim's skin and operates like a syringe. The venom is highly effective, and the inoculation system no less so. In many species, the bite is very dangerous and can be fatal for humans without rapid and adequate treatment. The primary function of the venom is to paralyse prey before it is swallowed.

The vipers are usually thickset snakes, sometimes quite squat; their body scales are keeled and the commonly triangular head is more often than not quite distinct from the body. In many of the species, especially the true vipers, the underside of the head is covered with small scales. The most visible difference between true vipers and pit vipers is that the latter have a specialised heat-sensing organ, the loreal pit, located on each side of the head between the nostril and the eye. These very sensitive heat detectors allow the pit vipers to locate and then catch warm-blooded animals even in total darkness.

The majority of Western Palaearctic vipers are terrestrial. However, in tropical forests, many species of true and pit vipers are arboreal and of a green colour that matches the foliage in which they conceal themselves.

True vipers only occur in the Old World (Europe, Asia and Africa), whereas pit vipers are widespread in the New World as well as in Asia. Halys Pit Viper is the only pit viper to occur in Europe, in Kazakhstan and Azerbaijan.

From a phylogenetic point of view, the members of the Viperinae are very divergent from the members of the Crotalinae, but both are retained as sub-families within the Viperidae, as they undoubtedly have a common ancestor that is more recent than those of other groups of snakes, and they share morphological

*A viper's skull showing its well-developed tubular venom fangs, located at the front of the maxillae on a mobile bone. In this example they are in the resting position, folded backwards.*

characteristics that are evidently ancestral and not the result of convergent adaptive evolution giving rise to the same morphology (e.g., the venom apparatus).

Six large clades are recognised within the sub-family Viperinae:

► genus *Bitis* (17 species in Africa, and reaching the southern part of the Arabian Peninsula, 1 species in the Western Palaearctic, the Puff Adder);

► genus *Atheris* (15 species, all in sub-Saharan Africa);

► genus *Causus* (7 species, also all in sub-Saharan Africa);

► genus *Echis* (some 10 species, in Asia and Africa);

► genera *Cerastes* (3 species) and *Proatheris* (just 1 species, in East Africa);

► all the other genera in the Viperinae (*Vipera, Montivipera, Macrovipera, Daboia, Pseudocerastes* and *Eristicophis*).

## Sub-family Viperinae (true vipers)

### Genus *Vipera*

Quite a few small or medium-sized viper species are placed in this genus, which is distributed throughout Europe and central Asia. This genus has reached Africa but only the north, the Maghreb. All are associated with Mediterranean or mid-European steppe habitats, even boreal habitats for the Adder (Common viper). From a phylogenetic point of view, the *Vipera* form a large clade along with the genera *Macrovipera, Daboia* and *Montivipera*. This clade contains all the species that until recently were placed in the genus *Vipera*; it is closely related to the genera *Pseudocerastes* and *Eristicophis*.

Three groups emerge from the genus *Vipera*, especially from a morphological perspective: the *V. ursinii/kaznakovi* group, the *V. berus* group and the *V. aspis/latastei/ammodytes* group; these first two are genetically closer to each other than they are to the third, *V. aspis, V. latastei* and *V. ammodytes*. For many authors, they constitute a sub-genus apart, the sub-genus *Pelias*, the other *Vipera* being part of the sub-genus *Vipera*. Other authors consider *Pelias* to be a genus in its own right and use, as an example, the designations *P. berus* and *P. ursinii* for the Adder and the Orsini's Viper. Our position is not to recognise the sub-genus *Pelias* as a genus, as hybrids exist, even in the wild (e.g., between *V. berus* and *V. aspis* in France). And recognising *Pelias* raises another problem: *V. ammodytes* (which is not in this group) is less related to the pair *V. aspis/latastei* than it is to the sub-genus *Pelias* pair. And here as well the existence of natural hybrids between *V. ammodytes* and *V. aspis* goes against the recognition of a separate genus for *V. ammodytes*.

In conclusion, to us it would appear preferable not to accept using the genus *Pelias* to denominate vipers of the *V. berus* and *V. ursinii/kaznakovi* groups.

## The Vipera *ursinii* group

This group contains numerous taxons whose systematics are argued and controversial today, despite the advent of molecular (genetic) techniques. Once grouped into their own sub-genus, *Acridophaga* (in reference to their preference for crickets and grasshoppers), a group very closely related to *Vipera berus*, these two groups were united in the sub-genus *Pelias*, which some authors considered was a genus in its own right. The *V. ursinii* groupe is closely related to the *V. kaznakovi* group, which is referred to as the *V. ursinii/kaznakovi* complex. Going into more

*A very good example of hybridisation in the wild: a hybrid between the Adder and Asp Viper (Vipera berus x V. aspis). Auvergne (central France). Note the general appearance and head pattern of the Adder, but also characters typical of the Asp Viper: slightly upturned snout, brown (not red) iris and the thin and more or less separated denticles of the dorsal zigzag.* Teynié.

*Hybrid Nose-horned Viper x Asp Viper* Vipera ammodytes x V. aspis. *Northeastern Italy.* P. Geniez and J. Garzoni.

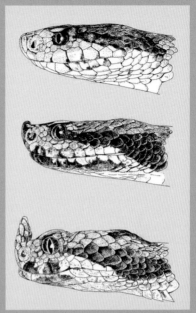

*Head shapes of Adder (above), Lataste's Viper (middle) and Nose-horned Viper (below), showing different snout shapes and number of rows of scales between the eye and the supralabials. Note that there are many more scales on the horn of the Nose-horned Viper than on that of Lataste's Viper.*

detail, the *V. ursinii* group is largely split into western populations that have 19 rows of scales around the body (*V. ursinii* in the strict sense) and eastern populations with 21 rows of dorsal scales (the *V. renardi* sub-group). From the basis of these two groups arise apparently two other very distinct lineages: *V. graeca* and *V. anatolica*, both with 19 rows of dorsal scales. Finally, within *V. ursinii* species, there is an exclusively western mountainous population and an eastern steppe plain population (called by some authors Grassland Viper) that genetically are within the mountain forms. All species of the *V. ursinii* group are small, slim vipers, with a small oval, patterned, only slightly triangular head that is hardly distinct from the body, with shiny scales, the continued existence of a frontal plate and parietal scales: the dorsal scales are arranged in 19 or 21 rows. Their vast distribution approximately covers areas of densely vegetated steppe with a continental climate and a large range of temperatures. Several populations or species have become very small and are on the verge of becoming extinct.

# Orsini's Viper

*Vipera ursinii* (Bonaparte, 1835)
**Family:** Viperidae (vipers)
**Sub-family:** Viperinae (true vipers)
**F.:** Vipère d'Orsini
**G.:** Wiesenotter

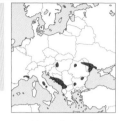

**Identification** Very small, quite slender to quite thickset viper, with a small oval head only slightly distinct from the neck. Length 30–40 cm, but farther east in its range it may attain 66 cm. Eyes with vertical pupils, golden-orange iris, darker at the bottom. More often than not, just 1 apical scale (but 2 in some individuals from certain populations – e.g., those of Mont Ventoux, southern France). Well-developed supralabials, with a thin black border on their outer edges, slightly curved above the eyes, which tends to give the Orsini's Viper a characteristic "severe" expression. The frontal and parietal plates are normally fully developed. The upper preocular usually touches the nasal scale. The nostril is situated below

*Orsini's Viper* Vipera ursinii ursinii, *adult. Alpes-de-Haute-Provence (southeastern France)*. Cluchier.

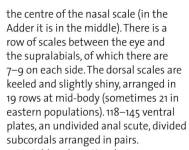

*Left: Orsini's Viper* Vipera ursinii ursinii, *adult male. Mont Ventoux (southern France). Évrard. Above: close-up of Orsini's viper, Mont Ventoux. P. Geniez.*

the centre of the nasal scale (in the Adder it is in the middle). There is a row of scales between the eye and the supralabials, of which there are 7–9 on each side. The dorsal scales are keeled and slightly shiny, arranged in 19 rows at mid-body (sometimes 21 in eastern populations). 118–145 ventral plates, an undivided anal scute, divided subcordals arranged in pairs.

Variable colouration, but never spectacular. A tight and quite wide zigzag on the back, often with small and blunt denticles; sometimes the zigzag is fragmented into marks separated one from the other. Males are often quite a bright pale grey or yellowish-grey with a black or red-brown with black-bordered zigzag; females are beige or pale brown with a dark brown zigzag. In certain populations, the ground colour of the back is paler than that of the flanks, thus evoking a pale line on each side of the zigzag: in other populations, the back is darker. There is a series of dark markings along the flanks that may be partially, but never completely, fragmented. The top of the head has several dark marks with 2 larger, elongated marks forming a forwards-pointing, very open V. A dark temporal band goes from the eye to the corner of the mouth and may continue along the sides of the neck. There is a small vertical dark mark on the edges of some labials; others are entirely pale, giving the snake a very pale face. The belly is mainly white or dark, depending on populations and subspecies, mottled with black, dark grey or brown, more densely towards the rear of the body than the front.

**Venom** Solenoglyphous fangs. Orsini's Viper is not particularly dangerous to humans, not that it cannot be aggressive if provoked: when cornered or captured, it tries to bite with incredible persistence. Cases of bites, common enough and well-documented, as they often concern

*Orsini's Viper* Vipera
ursinii macrops.
*Coatia*. Riegler.

manipulations by herpetologists, for
the most part result in localised pain
and slight swelling of the bitten limb;
these symptoms disappear after a few
days. Very rarely do bites result in a
general malaise.

**Habitat** Totally avoids forest. In
mountains, occupies short herbaceous
steppe with abundant dense woody
plants, with or without small rocks. On
plains, occupies natural meadows and
grassy steppe with damp depressions
or watercourses. Readily uses rodent
burrows, especially on plains and scree.
In mountains, occurs at about 1000–
2400 m elevation.

**Habits** A diurnal snake that avoids
extreme heat and especially strong
sunshine. Basks in the sun mainly in
the morning, but also in late afternoon,
especially in inclement weather or just
after rain. Orsini's Viper is very wary, but
as it has very poor eyesight, it can be
watched coiled up, basking in the sun,
if approached very carefully. Has a long
dormant period in winter that may last
six months.

**Diet** Especially large insects (crickets
and grasshoppers); sometimes also

small lizards and, when the occasion
arises, the young of small mammals.
Generally does not use venom on its
prey, which is swallowed live, especially
grasshoppers and crickets.

**Reproduction** Ovoviviparous. Gives
birth to live young. Mating usually
occurs in April or May. 4–8 (sometimes
as many as 12–15) minuscule young are
born in August or September.

**Range** Broken distribution of relict
populations from Mont Ventoux in
southern France to the west; east
as far as Moldavia: mountains of
southwestern France, central Italy
and the western Balkans (Croatia,
Bosnia-Herzegovina, Montenegro,
northern Albania and northwestern
Macedonia): plains of eastern Austria
(population extinct), Hungary,
western Romania (probably extinct),
northeastern Bulgaria and Moldovia.
More eastern viper populations
concern other species, especially
*Vipera renardi*.

**Geographic variation**
▶ *Vipera ursinii ursinii* (Bonaparte,
1835): southeastern France,
fragmented distribution, 21 sites

between 900 and 2200 m; central Italy, 5 groups of sites (central Apennines, the Marches, Ombrie and Abruzzes region). Small size (maximum 43.7 cm), 19 rows of dorsal scales at mid-body, generally has small numbers of scales, upper preocular often in contact with the nasal; the parietals are often fragmented, few supralabials (usually 8), there are small dark marks on the edges of the rear labial scales; the denticles of the zigzag are quite rounded; the sides of the back are not much paler than the flanks; there are spots on the flanks that are often separated into 2 series and that can be quite indistinct, with the ventral surface predominantly pale.

▶ *Vipera ursinii macrops* Méhely, 1911: mountains of the northwestern Balkans (Bosnia-Herzegovina, Montenegro, southwestern Serbia, western Kosovo, western Macedonia and northern Albania), between 1000 and 2300 m elevation. Populations in Croatia, traditionally considered to be *V. u. macrops*, in fact constitute a particular lineage closer to *V. u. ursinii*. Small size (maximum 47.4 cm), 19 rows of dorsal scales at mid-body, generally has small numbers of scales, upper preocular often in contact with the nasal scale, parietals often fragmented, few supralabials (5–9, usually 8 on each side), generally small dark vertical markings on the edges of the labials, the zigzag denticles quite triangular; the sides of the back a little paler than the flanks, the flank markings often distinct, not always split into 2 series, but that may be quite indistinct in females; ventral surface predominantly pale or dark.

▶ *Vipera ursinii rakosiensis* Méhely, 1893: steppe prairies of central European plains where just a few and rare very small populations subsist: eastern Austria (extinct), Hungary, northeastern Serbia (presumed extinct) and Romania (Transylvania). Large size (maximum 66 cm), 19 rows of dorsal scales at mid-body; generally has large numbers of scales; upper preocular often in contact with the nasal; parietals often not fragmented;

Vipera ursinii rakosiensis, *Bugac (southern Hungary). Note the sides of the back, the sides of the zigzag are paler than the flanks. This difference in colour does not normally occur in* V. u. ursinii, V. u. macrops, V. graeca *or* V. anatolica. Lyet.

generally 8 or 9 upper labials on each side that are of about the same size; the labials are nearly uniform, the zigzag denticles somewhat rounded; the sides of the back are much paler than the flanks, flank markings distinct and often arranged in 2 series; ventral surface predominantly dark.

▶ *Vipera ursinii moldavica* Nilson, Andrén and Joger, 1993: steppe prairies of southeastern central Europe: essentially the Danube Delta in Moldavia; also a small part of the Carpathians in Romania (at 950 m elevation, to the south of Mount Rarau), one site in extreme southwestern Ukraine, and a presumably now extinct population in an area in northeastern Bulgaria.

Morphologically and genetically very close to *V. u. rakosiensis*, but a little smaller (maximum 55.6 cm); slightly larger number of scales in certain parts of the body (in particular, the ventral plates: 130–145 compared to a maximum of 139 in *V. u. rakosiensis*, less in other subspecies of *V. ursinii*); upper preocular touches the nasal in about 50% of cases; parietals quite often fragmented (same proportion); the edges of the labials are normally dark (pale in *V. u. rakosiensis*).

The subspecies *Vipera ursinii graeca, V. u. anatolica, V. u. renardi* and *V. u. eriwanensis* are considered to be separate species in this book (see below).

## Steppe Viper

*Vipera renardi* (Christoph, 1861)
**Family:** Viperidae (vipers)
**Sub-family:** Viperinae (true vipers)
**F.:** Vipère des steppes or Vipère de Renard
**G.:** Steppenotter

**Identification** Small viper of the *Vipera ursinii* group, slightly larger and more slender, most populations characterised by having white labials with dense purple-black colouring. Total length around 40–50 cm, to 63.7 cm. Small oval, slightly triangular head, only slightly distinct from the body. Top of head flattened, with 5 often well-differentiated scales: 2 supraoculars, the frontal and 2 parietals, the last 2 fragmented in almost 50% of individuals. Normally only 1 apical scale, 1 row of scales between the eye and the supralabials, 7–10 (9 more often than not) supralabials on each side. Keeled body scales, quite shiny, arranged in 21 rows (rarely 19 or 20) at mid-body. Quite a large number of ventral plates, normally 129–151, slightly more than *V. ursinii*. The zigzag has various rounded or blunt-tipped denticles. The zigzag is generally brown with a black or dark brown border. The ground colour of the back is pale grey or brown, paler than the flanks, showing up the zigzag, especially in males, which have more contrasting colours than females. The flanks are decorated with dark vertical oval marks, sometimes arranged into 2 series. There are 2 dark bars forming a rear-opening V on the top of the head, unconnected to the first dorsal mark; thus there is a very pronounced V on the nape. On each side of the head, a sharp dark band extends from the rear of the eye to the sides of the neck. The labials and the scales under the head often have reddish-black-brown borders. The belly is dark grey, almost

black (darker than in *V. ursinii* and *V. eriwanensis*, in which white dominates) with white mottling.

**Venom** Solenoglyphous fangs. As in all members of the *Vipera ursinii* group, the venom is considered to be relatively harmless to man, not at all life-threatening. Symptoms are normally limited to local pain and inflammation of the bitten limb. The Steppe Viper is timid and flees at the slightest sign of danger. Cornered, it swells its body and hisses whilst flattening its head and neck.

**Habitat** Areas of grassy steppe with damp depressions and watercourses; also wooded steppe or slopes with scree and sparse vegetation. It occurs on lowlands as well as mountain areas to 2000–2500 m.

**Habits** Timid and always ready to flee; readily hides in rodent burrows. Diurnal; basks in the sun in the morning and late afternoon, but also during the middle of the day after rain. Inactive in winter for up to six months.

**Diet** Large insects such as crickets and grasshoppers; also spiders, lizards

*Steppe Viper* Vipera renardi renardi. *Northern Ukraine.* Crochet.

273

and sometimes young rodents. Kills larger prey by injecting its venom but swallows insects alive.

**Reproduction** Ovoviviparous. Young are born live. Mating occurs after winter inactivity and the first moult of the year, in April or May, depending on elevation. Young (4–8, sometimes as many as 15) are born in August or September.

**Range** Very vast range in central Asia that extends into southeastern Europe: Ukraine, southwestern Russia, foothills and north-facing slopes of the Greater Caucasus as far as northern Georgia, northern Azerbaijan, northern and eastern Kazakhstan and from there as far as Kirghizstan, Uzbekistan and northwestern China.

**Geographic variation** Across its vast range *Vipera renardi* is represented by many populations that are often isolated one from the other. If at present most authorities agree in considering the Steppe Viper to be a

*Close-up of* Vipera renardi renardi. *Note the motley coloured labials typical of the nominative sub-species of* Vipera renardi. Kreiner.

separate species from Orsini's Viper, a position supported by different genetic studies, the situation within this species proves to be very complicated; some specialists regroup all the different populations into one species, while others accord species status to these populations. Here we propose an intermediate position that not only takes into account published phylogeographic differences but also morphological characters.

▶ *Vipera renardi renardi* (Christoph, 1861): the major part of the species range except Crimea and the eastern part of central Asia. Amongst other characteristic features, the labials and the underside of the head are patterned with chestnut.

▶ *Vipera renardi puzanovi* Kukuskin, 2009: an isolated population endemic to the southern Crimean mountains, between 500 and 1000 m elevation. Very similar to *V. r. renardi*, differentiated by having fewer scales on many parts of the body.

▶ *Vipera renardi lotievi* Nilson, Tuniyev, Orlov, Hoggren and Andrén, 1995: dispersed over the central part of the Greater Caucasus (southern Russia); look for them in Georgia and in Azerbaijan near the frontier with Daghestan. This is a paraphyletic taxon (that is, a taxon that has several different lineages between which are inserted lineages attributed to other subspecies), originally described as a separate species and still considered as such by many authors. Thick, angular head, the edges of the labials and scales on top of the head are very pale and without, or nearly so,

Vipera renardi lotievi. *A very distinctive subspecies, recognised by a relatively thick head and commonly an absence of patterning on sides of the head.* P. Geniez and J. Garzoni.

any dark (but the supralabials are sometimes quite orange, as in *V. eriwanensis*); denticles of the dorsal zigzag very rounded, belly dominantly white with dense dark spotting and mottling. There exist uniform bronze-brown individuals.

► *Vipera renardi shemakhensis* (Tuniyev, Orlov, Tuniyev and Kidov, 2013), a new combination (taxon initially described as a species and in another genus: *Pelias shemakhensis* Tuniyev, Orlov, Tuniyev and Kidov, 2013). For the present, known from the Shemakha region in northern Azerbaijan. Closely related to *V. r. lotievi*, with which

is shares numerous colouration characteristics; the zigzag has fewer denticles and fewer ventral plates, flank marks are arranged in 2 or 3 lines, as in *V. eriwanensis*.

In the eastern part of its range, outside the area covered by this book, there are other subspecies of *V. renardi* (maybe even species, as the systematic situation is not yet clear): *V. r. tienshanica* Nilson and Andrén, 2001; *V. r. parursinii* Nilson and Andrén, 2001; *V. r. bashkhirovi* Garanin, Pavlov and Bakiev in Bakiev et al., 2004 and *V. r. altaica* Tuniyev, Nilson and Andrén, 2010.

# Armenian Mountain Steppe Viper

*Vipera eriwanensis* (Reuss, 1933)
**Family:** Viperidae (vipers)
**Sub-family:** Viperinae (true vipers)
**F.:** Vipère d'Arménie
**G.:** Armenische Wiesenotter

**Identification** A viper of the *Vipera ursinii* group and *V. renardi* subgroup. Smallish size compared to other taxon in the *V. ursinii* group, usually 35–40 cm, males to 45 cm, females to 50 cm. Resembles an Orsini's Viper but appears slightly more slender. Head slightly longer, a little squarer from above (this means there is a higher proportion, as many as 25% of individuals, having 2 apical scales instead of just 1, but not in Iran). The rostral is much higher than wide; the parietal scales are usually undivided, the upper preocular scale does not normally touch the nasal, except in Iranian populations (in all other members of the *V. ursinii* group, with the exception of *V. darevskii*, the upper preocular is nearly always in contact with the nasal). A relatively large number of scales overall: 9–10 supralabials on each side. Keeled dorsal scales arranged in 21 rows at mid-body; 123–143 ventral plates.

The underside of the head often has many darker markings, and there is often a pale transverse band between the eyes. The labials are a pale yellow without dark markings (in Iran) or frequently uniform pale grey, pinkish or reddish with a few small dark markings. The brown or dark grey dorsal zigzag with a dark border has many quite rounded denticles; the zigzag is often fragmented on part of the body; the sides of the back are slightly paler than the flanks. There are 2 series of generally fragmented, often indistinct, spots on the flanks. The belly has a dominantly pale colour. Males are paler grey with a black and brown zigzag, females browner and less contrasted.

**Venom** Probably similar to that of *Vipera ursinii*, thus presumed to be only slightly dangerous for humans.

**Habitat** Basaltic plateaus and hillsides densely covered with herbaceous vegetation, with scattered rocks, or drystone walls, dotted or not with bushes (e.g., dwarf junipers). Also in alpine meadows and extensively exploited agricultural plateaus. Found from 1000 to 3000 m; can be quite abundant locally, especially between 1900 and 2500 m.

**Habits** Quite similar to those of other mountain vipers of the *V. ursinii* group. Can be seen basking in the sun in the morning or during periods of overcast skies and especially after rain.

**Diet** Probably crickets and grasshoppers and lizards.

**Reproduction** Ovoviviparous. Gives birth to live young.

**Range** Distribution centred on the Armenian plateau: Armenia, western Azerbaijan and eastern Turkey, and farther east in isolated populations in the Elbourz chain in northern Iran and the Talysh mountains in extreme southeastern Azerbaijan.

**Geographic variation** From a genetic point of view, *Vipera eriwanensis* forms a subgroup when attached to *V. renardi*. The *eriwanensis-renardi* subgroup forms a sister-group with *V. ursinii* in the strict sense. *V. eriwanensis* is itself split into two lineages that correspond to two subspecies, according to the author, but they are considered as separate species by other authors or, on the contrary, as subspecies of *V. renardi* or even *V. ursinii* by yet others.

▶ *Vipera eriwanensis eriwanensis* (Reuss, 1933): the western part of the species' range. Characterised, among other features, by the upper preocular and nasal scales not touching, the relatively frequent presence of 2 apical scales (in 25% of the population), by a large number of ventral plates (133–143) and of denticles on the dorsal zigzag (54–79); the labials have few dark markings, and often a pink or brick-red hue.

▶ *Vipera eriwanensis ebneri* Knoepffler and Sochurek, 1955: the Elbourz chain (northern Iraq) and from there as far as the Talysh mountains in extreme southeastern Azerbaijan (a taxon that occurs in the Western Palaearctic). Averages smaller (maximum 43.8 cm) always a single apical scale, the upper preocular more often than not touching the nasal scale, the labials always without dark markings and

Vipera eriwanensis eriwanensis. *Northeastern Kömürlü (northeastern Turkey).* P. Geniez and A. Teynié.

often of a yellowish colour (sometimes an orange-pink as in *V. e. eriwanensis*); often fewer denticles on the zigzag

(54–65) and also fewer ventral plates to an extent that this feature is almost diagnostic (123–134).

*Close-up of* Vipera eriwanensis eriwanensis. *Note the reddish tint of the labials, a common trait of this taxon.* P. Geniez and J. Garzoni.

Vipera eriwanensis ebneri. *Lar national park, at an elevation of 3000 m (the Elbourz chain, northern Iran).* Behrooz.

# Greek Meadow Viper

*Vipera graeca* Nilson and Andrén, 1988
**Family:** Viperidae (vipers)
**Sub-family:** Viperinae (true vipers)
**F.:** Vipère grecque
**G.:** Griechische Wiesenotter

**Identification** Small viper very morphologically similar to mountain forms of Orsini's Viper (*Vipera ursinii ursinii* and *V. u. macrops*), but genetically very distinct, as shown by the work of Anne-Laure Ferchaud and colleagues. Characterised by its small size (maximum 43.5 cm, but can be to 60 cm in Albania); the sides of the back are no paler than the flanks; the zigzag denticles are triangular (the zigzag may be reduced to a serrated dorsal band); flank spots are often absent; there are no dark marks on the edges of the labials; the upper preocular always touches the nasal scale; and, especially, there are generally a small number of scales (the least of all the *V. ursinii* group): 6–7 supralabials on each side, the 3rd directly under the eye (nearly always the 4th in other members of the *Vipera ursinii* group); only 20–27 pairs of subcordal plates in the male (always

27 or more in all the other species of the *ursinii* group, with the notable exception of *V. anatolica*), 18–21 in the female (always 20 or more in the other taxon apart from *V. anatolica*).

**Venom** Venom probably similar to that of *Vipera ursinii*, so presumed to be of no particular danger to humans.

**Habitat** Subalpine meadows with scattered rocks and piles of stones, often without shrubby vegetation (the opposite in *Vipera ursinii macrops*) and with abundant crickets and grasshoppers.

**Habits** Probably identical to that of *Vipera ursinii ursinii* and *V. u. macrops*.

**Diet** Almost certainly mainly grasshoppers and crickets.

**Reproduction** Ovoviviparous. Gives birth to live young.

**Range** *Vipera graeca* was of very recent (1982) discovery. Considered endemic to Greece (about 12 sites known in the Pindus mountain range), it was found in 2006 in southern Albania, where it is probably very rare (1 single individual observed and photographed).

**Geographic variation** *Vipera graeca* is a monotypic species. Anne-Laure Ferchaud and her colleagues have shown that the genetic divergence between *V. graeca* and other taxon in the group (excepting *V. anatolica*) is almost as great as the one that exists between, for example, *V. berus* and *V. seoanei*, the two species the most genetically distinct within the *V. berus* group. So these authors rightly propose that *V. ursinii graeca* should be given specific status with the nomenclature *V. graeca* Nilson and Andrén, 1988.

*Greek Meadow Viper*
Vipera graeca. *Note the small number of large labial scales.*
Franzen.

# Anatolian Mountain Steppe Viper

*Vipera anatolica* Eiselt and Baran, 1970
**Family:** Viperidae (vipers)
**Sub-family:** Viperinae (true vipers)
**F.:** Vipère d'Anatolie
**G.:** Anatolische Wiesenotter

**Identification** Small viper similar to *Vipera ursinii*. Very localised in southern Turkey. Maximum size 43.4 cm, often less. Relatively large head, high and angular (more so than in *V. ursinii*); relatively thickset body (more so than in other members of the *V. ursinii* group); the rostral plate is higher than it is large (nearly as high as it is large in *V. ursinii*); the parietal plates are sometimes fragmented; the upper preocular touches the nasal scale, the keeled dorsal scales are arranged in just 17 rows at mid-body (it is the Western Palaearctic viper with the smallest number of dorsal scales); few ventral plates (114–124 in females) and subcordal plates (19–23), similar numbers to that of *V. graeca*. The dorsal zigzag denticles are fairly rounded or slightly triangular, and are less numerous than in other species of the *V. ursinii* group: 34–42 (compared to more than 43 in the other taxa except for *V. renardi tienshanica*, which rarely has only 41). The general colour is grey or beige with a blackish or chestnut dorsal zigzag with large dark edges, of which the denticles reach the top of the flanks; the flanks have 2 series of dark vertical markings that alternate with the zigzag's denticles. The edges of the labial plates are plain or have dark markings; the underside is primarily pale.

**Venom** Venom probably similar to that of *Vipera ursinii*, so presumed to be of no particular danger to humans.

**Habitat** Herbaceous hillsides dotted with limestone outcrops, with sparse small bushes around the rocks; edges of sinkholes.

**Habits** Probably identical to that of *Vipera u. ursinii, V. u. macrops* and *V. graeca*. A very rare and little known viper, only ever seen by a few people; considered to have been extinct, it has recently been rediscovered.

**Diet** Probably mainly grasshoppers and crickets.

**Reproduction** Ovoviviparous. Gives birth to live young.

**Range** *Vipera anatolica* is known only from the Kohu Dag mountains, near Çiglikara Ormanlari, southwest of Elmali (southern Turkey in Antalya province), between 1650 and 1750 m elevation.

**Geographic variation** *Vipera anatolica* is a monotypic species. Recent genetic studies have shown that the Anatolian Mountain Steppe Viper should be placed at the base of all vipers of the *V. ursinii-kaznakovi* group and that it is the most distinct species within the group. Their separation occurred some 5 million years ago. Thus the genetic distance between the Anatolian Mountain Steppe Viper (*V. anatolica*) and Orsini's Viper (*V. ursinii*), for example, is greater than that between Orsini's Viper and Caucasus Viper (*V. kaznakovi*), although they are visually very distinct (see page 284). As a comparison, *V. graeca* and *V. ursinii* diverged about 3 million years ago, *V. kaznakovi* and *V. berus* about 5.5 million years ago and *V. monticola* and the *V. berus-kaznakovi-ursinii* grouping some 12 million years ago.

## Darevsky's Viper

*Vipera darevskii* Vedmederja, Orlov and Tuniyev, 1986
**Family:** Viperidae (vipers)
**Sub-family:** Viperinae (true vipers)
**F.:** Vipère de Darevsky
**G.:** Darevskis Kreuzotter

**Identification** Small viper with appearance similar to *Vipera ursinii*, but with colouration characteristics particular to the *V. kaznakovi* group. Length 26–48 cm, maximum 51.7 cm. Narrow, oval or slightly triangular head, a little more pointed, flatter above, a little larger and higher and slightly more distinct from the body than other members of the *V. ursinii* group. Rostral plate only slightly higher than wide; well-developed frontal,

*Darevsky's Viper
Vipera darevskii.
Gukasyan, type
locality of the species
(northwestern
Armenia). Orlov.*

parietal and supraocular scales; all the other scales on the top of the head are subdivided. The head is slightly angular at its front, but less so than in the Adder or Caucasus Viper. The front edge of the snout is rounder than in the Caucasus Viper. Nostrils are in the lower part of the nasal scale. Either 1 or 2 apical scales, depending on the individual, even within the same population. The upper preocular scale is nearly always separated from the nasal scale. There is a row of scales between the eye and the supralabials (8–11 on each side). There are 8 or 9 scales around the eye. The keeled dorsal scales are arranged in 20 or 21 (sometimes 19) rows at mid-body. 125–144 ventral plates; a divided anal scute; subcordals divided and arranged in pairs. The ground colour of the back is paler than the flanks, pale grey or yellowish in males (this yellowish is typical of *V. darevskii* within the

*V. ursinii* group), beige or pale brown in females. The back zigzag is quite narrow, with quite a large central band but small denticles. The males have a black zigzag that has a tendency to be fragmented into transverse bars; in females it is brown, sometimes with a more reddish centre. There are 2 series of dark marks along the flanks that may merge to form a single series of larger vertical, oval markings (in *V. eriwanensis* the flank markings are separated into 2 indistinct series). The top of the head is very dark, largely black in the male, paler with large dark brown markings in females. The first dorsal mark, on the nape, is elongated and bordered on each side by a quite contrasting pale bar that ends abruptly against the 2 dark bars in an open V at the rear of the top of the head. This same character occurs in the Caucasus Viper *V. kaznakovi* and related taxa and suggests, along

Vipera darevskii, *two males of the population discovered near Zekeriya (northeastern Turkey) in September 2000. Note the very dark top of the head, more apparent than in other species of the V. ursinii* group. *P. Geniez and A. Teynié.*

Vipera darevskii, *a female from Zekeriya. Note the general appearance, similar to that of Orsini's Viper, but the lateral dorsal lines when they exist in that species are not straight or neatly cut off on the nape, nor so obvious. In this species, they are similar to those of* V. kaznakovi. *P. Geniez and A. Teynié.*

with other morphological characters and colouration, that *V. darevskii* could be a species of hybrid origin between *V. kaznakovi* and a species in the *V. ursinii* group. A black band in males, brown in females, extends from the eye to the angle of the mouth, sometimes continuing along the side of the neck. The belly is white with black speckling under the neck, progressively changing to dark grey with dense black and white speckling. Contrary to Orsini's Viper and the closely related taxon *V. eriwanensis*, the belly appears more black than pale. The underside of the end of the tail is orange-yellow. Melanistic individuals are unknown.

**Venom** Solenoglyphous fangs. Venom unknown but probably similar to that of *Vipera ursinii*, thus probably less potent than in most vipers.

**Habitat** High mountainous areas, between 2000 and 3000 m elevation. Subalpine mountain meadows, hillsides with stony moraine. In Turkey,

occurs exclusively on limestone, whereas its close relative, *Vipera eriwanensis*, has only been found on basalt areas. On Mount Legli, in Armenia, occurs on volcanic scree slopes with many rocky boulders as well as in alpine meadows.

**Habits** Little known, probably very similar to that of Orsini's Viper. Diurnal, basks in the sun in the morning or after rain. Long inactive period in winter, probably 6–7 months.

**Diet** Appears to be unknown; crickets and grasshoppers probably constitute the main part of its diet, and it probably also takes small lizards.

**Reproduction** Ovoviviparous. Gives birth to live young. Number of young unknown but probably small.

**Range** Until recently considered endemic to Mount Legli (Mount Achkasar), a high mountain in northwestern Armenia close to the Georgia border. It was discovered in the 2000s in two new sites in extreme northern Turkey, near Zekeriya and

Posof, in Ardahan province. Look for it in southern Georgia, close to Mount Legli and to the north of Posof.

**Geographic variation** There are only three known populations of *Vipera darevskii* in the world, each one isolated from the others and each with slight scale and colour differences. Some authors have not hesitated in describing a species new to science: *V. olguni* (S. B. Tuniyev, Avcı, B. S. Tuniyev,

A. L. Agasian and L. A. Agasian, 2012), known from the area of Posof, in extreme northern Turkey, near the Georgian border. In this taxon the zigzag denticles are close together and more or less quadrangular (and not rounded). In the absence of more information, especially on its genetics, we consider this population as a subspecies of Darevsky's Viper *V. darevskii olguni*.

## Caucasus Viper

*Vipera kaznakovi* Nikol'skij, 1909
**Family:** Viperidae (vipers)
**Sub-family:** Viperinae (true vipers)
**F.:** Vipère du Caucase
**G.:** Kaukasusotter

**Identification** A magnificent viper, characterised when adult by its bright black colouration marked with bright red or orange or, less often, lemon-yellow or whitish lateral dorsal lines. Juveniles are reddish with orange bands (never grey, as those of *Vipera dinniki* can be). The 2 longitudinal coloured bands correspond to the ground colour of the back, which has been highly reduced by the size of the zigzag, which, in this species, is in the form of a large band with undulating edges that covers almost the entire back. The same goes for the flank markings, which, in juveniles, are large are very close to each other and which come together to form a wide

black band when adult. The dorsal bands stop abruptly on the nape in the form of a wedge (a character also found in *V. darevskii*, but less clear-cut). Adults have a black belly. The sides of the snout are the same colour as the dorsal bands; the labial plates are paler, pinkish or yellowish-white. The iris is almost the same colour as the dorsal band, with a darker bottom half. Melanistic individuals are not rare.

Medium-sized snake (50–55 cm, to 65 cm), with quite high and thick triangular head. Snout triangular from above, square and angular in profile. The underside of the head is very flat, slightly concave. The relatively few scales on the top of the head are

*Caucasus Viper* Vipera kaznakovi kaznakovi, *young adult male. Hopa (northern Turkey).* P. Geniez and A. Teynié.

irregular, the frontal and parietals well-developed but often asymmetric, sometimes with small scales inserted between the frontal and the parietals. 1 or 1½ rows of scales between the eye and the supralabials, of which there are 8 or 9 each side. The dorsal scales are well keeled, arranged in 21–23 rows at mid-body. 126–143 ventral plates.

**Venom** Solenoglyphous fangs. Venom reputed to be more virulent than that of the Adder. Shy and wary but like all vipers tries to bite when caught.

**Habitat** Well-vegetated, hot, damp areas. In unkempt gardens and hillside tea plantations, damp alpine meadows and open mountain forest with much undergrowth; also on slopes with scree, scrub and shrubs. In mountainous areas rarely over 800 m (to 1000 m in Turkey), replaced higher up and farther north by *Vipera dinniki*.

**Habits** Very shy, lives a hidden life. Mainly diurnal. Readily basks in the sun along clearing edges or at the

Vipera kaznakovi kaznakovi, *adult male. Hopa (northern Turkey).* P. Geniez and A. Teynié.

base of brambles or low bushes – for example, tea plants – disappearing into cover at the slightest sign of danger. There have been few observations of this species' biology. This snake is quite difficult to procure, and is very much prized by collectors due to its rarity and bright colours; for these reasons, it is considered to

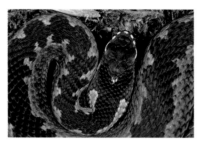
Vipera kaznakovi kaznakovi, *adult male. Near Sochi (Russia, Krasnodar territory).* Orlov.

be threatened with extinction. Today it is totally protected; its capture and transport may be severely punished by the authorities of the countries in which it occurs.

**Diet** Small mammals (rodents and shrews), lizards, occasionally small snakes (e.g., grass snakes) and, when the occasion arises, young birds on the ground.

**Reproduction** Ovoviviparous. Each year gives birth to 3–5 live young. Mating occurs in March or April; young are born in late August. The birth process may take 2 hours with 20 to 40 minutes between each birth.

**Range** Species endemic to southern slopes in the western Caucasus; as far as the shores of the Black Sea: northeastern Turkey, in the region of Hopa, western Georgia, the Abkhazia republic and southwestern Russia.

**Geographic variation** The systematics of the *Vipera kaznakovi* group is still uncertain and controversial. Two species have recently been described:

▶ *Vipera magnifica* Tuniyev and Ostrovskikh, 2001. Known from a small area in the northwestern Caucasus, in southern Russia (Adygea republic and Krasnodar territory), at an elevation from 700 to 1000 m. Resembles *V. kaznakovi*, with the red colouring, and has a very serrated zigzag recalling that of *V. darevskii* or *V. dinniki*; also characterised by a larger number of ventral and subcordal plates and by more fragmentation of the head scales.

▶ *Vipera orlovi* Tuniyev and Ostrovskikh, 2001. The most western population of the *V. kaznakovi* group: slopes overlooking the Black Sea, in Krasnodar territory (southern Russia), between 200 and 950 m elevation. Often paler than *V. kaznakovi*, with a more slender zigzag on the back.

Most authors, without supplementary information, especially concerning their genetics, consider these two "species" to be subspecies of *Vipera kaznakovi* or *V. dinniki*.

Vipera kaznakovi kaznakovi, *adult female. Hopa (northern Turkey).* P. Geniez and A. Teynié.

*Very young Caucasus Viper* Vipera kaznakovi kaznakovi. *Hopa (northern Turkey).* P. Geniez and A. Teynié.

Vipera kaznakovi magnifica. *Holotype. Krasnodor territory (southern Russia).* Orlov.

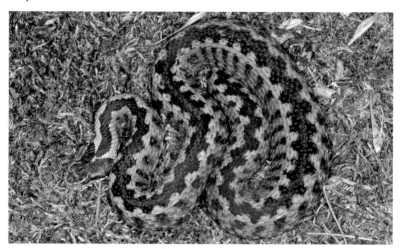

Vipera kaznakovi orlovi, *Female from the type locality, Sochi area (southern Russia).* Orlov.

Vipera kaznakovi orlovi. *Subadult male and juvenile photographed in captivity.* P. Geniez.

# Dinnik's Caucasus Viper

*Vipera dinniki* Nikol'skij, 1913
**Family:** Viperidae (vipers)
**Sub-family:** Viperinae (true vipers)
**F.:** Vipère de Dinnik
**G.:** Westkaukasische-Kreuzotter,
Dinnikis Kaukasusotter

**Identification** A high-elevation viper, closely related to *Vipera kaznakovi* but slightly smaller on average (maximum size 59 cm) and a much more variable colouration, the most variable of all Western Palaearctic vipers. Typically is similar in appearance to *V. kaznakovi* with a slightly less thick, less angular head, its underside nearly or not at all concave and a more slender body. Typically similar colouration to *V. kaznakovi* but less bright, with a wide dark zigzag on the back that lets more of the ground colour appear than in that species. The ground colour on the back varies from bluish-grey to orange-red, including all shades

of brown, bright yellow and quite whitish. The zigzag markings are often black in males, brown bordered dark in females. Some individuals have the zigzag markings separated one from the other, forming rounded transverse bars that are close together, giving a tiger-stripe appearance (indeed, this morph is termed "tigrina" by specialists). In other individuals, the zigzag is very wide, restricting the ground colour on the back to 2 pale zigzag bands ("kaznakovi" morph). Inversely and especially in the eastern part of the species' range, the colours are less bright, and the dorsal markings form a zigzag similar to that

*Dinnik's Viper* Vipera dinniki. *Greater Caucasus.* Schulz.

*Dinnik's Viper* Vipera dinniki, *Male. Mount Ficht, north of Sochi (Krasnador territory, southern Russia).* Orlov.

in *V. ursinii* ("ursinii" morph). There are also some strange individuals, uniform pale brown with a bronze sheen and dark, ill-defined dorsal band ("bronzed" morph). In all cases, *V. dinniki* is distinguished from all other members of the *V. ursinii* group by having a large, flat head, the frontal being the same size as, or only slightly larger than, the parietals (obviously larger in the *V. ursinii* group, including *V. darevskii*) and in most cases a scale separates the upper preocular from the nasal scale; in individuals of eastern populations, the underside of the head is dark with a pale border and a pale mark on each parietal scale; the head pattern may be reminiscent of that of *V. renardi*. And there exist entirely melanistic individuals without the slightest pale marking (contrary to melanistic *V. kaznakovi* individuals that always retain a few pale markings, such as the labials or around the chin).

Morphological characters may also help in identifying this species: a thin rostral plate, touching only 1 apical scale in 50% of individuals (the rostral is large and touches 2 apicals in 90% of *V. kaznakovi*). Other characters are similar to those of *V. kaznakovi*: black belly (but black with white mottling in eastern populations), 1 or 2 apical scales, 21 (sometimes 23) rows of dorsal scales at mid-body, 126–141 ventral plates.

**Venom** Probably similar to that of *Vipera kaznakovi*.

**Habitat** At high elevations, in more variable and cooler climates than *Vipera kaznakovi*: the upper stages of forest, subalpine and alpine meadows,

Vipera dinniki, *male of "tigrina" morph (above) and "bronzed" morph. Along the Mzymta River, east of Sochi (Krasnodar territory, southern Russia).* Orlov.

moraines, rocky hillsides. Prefers south- and southeast-facing situations.

**Habits** Diurnal, spends much time in the sun, especially in the morning but also the afternoon; shelters in a burrow during the hottest parts of the day. In inclement weather, it may well remain active in the middle of the day. As with the Adder, Dinnik's Viper increases its body temperature to well above that of the air: at 10°C, its body temperature may approach 30°C. It is inactive when ambient temperatures are below 9°C.

**Diet** Lizards, small mammals and the young of ground-nesting birds (e.g., Water Pipit). Juveniles consume crickets, grasshoppers and small lizards.

**Reproduction** Ovoviviparous. Gives birth to live young. Mating in late April and May, young born in August or September. The young wait until the following spring before feeding (in *V. kaznakovi*, juveniles capture prey as soon as their first autumn).

**Range** *Vipera dinniki* occupies the central part of the Greater Caucasus, from Mount Fisht (about 50 km to the north of Sochi, in Russia) to the northwest, as far as Mount Shkhara (Dagestan) in the southeast. Reaches the south of northwestern Georgia and extreme northwestern Azerbaijan (above Lgodekhi and Zakatala). The distribution of *V. dinniki* thus lies to the north and east of that of *V. kaznakovi*, at a greater elevation: between 1500 and 3000 m, sometimes lower.

**Geographic variation** *Vipera dinniki* is a paraphyletic taxon – that is, it is represented by several divergent

Vipera dinniki, *juvenile photographed in captivity.* P. Geniez and J. Garzoni.

lineages within which other lineages of other species (*V. kaznakovi* in this case) are mixed. It is provisionally considered to be a monotypic species by most authors. However, we know that less colourful populations from the east probably represent a separate subspecies and may even deserve species status. As well, some authors consider the taxons *magnifica* and *orlovi* to be subspecies of *V. dinniki*, others that *dinniki* is a subspecies of *V. kaznakovi*, along with *magnifica* and *orlovi*.

## Adder or Common Viper

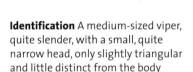

*Vipera berus* (Linnaeus, 1758)
**Family:** Viperidae (vipers)
**Sub-family:** Viperinae (true vipers)
**F.:** Vipère péliade
**G.:** Kreuzotter

**Identification** A medium-sized viper, quite slender, with a small, quite narrow head, only slightly triangular and little distinct from the body

compared to other vipers. Average length 50–70 cm; may sometimes reach 85 cm, with extraordinary records from Finland of 94 and 104 cm.

Females are often a little larger than males. Snout appears square from above, slightly angular in profile but not upturned (contrary to the Asp Viper). Eyes with reddish or dark red irises, contrasting with the animal's general colour. Body scales are keeled, slightly shiny and generally arranged in 21 (sometimes 19 or 23) rows at mid-body. The top of the head is covered in smooth, shiny, medium-sized scales contrasting with the 2 supraocular scales, the frontal and 2 parietals, which are small but well developed. On the top of the front of the snout there are generally 2 apical scales (compared to often just 1 in the Orsini's Viper), 1 row of subocculars between the eye and the supralabials, which normally number 9 on each side. 136–155 ventral plates, anal scute undivided, subcordales divided and arranged in pairs.

Variable in colour but less so than the Asp Viper. 2 dark bars on the back of the head are arranged obliquely and open at the rear, sometimes extending forwards to form a vague X (hence its German name Kreuzotter, "Cross Viper"). There is a dark band along the side of the head from the eye to the rear of the temple, sometimes extending along the side of the neck. The tight zigzag on the back is tighter than that of the Asp Viper. On the flanks, its ground colour is a little paler than that of the back, and there is a series of often ill-defined markings. Dark underside is often black, with feeble pale mottling on the sides. The rear of the underside of the tail is orange-yellow to brick-red. In most populations (*Vipera berus berus*) there is quite a marked sexual dimorphism: the males are generally a bright pale grey (especially just after the moult), sometimes pale brown, with a black zigzag; females are more brown or yellowish-brown with a chestnut or rufous zigzag. In certain populations, there is a high proportion of melanistic individuals, a trait that helps absorb the sun's heat in cold habitats, characteristic of this species of boreal affinity.

*Adder or Common Viper* Vipera berus berus, *adult male. Puy-de-Dôme (central France). Note the contrast in colour between the black and very pale grey colouration characteristic of males and the dark red iris, typical of* V. berus. Teynié.

*Female* Vipera berus berus. *Mont Mézenc (Haute-Loire, France). Note the colour, browner than that of the male. P. Geniez.*

**Venom** Has solenoglyphous fangs. The venom acts on the blood and vascular system (essentially a cytotoxic effect). The bite is painful. Attacks on humans are rare; death, as a result, is rare. In the case of an Adder bite, It is necessary to seek medical advice, which nowadays normally does not involve using an antivenin. An unaggressive and shy species, it will try to bite if captured (beware!).

**Habitat** Frequents a wide variety of habitats, but with wide daily fluctuations in temperature or, as in England and Brittany, high relative humidity. It is the viper that is most adapted to living in wet habitats such as marshes, wet meadows and peat bogs. At high elevations or in areas without competition from other species (e.g., the Asp Viper or the Nose-horned Viper), it also occupies alpine meadows, forest edges and clearings, quarries, railway embankments, heath, field edges and open conifer or birch forest. May locally occur to 3000 m elevation. It finds refuge in dense shrubs, under tree roots, in piles of stones or rodent burrows.

**Habits** Diurnal, active at dusk only in very hot weather. Basks in the sun in the early morning and late afternoon, even in cold, rainy conditions; its body temperature is then much higher than that of the air, up to 30–33 °C, which corresponds to its optimal body temperature when active. In the northern part of its range or at a high elevation, it is especially active during warm, humid weather and after long periods of rain; it avoids wind. Swims well. Emerges early from winter inactivity, in some regions as early as March or April, when it might be seen sunning itself even between patches of snow. In general, shy and discreet but quite easy to find in the morning when it will be coiled up, head in the centre of its body, basking in the sun. When threatened, moves away silently and quickly into vegetation or under stones. Winter rest of 5–7 months, even up to 8° north of the Arctic Circle. Often passes the winter in communal sites.

**Diet** Rodents, young birds, lizards and frogs, especially "brown" frogs. Kills prey with its venomous bite. After injecting its venom, it waits until the prey is dead and then swallows it headfirst. Whilst digesting prey, moves less quickly, even sometimes heavily and uneasily.

**Reproduction** Ovoviviparous. Gives birth to live young. Mating occurs in April or May. Before copulation, which may take hours, males perform mating displays. They often demonstarte their force in ritual fights (coiling together with the front of the body raised and attempting to push the opponent to the ground). Between late August and early October, the female gives birth to 5–15 (rarely 20) young that are born within a transparent egg membrane from which they free themselves immediately. After giving birth, the female's flanks appear flattened and wrinkled. After their first moult, the young snakes immediately start hunting young lizards or young Common Frogs. They are able to start breeding at 3 or 4 years of age.

**Range** This is the terrestrial snake with the largest range in the world: from Britain and Brittany (northern France) in the west as far as eastern Siberia and Sakhalin Island in Russia in the east; to the north it occurs slightly above the 70th Parallel in Finland; its southern limit passes via Mont Lozère in France, the Pô plain in Italy (where it is almost extinct), the mountains of northern Greece, northern Turkey, extreme northern Kazakhstan, Mongolia, northeastern China and northern Korea.

**Geographic variation** This is a slightly variable species in northern and the mid-latitude parts of its range, but is highly variable in the south, with more or less isolated populations, some of which are considered to be separate species by certain authors.

▶ *Vipera berus berus* (Linnaeus, 1758): the greater part of its range. Dorsal zigzag quite wide, with short denticles (no separated dorsal bars).

▶ *Vipera berus bosniensis* (Boettger, 1889): Balkan mountains from southern Slovenia as far as northern Greece, including Albania, Macedonia, and western Bulgaria. Zigzag formed of transverse bars in the form of narrow lozenges often separated one from the other (from a distance resembles an Asp Viper).

▶ *Vipera berus nikolskii* Vedmederja, Grubant and Rudajewa, 1986: the Ukraine and southern European Russia, farther west extends as far as Moldavia and probably northeastern Romania. A subspecies of which most populations are entirely melansitic; 1, 1¹/₂ or 2 rows of scales between the eye and the supralabials (always 1 in *V. b. berus*), more often than not 23 rows of dorsal scales at mid-body, on average more ventral plates (to 155 compared to a maximum of 149 in *V. b. berus*); non-melanistic individuals have a dorsal pattern more like the "*bosniensis*" type, juveniles are typically red-brown (more grey-brown in *V. b. berus*), it has clear venom (yellow in *V. b. berus*). Considered by certain authors to be a separate species, Nikolsky's Adder *V. nikolskii*

*Melanistic Adder. Lioran, in Cantal (France). Note the dark red iris colour. In melanistic specimens of the Asp Viper, the eye is entirely black. Teynié.*

Vipera berus, *subadult female in Finland (18 km east of Kajaani). Note the large reddish band with small, compact denticles and the pale V that separates the brown of the head from the zigzag. This type of colouration is common in Finland but apparently unknown in France. It is probable that Finnish populations belong to* V. b. sachalinensis. *Crochet.*

Vipera berus bosniensis, *adult female in Bosnia. This subspecies, genetically distinct, can be recognised by the zigzag markings being transformed into transverse bars, a little like the Asp Viper. Note the dark red iris, the square but not upturned snout and three large plates on the top of the head, all characters that show that* bosniensis *is a subspecies of* V. berus. *P. Geniez.*

Vedmederja, Grubant and Rudaeva, 1986, despite the fact that populations are morphologically intermediate with *V. b. berus.*

► *Vipera berus barani* Böhme and Joger, 1984: northern Turkey, close to the Black Sea shores, from sea level to 1500 m elevation. A subspecies considered by several authors to be a species in its own right, Baran's Viper, *V. barani* Böhme and Joger, 1984. Melanistic individuals dominate in certain populations. Non-melanistic individuals have a colouration similar to *V. b. bosniensis*, always with dark dorsal bars on a greyish background,

finely connected one to the other; top of the head covered in small irregular scales and generally no frontal or parietal scales; always 2 complete vertical rows of small scales between the nasal scales and the eye (more commonly $1\frac{1}{2}$ in the other subspecies), generally 2 rows of scales between the eye and the supralabials, 9 or 10 supralabial scales on each side (8 or 9 in the other subspecies).

*Nikolsky's Viper* Vipera berus nikolskii, *taxon from southern Russia considered to be a full species by most authors, despite being genetically very close to* V. b. berus. *Certain populations of* V. b. nikolskii *are entirely melanistic, others not. P. Geniez.*

► *Vipera berus sachalinensis* Zarevskij, 1916: subspecies morphologically similar to *V. b. berus* but genetically very distinct, that occurs from eastern Russia to the Sakhalin and Chantar Islands. Considered to be a full species by certain authors: *V. sachalinensis* Zarevskij, 1916. Depending on populations, the dorsal zigzag is composed of quite narrow denticles that are sometimes almost separate one from the other, or it is quite wide with many close denticles that only just reach the side of the back. This wide zigzag is often mahogany brown on a grey background; the first dorsal mark, on the neck, is often separated from the dark pattern on the head by a pale V that is open at the rear. It is considered to be absent from the Western Palaearctic despite the fact that the work of Sylvain Ursenbacher has shown that the genetic lineage "*sachalinensis*" extends much farther west, reaching western Russia, Estonia and Finland.

**Remarks** Another species of viper has been described from northeastern Turkey, in the Small Caucasus chain, in the upper valley of the Çoruh, *Vipera pontica* Billing, Nilson and Sattler, 1990. The holotype (the specimen collected to describe the species) has been shown genetically to be a hybrid between *V. ammodytes transcaucasiana* and *V. kaznakovi*. Its general morphology is absolutely intermediate between these two species and the locality of its discovery is exactly in the zone of contact between them. Since then various work based on the discovery of new sites for *V. pontica*, has tried to show that it is merely a synonym for ... *V. (berus) barani*! Photos published on these occasions clearly show examples of *V. berus barani* that are quite dissimilar to *V. pontica*.

*Baran's Viper* Vipera berus barani, *a rare subspecies of northern Turkey considered by most authors to be a separate species, despite the fact that it is closely related to* V. b. bosniensis, *Çarsamba Ovasi (Turkey, banks of the Black Sea).* P. Geniez and A. Teynié. *Right: newborn young, Çarsamba Ovasi (Turkey). Note the dorsal pattern recalling that of* V. b. bosniensis. P. Geniez and A. Teynié.

# Iberian Viper

*Vipera seoanei* Lataste, 1879
**Family:** Viperidae (vipers)
**Sub-family:** Viperinae (true vipers)
**F.:** Vipère de Seoane
**G.:** Nordiberische Kreuzotter

**Identification** A medium-sized viper appearing at a quick glance to be intermediate between the Adder, to which it is most closely related, and the Asp Viper. Average length 45–60 cm, with a maximum of 72 cm. The head is slightly triangular and quite distinct from the body. Slightly angular snout, flat above, sometimes slightly lifted (but never obviously upturned, as in the Asp Viper). On the head pattern, the supraoculars are well differentiated; the frontal and the parietals are quite visible in most cases, but are sometimes fragmented. There is 1 row of scales between the eye and the supralabials, which generally number 9 on each side (rarely 8 or 10). The dorsal scales are clearly keeled, arranged in 21 rows (rarely 19 or 23) at mid-body, and there are 129–150 ventral plates. Very variable colouration, often very similar to Asp Vipers of the Pyrenees (*Vipera aspis zinnikeri*). The large zigzag on the back has small, close denticles (closer together than in *V. aspis*), often chestnut in the centre with darker sides. The background colour of the back is often paler than that of the flanks and extremely variable, ranging from pale grey to reddish-brown, passing through beige, pale brown and yellowish-brown. A series of dark, vertical oval markings on the flanks is sometimes fragmented. In

*Iberian Viper* Vipera seoanei. *Note the relative similarity with the Adder but that the snout is slightly upturned (straight in the Adder) and a somewhat different colouration.* Kreiner.

*Above:* Vipera seoanei seoanei, "bilineata" *form. Note the considerable resemblance to V. aspis zinnikeri, page 303. Mendive, Pyrénées-Atlantiques (southwestern France). Pottier.
*Right:* V. s. seoanei, *an almost uniform individual. South of Oviedo (northwestern Spain).* Évrard.

some individuals the zigzag and the flank markings cover the whole body, leaving the ground colour to appear only in the form of 2 magnificent yellow, orange or reddish lines, exactly as in the Caucasus Viper *V. kaznakovi*. Rarely the back is a uniform pale brown without any markings. Usually few ill-contrasting markings on the top of the head; nevertheless, 2 oblique dark bars outline a pale V on the neck. There is a dark temporal band from the eye to the neck. The iris is orange or bronze, with the bottom part slightly darker. Belly dark grey to black. Entirely melanistic individuals occur but are rare.

**Venom** Solenoglyphous fangs. Venom quite or highly potent depending on region. It is reputed to be highly toxic in the subspecies from the centre of the Cantabrian mountains, *Vipera seoanei cantabrica*, and of a different composition than that of other Iberian Vipers. A shy and retiring snake, unaggressive unless caught.

**Habitat** Warm, humid areas of high rainfall, with an oceanic climate. Occupies heath with bracken, broom and heather, rocky slopes with plentiful scrub and ground vegetation, areas of scree, piles of vegetation-covered stones, drystone walls between meadows, open forest, abandoned gardens. It particularly occurs in thick gorse hedgerows that separate hay meadows. From sea level to about 800 m elevation, higher locally, to 1900 m in the Cantabrians.

**Habits** Shy and discreet, often hidden. Principally diurnal; avoids brightest sun but basks in the morning or after rain. Inactive winter period quite short, 3 or 4 months.

**Diet** Small mammals, lizards, frogs and salamanders. The young feed

Vipera seoanei cantabrica. *Note the dorsal pattern of blackish transverse bars that is similar to the colouration of* V. aspis *or* V. berus. *The Cantabrian Mountains, between Valdeteja and Genicera (northern Spain).* P. Geniez.

principally on small lizards and young "brown" frogs, whereas small mammals form the majority of the adults' prey. Kills prey by a venomous bite.

**Reproduction** Ovoviviparous. Gives birth to live young. Males moult in the spring after mating. Low reproductive rate, 3–10 young each litter. Many females apparently breed only once every other year.

**Range** Endemic to northwestern Iberia: the whole of northern Spain, extreme northeastern Portugal, where it is quite restricted, and extreme southwestern France (a small part of the Basque country).

**Geographic variation** The Iberian Viper belongs to the sub-genus *Pelias*, as does the Adder, and the *Vipera berus* group, within which it is represented by the most divergent lineage (the first to be separated from the others); this testifies to a time in the long-distant

past when the ancestor of these two species occupied northern Iberia. There are two recognised subspecies of *Vipera seoanei*:

► *Vipera seoanei seoanei* Lataste, 1879: in the greatest part of the species' range.
► *Vipera seoanei cantabrica* Braña and Bas, 1983. Central part of the Cantabrian region of Spain: León hills and southwestern part of the Picos de Europas. Characterised by more fragmentation of the head scales, a greater number of ventral plates and a dark zigzag, often black, composed of only slightly touching, quite thin denticles; due to this feature, from a distance this subspecies appears similar to the dull-coloured *V. aspis*; finally, its venom is considered more potent than that of the nominative subspecies.

# Asp Viper

*Vipera aspis* (Linnaeus, 1758)
**Family:** Viperidae (vipers)
**Sub-family:** Viperinae (true vipers)
**F.:** Vipère aspic
**G.:** Aspisviper

**Identification** Medium-sized viper, quite slim, characterised by having a square, upturned snout forming a prominent transverse ridge: it is the only viper in the range covered by this book to show this character. Length 50–70 cm, to 85 cm in southern France, rarely to 94 cm in Italy. Fairly elongated head, quite triangular and somewhat enlarged at the rear, very distinct from the neck. Eyes with vertical pupils, iris almost the same colour as the animal and darker towards the bottom. The supraocular scales form a sharp ridge above the eye, giving the snake a "serious" expression. The top of the head is covered with small, only slightly keeled scales. There can be, especially in the southwestern part of its range, 1–3 larger scales in the position of the frontal and parietals. Generally 2 rows of scales between the eye and the supralabials, rarely 3, but quite often only 1½ in southwestern France and in Spain; 9 or 10 supralabials on each side. Keeled dorsal scales, arranged in 21 (rarely 23) rows at mid-body. 136–160 ventral plates, an undivided anal scute. Extremely variable colouration, males often more contrasting than females. The zigzag on the back formed of generally alternating transverse half bars, sometimes joined, in which case they form rings. These bars may be joined by a thin or, on the contrary, a very large, dorsal band, reducing the dorsal bars to denticles on each side of this central bar. The ground colour is often pale brown or yellowish-brown in the female, sometimes grey. In the male it is brighter, varying from an almost white pale grey to brick-red, passing through all sorts of shades of brown. The dark bars in a series on the flanks alternate with the dorsal zigzag bars. The pattern on

*Magnificent female* Vipera aspis aspis. *Le Gâvre, Loire-Atlantique, western France. The thin zigzag is characteriestic of the nominate subspecies in France.* Évrard.

Vipera aspis aspis, *close-up of a male of the so called "garigue" form. Hortus cliffs, north of Montpellier (southern France). Note the upturned snout (but not forming a horn) and the two rows of scales between the eye and the supralabials, distinctive characters of* V. aspis. P. Geniez.

the back of the head is variable; it is sometimes pronounced, sometimes hardly present, is generally formed of 2 oblique bands open at the rear but sometimes also, especially in the Alps, in lyre form, with tips pointing forwards. A dark band on the side of the head above the pale labials goes from the eye as far as the sides of the neck. The belly is dark, from yellowish-brown or reddish-buff to grey or black, with much black and white mottling. The underside of the tip of the tail is often orange-yellow. In certain mountain populations, especially in Switzerland, there are entirely melanistic individuals (which sometimes represent as much as 50% of the population). Elsewhere melanism is rare.

**Venom** Solenoglyphous fangs. Quite potent venom, but less so than vipers of the *Echis*, *Bitis*, *Daboia* or *Macrovipera* genera. However, a bite on a human requires medical attention as symptoms can be quite alarming (fall in blood pressure, giddiness, vomiting, excruciating pain, often extensive swelling, dead tissue). Cases of death are very rare today. The Asp Viper is a shy snake that prefers to flee rather than defend itself, but if cornered, and especially handled, it does not hesitate to bite. The venom of *Vipera aspis zinnikeri* is known to be four times as potent as that of *V. a. aspis* (personal communication with Philippe Golay).

**Habitat** From plains to high mountains and all elevations between in various habitats but avoiding urban or intensively cultivated areas. Dry slopes with scrub, drystone walls, hedgerows, scree, rocky areas. Equally in open deciduous forest. Locally in limestone garigue at low elevation, with rocks and many shrubs. Farther north, abundant in extensively farmed countryside, where it occupies hedgerows. Also present in peat bogs and wet meadows (especially when the Adder is not present) and rocky alpine meadows or those with drystone walls above the tree line. May locally reach 2500 m elevation, with records at 2930 m in the Hautes-Pyrénées (southern France) and 3300 m in the Valais (Swiss Alps), the elevational record for any European snake.

**Habits** A terrestrial snake, essentially diurnal in spring and autumn but also active at dusk in mid-summer or at night after warm rain. Avoids wind. Basks in the sun in early morning, then retires to broken shade. An inactive period in winter of up to 6 months at high elevations, far less at lower elevations, may even bask in the sun in

December or January in Mediterranean areas during fine weather.

**Diet** Small rodents, shrews, lizards, small birds, sometimes other snakes. Immobilises its prey with its venomous bite.

**Reproduction** Ovoviviparous. Mating soon after coming out of hibernation. Males then take part in ritual fighting. Females give birth in late August or September; the 5–15 young have to fend for themselves, the first stage being to eat enough to survive the winter hibernation successfully and to avoid predators – and humans.

**Range** Western Europe: northeastern Spain, middle and southern France, southern, central and western Switzerland; relict populations in Germany (southern Black Forest). Also present in Sicily, on Montecristo and Elba islands (Mediterranean Italy) and on a few French oceanic islands, including Oléron.

**Geographic variation** The Asp Viper is structured into four main quite divergent lineages and five subspecies:

▶ *Vipera aspis aspis* (Linnaeus, 1758): almost all of France except the southwest, northwestern Switzerland, southwestern Germany and northwestern Italy. Quite variable colouration, especially in males, which

may have a pale grey, yellowish-brown or brick-red ground colour; the dorsal zigzag is composed of thin transverse bars that may be alternate, unattached or attached; the females often have a diffuse brown line along the back.

▶ *Vipera aspis atra* Meisner, 1820: east of the French Alps, most of Switzerland and northwestern Italy. Closely related to *V. a. aspis*; characterised by having dorsal bars that descend lower on the flanks and are often connected, thus forming rings. In mountains they have larger, often rectangular dorsal bars, often larger than the spaces between them; in certain populations 50% or more of individuals may be melanistic. This subspecies is not recognised by certain authors due to it being genetically very close to *V. a. aspis*.

▶ *Vipera aspis francisciredi* Laurenti, 1768: the upper three-quarters of Italy, south as far as Naples and Mount Gargano, to the northeast as far as southeastern Switzerland and northeastern Italy; also present on Elba island. A lineage very divergent from the others; morphologically close

*Vipera aspis atra, male. Switzerland. A subspecies that is not recognised by some authors. However, it does have specific characteristics: straight-sided denticles to the zigzag that descend lower on the flanks than in* V. a. aspis; *when they are wide, they are rectangular; there is often a lyre-shaped mark on the hind neck. P. Geniez and J. Garzoni.*

Vipera aspis atra, *male. Les Agites (Vaud canton, Switzerland). It is often thought that the subspecies* atra *is characterised by being dark. At mid-elevation this is not the case: the denticles can be slim. Note, however, that they descend onto the top of the flanks, and there is a well-marked lyre on the hind neck.* P. Geniez.

Vipera aspis francisciredi, *female. Massa-Maritima (Livorno province, Italy). This Italian subspecies resembles* V. a. aspis. *The dorsal zigzag is even thinner, the head wider at the rear, and the labials are white, contrasting well with the general colour of the snake.* Deso and Maran.

to *V. a. aspis*, the dorsal bars are very thin, often widely separated from one another; wide head; labials are very white.

▶ *Vipera aspis hugyi* Schinz, 1833: southern Italy (Basilicata, Puglia) and Sicily. Spectacularly coloured: the dorsal markings do not form bars but large rounded and oval spots, brown with black borders, sometimes separated. *V. a. montecristi* (Mertens, 1956), endemic to the island of Montecristo (between Corsica and mainland Italy) is exactly the same as *V. a. hugyi*. We now know from genetic studies that the Asp Vipers on Montecristo were introduced by humans from Sicily, between the 8th and 3rd centuries BC, so that *montecristi* is synonymous with *hugyi*.

▶ *Vipera aspis zinnikeri* Kramer, 1958: northern Spain and southwestern France, including the whole of the Pyrenees; northwest as far as the Gironde; to the east it occurs as far as the southwestern part of the

Massif Central, even reaching the western edge of the Puy-de-Dôme department (personal communication with Alexandre Teynié). In this subspecies the dorsal bars are reduced to denticles that do not reach the flanks, bordering a dorsal line of variable width, commonly chestnut or reddish with black edges; the snout is often less upturned than in the other subspecies, and there are often 3 larger scales, the frontal and 2 parietals, on the top of the head (as in the Adder). This subspecies is so divergent from *V. a. aspis* that some authors have proposed that it be a separate species despite the fact that there is a very wide zone of morphological integration of the two: *V. a. aspis* and *V. a. zinnikeri*. Furthermore, Spanish populations, often with a thinner zigzag, are also considered by several authors to be part of *V. a. aspis*, although from a genetic point of view they are clearly part of the subspecies *V. a. zinnikeri*. There exists a particular form in the

Mediterranean garigue of southern France: large size, a large number of ventral plates, always a pale chalky-grey colour and dorsal zigzag limited to thin black bars. Assumed to be a distinct subspecies by certain authors and named "Garigue Asp Viper", it is in fact just a form of *V. a. aspis* that is perfectly adapted to its calcareous, rocky habitat.

*Left*: Vipera aspis hugyi, *adult male. Godrano (Sicily). A spectacular subspecies characterised by its large dark dorsal markings with black borders.* P. Geniez.
*Right*: Vipera aspis hugyi, *subadult female. Southern slopes of Etna (Sicily).* M. Geniez.

Vipera aspis zinnikeri, *Beyrède-Humet (Haute-Pyrénées, southwestern France). This subspecies, very distinct genetically, is in part characterised by the more or less amalgamated markings of the zigzag and its being brown within the dark denticles.* Pottier.

# Lataste's Viper

*Vipera latastei* Boscá, 1878
**Family:** Viperidae (vipers)
**Sub-family:** Viperinae (true vipers)
**F.:** Vipère de Lataste
**G.:** Stülnasenotter

**Identification** A small to medium-sized viper, quite thickset, with a large, triangular, flattened head with a quite short horn on the snout composed of few scales; in general, the rostral scale touches the bottom of the horn (it is, however, partially fragmented in certain coastal populations, particularly in Algeria). Average length 40–55 cm, with a maximum of 72 cm. Males are a little larger than females. Most individuals resemble the Nose-horned Viper, due not only to the horn but also the dorsal pattern composed of lozenges; others more closely resemble the Asp Viper. The top of the head is covered in small scales, among which it is sometimes possible to distinguish the vestiges of the frontal and the parietals; on the contrary, the supraoculars are always well defined. There are 2 rows of scales between the eye and the supralabial plates, 9–10 on each side. The dorsal scales are keeled, arranged in 21 (rarely 23) rows at mid-body, there are 122–147 ventral plates.

Variable colouration. The zigzag on the back is composed of lozenge-shaped markings that are often connected; these marks, with angular or rounded edges, are normally bicoloured, brown, reddish or orange, with black or darker brown borders. Ground colour is dark grey, beige, pale brown or reddish-brown. In some individuals, especially in the

*Lataste's Viper* Vipera latastei, *subadult male of the western Iberian lineage. Serra da Estrela (Portugal).* P. Geniez.

northern part of the range, the zigzag markings can be thin, looking more like transverse bars than lozenges. The top of the head is often uniformly coloured except for 2 bars forming a V on the hind neck, which can be quite subdued. There is a dark temporal band from the eye to the corner of the mouth. Belly grey or blackish with lighter or darker mottling. The underside of the tip of the tail is often yellow in males, black in females. Melanistic specimens exist but are rare.

**Venom** Solenoglyphous fangs. The toxin is as potent as or less potent than that of *Vipera aspis*. Bites are rare, and cases of death almost unknown.

**Habitat** Sunny rocky slopes with scrub, open deciduous forest, alpine meadows with bushes and drystone walls, Mediterranean garigue, from plains to hills and mountains. Also present in extensive areas of coastal dunes in southern Spain and Portugal. It occurs to 3000 m in the Sierra Nevada (southern Spain) and 2000 m in the Serra da Estrela (Portugal).

**Habits** Partially diurnal terrestrial snake that is also active at dusk and on warm or humid nights, the Lataste's Viper is reputed to be hard to find. It has a localised distribution, is very discreet and usually occurs at low densities, avoiding urban areas or habitat modified by man. Inactive in winter for 2–4 months.

**Diet** Small mammals, lizards, young birds in the nest, sometimes amphibians or arthropods (molluscs, myriapods, beetles).

**Reproduction** Ovoviviparous. Gives birth to live young; generally 2–6

*Lataste's Viper* Vipera latastei, *adult female of the eastern Iberia lineage. Spain, Serranía de Cuenca. Note the very different pattern with far less compact zigzag denticles compared with those of the individual of western lineage.* P. Geniez.

young at a time, rarely up to 13 in the same litter. When born they measure 17–19 cm, and one week later, after their first moult, they feed on crickets, grasshoppers, centipedes and small lizards.

**Range** Occurs throughout most of the Iberian Peninsula (Spain and Portugal) except in the more humid north, where it is replaced by the Asp Viper. Also present in the northern part of North Africa, in Morocco and Algeria; it is controversial as to whether or not it occurs in Tunisia.

**Geographic variation** Recent genetic studies have shown there to be two main lineages in Lataste's Viper, one in Iberia and the other in North Africa, within which the Atlas Dwarf Viper *Vipera monticola* is a component. In Iberia there are three lineages: one western; one southern and one eastern. *V. latastei gaditana* occurs within this last lineage; it is the subspecies of low elevations that is

Vipera latastei gaditana, *adult male. Coto Doñana national park (southwestern Spain).* P. Geniez.

found particularly in the extensive vegetated dunes of the Coto Doñana national park in southwestern Spain. It is characterised by having more scales on the horn, a large fragmentation of head scales (no frontal or parietals),

fewer ventral plates and subcordals, and having the zigzag denticles that are more widely spaced, quite slender and can be rounded or triangular. It is possible that the nomenclature *V. latastei gaditana* applies to all populations of western Spain and Portugal given that they belong to a lineage that contains the "*gaditana*" population.

## Atlas Dwarf Viper

*Vipera monticola* Saint Girons, 1953
**Family:** Viperidae (vipers)
**Sub-family:** Viperinae (true vipers)
**F.:** Vipère naine de l'Atlas
**G.:** Atlas-Zergotter

**Identification** Very small viper with a total length of 25–35 cm, considered to be the smallest of the genus *Vipera*. Has the appearance of a small, dull Lataste's Viper with a reduced number of scales. The head is triangular and clearly distinct from the body, covered in small, irregular scales from which the supraoculars are distinct. The horn is a little less developed than in *Vipera latastei* but has the same

configuration; the rostral scale always reaches its tip. 1½ or 2 rows of scales between the eye and the supralabials, of which there are 9 or 10 on each side. The dorsal scales are keeled, arranged in just 19 rows at mid-body (compared to 21 in *V. latastei*).

The back is dull grey, yellowish-grey or grey-brown, with a dark zigzag formed of slightly alternating transverse markings that evoke denticles rather than the lozenge markings typical of *V. latastei*. There is a series of dark, faded spots along the flanks. The underside of the tail tip is yellow in the male.

**Venom** Solenoglyphous fangs. Considering the small size of this viper, any bites should have little effect on humans (especially compared to other venomous snakes of North Africa). A shy species, but not inclined to flee, relying on its mimetic colours as a best defence.

**Habitat** Slopes at high elevations in the Moroccan High Atlas that are covered with short, tufted (xerophytes) vegetation, above the tree line. Also occupies small areas of scree and locally slopes in steppe with sparse xerophytes vegetation. Occurring between 2000 and 3500 m, but could potentially be found as high as 4000 m elevation.

**Habits** Little known, probably similar to Lataste's Viper. Essentially active in the morning or during overcast weather. It is then possible to find it coiled up under a clump of spiny vegetation or at the foot of some scree. Avoids very hot situations. Passes much of its time coiled up under spiny bushes. Has an inactive period in winter.

**Diet** Young rodents, small lizards, grasshoppers and crickets.

**Reproduction** This species has a reputation for giving birth to one young at a time; it said that the young is large in comparison to the female, and that the female gives birth only once every four years. It is probable that it reproduces more often than that if conditions are favourable (e.g., snow cover is less significant in some years than others).

**Range** Considered to be endemic to the Moroccan High Atlas but, according to certain authors, also occurs in the Mid Atlas, at elevations between 1000 and 2300 m.

**Geographic variation** The vipers of the *latastei-monticola* group are separated into at least three main lineages in North Africa:

► One corresponds solely to *Vipera monticola* in the strict sense, such as was defined when it was described – that is, from the western Moroccan High Atlas;

► Another has a wider range (eastern High Atlas, the Mid Atlas and the Rif chain), which thus also comprises

*Atlas Dwarf Viper* Vipera monticola, *male. Oukaïmeden (Moroccan High Atlas). It is a replica in miniature of Lataste's Viper but with duller colours and only slightly triangular denticles on the zigzag.* García-Cardenete.

*Atlas Dwarf Viper* Vipera monticola, *male. Oukaïmeden (Moroccan High Atlas). P. Geniez.*

*Right: Small viper of the* Vipera latastei-monticola *group, juvenile female, north of Imilchil, in the eastern High Atlas. According to genetic data, this population is part of Lataste's Viper of northern and central Morocco, and is not a "true" V. monticola, as was thought until quite recently. P. Geniez.*

vipers assigned to *Vipera latastei* as well as others to *V. monticola*;
► A third has been proposed for the Lataste's Vipers of extreme northeastern Algeria. If it is considered that *Vipera monticola* is a species distinct from *V. latastei*, then this nomination should apply to the whole of North Africa, with distinct subspecies well-differentiated on genetic as well as morphological criteria.

## Nose-horned Viper

*Vipera ammodytes* (Linnaeus, 1758)
**Family:** Viperidae (vipers)
**Sub-family:** Viperinae (true vipers)
**F.:** Vipère ammodyte
**G.:** Sandotter, Europäische Hornotter

**Identification** Quite a long viper on average, thickset, often brightly coloured, has a noticeable horn on the snout, generally larger than the Asp Viper. Normally 50–70 cm but may reach 90 cm, even 110 cm in the northern part of its range. Very triangular head, distinct from body, with an obvious horn on the snout that is covered in small scales. Top of head covered in small scales; well-defined supraoculars, 2 rows of scales between the eye and the supralabials, generally 10 ( sometimes 9 or 11) supralabials on each side. Keeled dorsal scales, arranged in 21–23 rows at mid-body. 143–163 ventral plates, undivided anal scute, divided subcordals arranged in pairs.

Quite variable colouration, anything between whitish-grey to brick-red, including beige, browns, yellow and orangish. The wide zigzag on the back is composed of adjacent or joined lozenges, generally with angular sides. In *Vipera ammodytes transcaucasiana* of Turkey and Transcaucasia, the lozenges are extremely reduced into transverse bars widely separated from each other (reminiscent of the

colouration of certain *V. aspis*). There is a series of marks on the flanks, often smaller, duller and less geometric than those of the back. The top of the head may show no pattern or may have a V open at the back that may, when wide, encircle a double pale spot at the front of the nape. There is a dark band that may be very visible or nearly absent, from the eye to the sides of the neck. The iris appears to be almost the same colour as the background colour of the snake, the lower part darker. The belly is grey, reddish or yellow-grey with numerous dark flecks or mottlings. The underside of the tail can be

*Right: Nose-horned Viper* Vipera ammodytes ammodytes, *adult male. Ex-Yugoslavia.* P. Geniez and J. Garzoni.

Vipera ammodytes ammodytes, *subadult male. Montenegro.* Cluchier.

green, yellow or red. The Nose-horned Viper's colouration, often glossy, does provide for an excellent camouflage and it is very difficult to locate when immobile on a pile of dead leaves or on a rock covered with lichens and moss. Melanistic individuals exist but are rare.

**Venom** Quite virulent venom that acts especially on the blood and vascular system (essentially cytotoxic); solenoglyphous fangs. This viper is quite slow-moving, generally preferring to flee as the best defence or, on the contrary, relies on its remarkable camouflage. Cornered, it can be quite aggressive, especially during hot weather. Anyone who has been bitten should seek medical attention, but deaths are very rare.

**Habitat** Relatively dry areas, principally in the Mediterranean bioclimatic zone. Dry slopes with scrub, edges of maquis, garrigue, piles of stones, drystone walls, abandoned gardens, hedgerows on the borders of meadows or crops, forest edges, open oak or conifer forest, scrub, and also riverbanks and stream banks, sides of lakes and undisturbed beaches. Penetrates into mountain areas on rocky slopes, locally as high as 2000 m.

Vipera ammodytes ammodytes, *male ready to moult, which is the reason for the eye's blue colour. Lake of Niksic (Montenegro). P. Geniez.*

**Habits** Diurnal and partially nocturnal. Readily basks in full sun in the morning before retiring to shade. Hunts mainly in the evening and at dusk. Relatively slow. A period of inactivity in winter for 2–6 months, depending on region and elevation. Many Nose-horned Vipers, as many as several hundred, may come together in favourable over-wintering sites.

**Diet** Mice and other small mammals, birds, lizards and similar and small snakes. Kills prey with venomous bite.

**Reproduction** Ovoviviparous. Gives birth to live young. Mates in the spring, just after emerging from hibernation and the first moult of the spring. Ritual fights between males may occur. 4–20 young snakes are born between late August and early October. The young start by mainly feeding on small lizards and large insects.

Vipera ammodytes ammodytes, *subadult female. Lake of Niksic (Montenegro). Note the well developed, forwards-pointing horn, covered with numerous scales, characteristic of the nominate subspecies. P. Geniez.*

**Range** Quite an extensive distribution centred on the Balkans: from Austria (Carinthia, Styria) and northwestern Italy (southern Tyrol) to western Transcaucasia, passing through southern Central Europe and northern Turkey. Also present on several Greek islands.

**Geographic variation** Four main subspecies are generally recognised. Certain are, however, termed polyphyletic as they include several quite distinct lineages, of which some may be more closely related to lineages attributed to other subspecies.

► *Vipera ammodytes ammodytes* (Linnaeus, 1758): northwestern part of the species' range, south as far as northern Albania and southern Serbia. Large subspecies, extremely variable in colour with a very well developed forwards-pointing horn, covered in numerous scales (14–20) arranged in 4 or 5 rows, one above the other; the dorsal zigzag is often reduced, not really reaching the top of the flanks; underside of tip of tail often orange or orange-red. Two other localised subspecies, hardly distinct genetically, are grouped into *V. a. ammodytes*: *V. a. gregorwallneri* (Sochurek, 1974) in Austria and *V. a. ruffoi* (Bruno, 1968) in the Bolzano area (northeastern Italy).

► *Vipera ammodytes meridionalis* Boulenger, 1903: the southern part of the Balkans – that is, the whole of Greece, southern Albania and Macedonia, including the island of Corfu. Of medium size (maximum 65 cm); quite thickset; the snout horn usually high and straight, perpendicular on top of head of 14–20 scales arranged in 4 or 5 rows; zigzag often more extensive, descending onto the sides of the back, which may be angular or on the contrary very rounded; underside of tip of the tail often yellow or greenish.

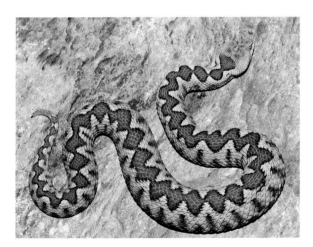

Vipera ammodytes meridionalis, *female. Greece. Note the yellow colour of the tail; in* V. a. ammodytes *it is orange or reddish.* P. Geniez and J. Garzoni.

► *Vipera ammodytes montandoni* Bolenger, 1904: countries to the west of the Black Sea: eastern Romania and Bulgaria, eastern European Turkey. Medium to large snake (maximum 89 cm); the quite short horn normally points forwards and has only 10–14 scales arranged in 3 (rarely 2 or 4) tiers; of variable colour, but especially brownish or greyish with little sexual dimorphism; underside of tail yellow or greenish.

► *Vipera ammodytes transcaucasiana* Bolenger, 1913: north of Turkey and Transcaucasia (the upper valley of the Kura in southern Georgia). Easily identified by the dorsal zigzag being reduced to well-separated lateral bars; only slightly variable colouration, males are usually pale grey, females brown; the horn is not well-developed, usually pointing forwards; underside of the tail yellow or greenish-yellow. Several authors have suggested raising this subspecies to species level, the Transcaucasian Nose-horned Viper, *Vipera transcaucasiana* Boulenger, 1913. However, this new arrangement is not supported by genetic studies,

Vipera ammodytes meridionalis. *Péloponnèse, in southern Kalamata (Greece). Note the wide, triangular horn that's vertical (and not angled forwards as in* V. a. ammodytes). P. Geniez.

which show that *transcaucasiana* is one of the lineages of an assemblage that regroups also *V. a. meridionalis* (from the east) and *V. a. montandoni*. Populations from the Greek islands in the Aegean Sea are traditionally assigned to *V. a. meridionalis*. They are however distinct particularly in their smaller size (30–50 cm), by having a slim but quite long horn, often pointing forwards and by an often thin zigzag; indeed, they resemble more a small *V. a. ammodytes* than *V. a. meridionalis*. Additionally, genetic work shows that they form a special lineage within an assemblage that regroups *V. a. meridionalis*, *V. a. montandoni* and *V. a. transcaucasiana*.

Vipera ammodytes montandoni, *two females of different colour. South of Dobroujda (Romania). Note the relatively short horn and very rounded zigzag denticles.* P. Geniez and J. Garzoni.

The populations of the Nose-horned Viper of the Greek islands are traditionally attached to Vipera ammodytes meridionalis. *They are, however, quite distinctive; in addition to their small size, they often have a forwards-pointing horn (as in* V. a. ammodytes) *and a much narrower dorsal zigzag, as opposed to the very wide one in* V. a. meridionalis. *They could represent one or several distinct island subspecies. Here an adult female on the Greek island, Santorini.* P. Geniez.

*Transcaucasian Nose-horned Viper* Vipera ammodytes transcaucasiana. *A male (left) and a female at Zara (northern Turkey) (right). In this subspecies, considered by some authors to be a separate species, the normal lozenges are replaced by dark transverse bars that are well separated from each other.* P. Geniez and A. Teynié.

### Genus *Montivipera*

There are some 10 species in the genus; they are characterised by being of medium to large size (60–116 cm), quite thickset and often very colourful; their range is centred on Anatolia and Iran. From a phylogenetic point of view the *Montivipera* is a sister-genus of *Macrovipera*. The *Montivipera* + *Macrovipera* is a sister-group of the assemblage *Vipera* + *Daboia*. The genus *Montivipera* consists of two distinct groups: the *Montivipera xanthina* and *Montivipera raddei* groups.

The *Montivipera xanthina* group has a western distribution, occurring in eastern Greece, much of Turkey and higher mountains in the Levant countries. The systematics of this group is still much debated and highly controversial, despite its phylogeny being well known. For one thing, there are species that have very distinctive colouration and are easily recognisable when seen; on the other hand, all are very close genetically. Moreover, *Montivipera xanthina* is structured in several lineages, more divergent one from another than they are from other species. Several authors suggest that they all be considered as subspecies of *M. xanthina*; others recognise five distinct species (*M. xanthina*, *M. bulgardaghica*, *M. albizona*, *M. bornmuelleri* and *M. wagneri*), whilst still others take an intermediate stand, grouping the taxa most genetically similar, *M. bulgardaghica* and *M. albizona*, into a single species, with the other three remaining.

Without further information and new work using different genes, we prefer to adopt this intermediate position and consider *M. bulgardaghica* and *M. albizona* as the same species (as they are the closest genetically, and also because there exist specimens of *bulgardaghica* that have a colouration intermediate between the two taxa), and keep *M. bornmuelleri* and *M. wagneri* as separate species.

Effectively, the species that are more closely related, *M. bulgardaghica* (including *albizona*), *M. bornmuelleri* and *M. wagneri*, are those that have the most restricted and isolated distributions. It can be easily conceived

that they evolved more slowly than *M. xanthina*. This last taxon has a much wider and more fragmented distribution and a more diversified ecology; it is better adapted to "modern" environmental constraints: climatic change, change in vegetation cover, competition with other snakes with the same ecological needs, anthropization. On the other hand, the other species have a reduced distribution and are confined to extremely limited dry mountain habitats: it is quite possible that they have evolved little, long ago, earlier than the diversification of *M. xanthina* in several lineages that are not all necessarily evolutionarily independent. On the contrary, we now know, and probably definitively, that the other species of the *M. xanthina* group are separate one from the other and thus represent independent evolving unities.

Within the area covered by this book, the *Montivipera raddei* group is represented solely by *M. raddei*. In Iran, apart from *M. raddei*, there are three other endemic species isolated one from another: *M. albicornuta*, *M. kuhrangica* and *M. latifi*. The systematic situation is identical to that of the *M. xanthina* group: one species with a wide range and quite distinct from a genetic point of view, *M. raddei*, while the others have quite restricted ranges and are genetically very similar to *M. raddei*, so much so that certain authors recognise only one species, *M. raddei*, despite their significant differences in colouration.

## Ottoman Viper

*Montivipera xanthina* (Gray, 1849) (formerly *Vipera xanthina*)
**Family:** Viperidae (vipers)
**Sub-family:** Viperinae (true vipers)
**F.:** Vipère xanthine or Vipère ottomane
**G.:** Kleinasiatische Bergotter

**Identification** The largest species of the genus *Montivipera*. Average length 80–100 cm, but can reach at least 116 cm for a diameter exceeding 5 cm. The wide, triangular head is distinct from the body with a large rounded snout. The eyes have pupils with vertical slits; the iris is silvery-grey and gold, darker towards the bottom. Very large nasal plates, as large as the eye, nostrils situated in their bottom half. The top of the head is covered with keeled scales, except for the supraoculars, which are well developed and go a little beyond the eye; curved slightly on the top of the eye, they give the Ottoman Viper with a "severe" or "worried" look; the supraoculars touch the upper edge of the eye, a feature characteristic of all species in the *M. xanthina* group (the eye is separated from the supraoculars by a row of small scales in members of the *M. raddei* group). 2 rows of scales between the eye and the supralabials. There are normally 10 (more rarely 9 or 11) supralabials on each side. The dorsal scales are keeled and arranged in 23 (sometimes 25) rows at mid-body. It has 170–185 ventral plates, an undivided anal scute, divided subcordals arranged in pairs. In males the back is pale or pearly grey; more beige or brown in females with a wide zigzag along its length composed of rhomboid or rounded markings that may be separated from each other. These markings are large and reach onto the top of the flanks (in other members of the *M. xanthina* group they are limited to the central dorsal band). And *M. xanthina* is the only species of the group in which these dorsal markings can be rhomboid in shape, forming angular triangles on the top of the flanks (in the others, the edges of the markings

*Ottoman Viper* Montivipera xanthina, *adult male. North of Alexandroúpoli (Greece).* P. Geniez.

*Close-up of Ottoman Viper* Montivipera xanthina, *adult. Tavçancil, to the east of Istanbul (northwestern Turkey).* P. Geniez and A. Teynié.

are rounded or quadrangular). In the male the markings are a uniform black, and in the females they are dark brown, uniform or with a paler orange-brown centre, when they resemble the colouration of *M. bulgardaghica albizona*. The ground colour of the dorsal area can be very pale, almost white, so that the zigzag can be very prominent. The dark flank markings may be a single row of vertical bars arranged alternately with the back markings, or in 2 series of separated markings. There are 2 long dark marks on the back of the head that come together at the front, preceded by 2 dark spots. There is a dark temporal band from the eye to the corner of the mouth and a vertical black mark below the eye; the belly is pale grey with dark mottling. The underside of the tail tip is often orange or yellow. Juveniles resemble adults, but the dorsal markings are more brown or olive-brown.

**Venom** The venom is very effective, essentially cytotoxic. A bite from a large individual can be fatal for a human. The Ottoman Viper is very shy and can flee rapidly despite its large size. Cornered, it hisses loudly, coils on itself, the neck bent in an S. Captured, it may bite violently and is even capable of lifting its body up when held by the tail. Solenoglyphous fangs.

**Habitat** Various types of hillside and mountains, often of a Mediterranean character. Open forest, valley bottoms with meadows and a stream, grassy slopes with scattered bushes and stones, garigue, orchard edges, drystone walls, areas of ruins, scree in mountains. From near the sea to 2500 m in mountains, at Erciyes Dagi (central Turkey).

**Habits** Essentially diurnal but also active at dusk and during the early nighttime hours during the hottest parts of the year. Winter inactivity can last up to 5 months in mountain areas but is much less at sea level, where individuals may even be seen basking in the sun during winter in periods of fine weather.

**Diet** Small mammals, birds, lizards. Kills its prey with its venomous bite.

**Reproduction** Ovoviviparous. Gives birth to 2–15 (occasionally 20) live young at the end of July or in August. The young do not moult until 2 weeks later, after which they start feeding.

**Range** The Alexandroúpoli region in eastern Thrace (extreme eastern Greece), European Turkey, the Greek and Turkish Dodecanese islands near the Mediterranean coast of Turkey, western and central Turkey to the southwest as far as Erciyes Dagi (Mount Erciyes) in southern Kayseri. Replaced farther east by other species of viper in the same group.

**Geographic variation** Some authors consider the OttomanViper to be polytypic (the species contains several

*A newly born Ottoman Viper* Montivipera
xanthina. *P. Geniez.*

subspecies) and within the species
include *Montivipera b. bulgardaghica*,
*M. b. albizona* and *M. b wagneri*.
Others consider these three taxa to be
separate species and thus consider the
last of these to be monotypic. However,
genetically, we know that *M. xanthina*
in the strict sense is structured into
three very divergent lineages, much
more so than are any other taxa
of the group within themselves
(*bulgardaghica, albizona, bornmuelleri*
and *wagneri*): a western lineage, a
southern lineage and an eastern
mountain lineage.

# Bolkar Viper

*Montivipera bulgardaghica* (Nilson and Andrén, 1985)
**Family:** Viperidae (vipers)
**Sub-family:** Viperinae (true vipers)
**F.:** Vipère du Bolkar
**G.:** Taurische Bergotter

**Identification** A species represented by
two subspecies that have such distinct
colouration that they are treated as
two distinct species by many authors.
To facilitate their description they are
separated in the text.

## *Montivipera bulgardaghica bulgardaghica* (Nilson and Andrén, 1985)

A very rare viper, with just a few
records, that is similar in appearance
to the Ottoman Viper. Smaller
(50–70 cm, may reach at least 78 cm).
Is readily recognised by the tight
dorsal zigzag that does not continue
onto the flanks, composed of black

Montivipera bulgardaghica bulgardaghica. *Note that the dorsal zigzag is much narrower than that of* M. xanthina. Teynié and Bettex.

or blackish markings, sometimes dark brown or with a slightly orange centre in females; the markings are placed alternately, sometimes joined, sometimes separated from each other; they are more numerous than in *M. xanthina*: on average 40 (range 35–46) as opposed to 30 (range 22–41). There are 2 or 3 series of dark spots along the flanks that can appear as vertical bars. In males the back and flanks have a shiny pale grey ground colour that is duller grey or pale brown in females. The belly is covered with fine dark speckling or ill-defined spots (well-pronounced black markings in *M. xanthina*). In some individuals, there is only 1 row of scales between the eye and the supralabials (always 2 in *M. xanthina*). On average, there are fewer ventral plates than in *M. xanthina*: 145–154 compared to 148–169. Compared to the Lebanon Viper *M. bornmuelleri*, which it resembles, the zigzag has diagnostically less markings: 40–46 compared to 47–64.

**Venom** Venom probably cytotoxic and quite dangerous for humans. Solenoglyphous fangs.

**Habitat** Subalpine zones with snow in winter; abandoned grassy slopes covered with tussocks of spiny, woody vegetation and rocks; vegetation-covered scree, base of cliffs. The bedrock where the species occurs is a very pale marble, nearly white, on which the Bolkar Viper is well camouflaged. Reputed to occur between 2000 and 2700 m, but in reality more likely to be limited (the rare sites for this species are kept secret).

**Habits** At the elevations at which it occurs, it is probably diurnal. Winter inactivity may well last for several months.

**Diet** Probably small mammals, birds, lizards and large grasshoppers and crickets.

**Reproduction** Almost unknown, but certainly ovoviviparous like other *Montivipera*.

**Range** Known from a single group of sites on Bulgar Dagh (in the Bolkar massif) that are very close to each other. The type locality is situated at Kar Boghaz.

**Geographic variation** *Montivipera bulgardaghica* is considered to be either monotypic, or composed of two subspecies if it is considered that *M. albizona*, the genetically closest taxon, is a subspecies of *M. bulgardaghica*.

## *Montivipera bulgardaghica albizona* (Nilson, Andrén and Flärdh, 1990)

**Identification** A very rare viper, genetically very close to *Montivipera b. bulgardaghica*, but with a general appearance similar to that of *M. wagneri*. Also similar in appearance to the Ottoman Viper but smaller (50–70 cm, reaches at least 78 cm). Ground colour bright pale grey to bluish-grey in males, browner grey or pale beige in females, with 2 series of large, rounded, half-moon markings on the back, in most cases joined in pairs forming beautiful round orange marks edged with black and separated from each other by the dorsal ground colour that is almost white along the spine. The

marks are sometimes relatively square; but in such a case they are much larger than those of *M. wagneri*. There is a series of poorly defined black markings on the flanks that alternate with the back markings. The head is pale grey with the black markings typical of members of the *M. xanthina* group. There are generally 9 supralabials on each side (as in other taxa of the group, excepting *M. xanthina*, which nearly always has 10). A relatively small number of ventral plates (151–155), on average less than *M. xanthina* (148–169) and especially *M. wagneri* (161–168).

**Venom** Solenoglyphous fangs. A cytotoxic venom quite dangerous for humans. I was witness to a bite inflicted by a large male *Montivipera albizona*. For the first $5\frac{1}{2}$ hours after the event, the only symptom was the swelling of the bitten finger, accompanied by increasing pain that became almost unsupportable; after $5\frac{3}{4}$ hours there was a brutal

Montivipera bulgardaghica albizona. *Adult male 78 cm long. Osmandere (Turkey). Note the beautiful colouration, completely different from the subspecies* bulgardaghica. P. Geniez and A. Teynié.

Montivipera bulgardaghica albizona. *Adult female. Osmandere (Turkey).* P. Geniez and A. Teynié.

drop in blood pressure, resulting in extreme paleness of the face, abundant perspiring; the victim felt cold (hypothermia) despite the clement weather at the end of May and suffered partial loss of consciousness; after administration of a sodium chloride solution by a medical doctor (physiological salt solution, not an antivenin), the blood pressure rose and the patient recovered consciousness; injections of Heparin (an anticoagulant administered to avoid clotting), an antihistamine (to prevent an eventual anaphylactic reaction) an analgesic and antibiotics were administered as precautions; after two days the finger became necrotic, resulting about permanent and insupportable pain for a week; when possible during the journey, the victim was given morphine injections to alleviate the pain; the finger almost needed amputating but eventually only required two surgical interventions, nothing else.

**Habitat** Grassy plateaus and slopes on a basalt bedrock with rock crevices. At the foot of cliffs strewn with scree with dense bushes and scrub. Between 1300 and 1700 m elevation, even to 2200 m.

**Habits** Has been observed several times in the middle of the day in bright but not too hot weather, basking in the sun. Winter inactive period probably of several months.

**Diet** Probably small mammals, birds, lizards and large grasshoppers and crickets.

**Reproduction** Almost unknown; almost certainly ovoviviparous like other members of the genus *Montivipera*.

**Range** Known from several sites along the "Anatolian diagonal", a long, large mountain that separates Turkey from the Gulf of Annaba at Erzincan. The sites are more or less staggered from Kahramanmaraş in the southwest to Erzincan. A new locality, which needs confirmation, has been published: near Mount Bozdağ in the Anamos massif, Hatay province, at about 2200 m elevation.

## Wagner's Viper

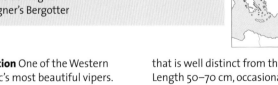

*Montivipera wagneri* (Nilson and Andrén, 1984)
**Family:** Viperidae (vipers)
**Sub-family:** Viperinae (true vipers)
**F.:** Vipère de Wagner
**G.:** Wagner's Bergotter

**Identification** One of the Western Palaearctic's most beautiful vipers. Of medium to quite large size, quite thickset with a wide triangular head that is well distinct from the body. Length 50–70 cm, occasionally to 95 cm. Top of the head covered in small, keeled scales, excepting the well-defined

*Wagner's Viper* Montivipera wagneri, *female. Northeastern Turkey. Note the shape of the orange markings at the front of the body, which are more or less quadrangular: they are very rounded in half-moons in* M. bulgardaghica, *which it otherwise very much resembles.*
P. Geniez and A. Teynié.

supraoculars; they go slightly beyond the eyes and touch the eye. There are 2 rows of scales between the eye and the supralabials, which normally number 9. The keeled dorsal scales are arranged in 23 rows at mid-body.

The colouration is extremely bright and shimmering: males have pearl to bluish-grey backs; females are brownish beige, covered with 2 series of long, more or less quadrangular markings that are bright orange with black edges; these markings are either staggered and partly merged two by two, or completely merged, forming wide, separated patches; the colour of the back between the markings is brighter and paler, almost white in some individuals. In aged males, the dorsal markings become totally blackish through elimination of the orange part. Large black (males) or dark brown (females) markings on the flanks form vertical bars that are sometimes attached to the markings on the back and those of the underparts. There are hardly any

markings at the front of the head, but 2 dark bars form a V, open at the rear and sometimes connected on the nape by a black transverse bar; there are 2 small black spots in front of this pattern. There is often, but not always, a vertical black bar below the eye that does not touch the eye. A wide black band runs from the eye to the corner of the mouth. The belly is grey, more or less marbled and finely speckled with black. The underside of the tip of the tail is orangey. Overall, it can be said that *Montivipera wagneri* much resembles *M. albizona*, but that the latter has rounded, half-moon-shaped orange and black markings on the back, whilst in *M. wagneri* these are quadrangular and elongated.

**Venom** The venom is supposedly as virulent as that of the Ottoman Viper, with an essentially cytotoxic action; Wagner's Viper is shy and inclined to flee but cornered or captured it rapidly tries to bite. Solenoglyphous fangs. Apparently no known case of a human having been bitten has been recorded.

**Habitat** Stony mountain slopes with a little short vegetation; often near mountain streams; also at the foot of scree and rocks. Occurs only on basalt ground. Known sites are at between 1200 and 2146 m elevation.

**Habits** Occurs exclusively at high elevations and thus is probably diurnal. Its biology is still largely unknown but is probably very similar to that of the Ottoman Viper.

**Diet** Probably small mammals, lizards, ground-nesting birds and large crickets and grasshoppers.

**Reproduction** Ovoviviparous. Gives birth to live young.

**Range** Only a few known sites in northeastern Turkey, between Dolabaş to the south, Karakurt to the northwest and Camuşlu to the northeast. Described as recently as 1984 and then

*Close-up of a male* Montivipera wagneri. P. Geniez and J. Garzoni.

considered as of Iranian origin, but it now almost certainly no longer exists in Iran.

**Geographic variation** *Montivipera wagneri* is a monotypic species.

---

## Lebanon Viper

*Montivipera bornmuelleri* (Werner, 1898)
**Family:** Viperidae (vipers)
**Sub-family:** Viperinae (true vipers)
**F.:** Vipère du Liban
**G.:** Libanesische Bergotter

**Identification** A medium-sized viper, quite thickset, endemic to Mount Lebanon and Mount Hermon. Average length 50–60 cm but can attain 74 cm, rarely 80 cm. Triangular head very distinct from body, covered in small keeled scales, except for the supraoculars, which are very distinct and slightly protrude above the eyes. Eyes with vertical pupils, and golden-silver iris that is darker towards the bottom. Keeled dorsal scales, arranged in 23 (rarely 21) rows at mid-body. 2 rows or scales between the eye and the supralabials that normally number 9 on each side.

General colour grey, grey-brown or pale brown, which is a good camouflage against the calcareous, rocky substrate of its habitat. Its dorsal pattern is difficult to describe: 2 series of more or less rectangular and partially joining markings form a vertebral dark band; these markings are partly dark brown but have wide blackish borders; the very pale grey ground colour of the back appears between these markings, forming small, almost white squares. In most adults the brown of the dorsal markings disappears, leaving just the dark borders, which form transverse bars. The flanks are dotted with smaller dark markings, sometimes very subdued. Males are generally paler, females browner; in some aged females the markings disappear almost completely. There is a dark temporal band from the eye to the corner of the mouth. Belly grey often with fine speckling. Yellow extremity to tail.

**Venom** Bite can be quite dangerous for humans, causing decrease in blood pressure, much necrosis of the bitten area followed by the cytotoxic effects of the toxin. Necessitates treatment with use of an antivenin. Normally a slow, shy snake but if caught will bite rapidly. When cornered it hisses loudly. Solenoglyphous fangs.

**Habitat** Mountain slopes with much low, dense vegetation and scattered scree and rocks, especially near remains of Cedar forest, between 1600 m and 2000 m elevation.

**Habits** Little known. Although apparently living in dry habitats the Lebanon Viper needs a certain level of atmospheric humidity. Probably has an inactive period in winter of several months.

**Diet** Especially small mammals.

**Reproduction** Ovoviviparous. Gives birth to 5–18 live young; in their first reproductive year, young females have just 2 young.

**Range** An endemic strictly limited to Lebanon (Mount Lebanon) and the frontier zone between Lebanon, Syria and occupied territories of Israel (Mount Hermon).

**Geographic variation** *Montivipera bormuelleri* is a monotypic species.

*Male Lebanon Viper* Montivipera bornmuelleri *in Lebanon.* Cluchier.

# Radde's Rock Viper

*Montivipera raddei* (Boettger, 1890)
**Family:** Viperidae (vipers)
**Sub-family:** Viperinae (true vipers)
**F.:** Vipère de Radde
**G.:** Armenische Bergotter

**Identification** Medium-sized viper, quite slender when young, becoming quite massive and thickset when adult, especially females. Total length 60–90 cm, occasionally to 99 cm. Males are a little larger than females. Wide, triangular head that is quite distinct from body covered with small keeled scales; well-developed supraoculars protrude above each eye, forming a small, pale triangular horn, typical of this species in the area covered by this book. Eyes with vertical pupils and silver-grey or golden iris, almost black at the bottom. There is a complete ring of scales around the eye, separating it from the supraoculars (in members of the *Montivipera xanthina* group, the supraocular touches the top of the eye, and thus the ring of scales is broken). Well-developed nasal plates, as large as the eye. 2 rows of scales between the eye and the supralabials, which number 9 (sometimes 10) on each side. Keeled dorsal scales, arranged in 23 or 24 (more rarely 21 or 25) rows at mid-body. 163–181 ventral plates, an undivided anal scute, divided subcordals arranged in pairs.

*Male Radde's Rock Viper* Montivipera raddei. *Mount Ararat (northeastern Turkey). Note the remarkable colouration, with orange markings on the lead-grey back.* P. Geniez and A. Teynié.

*Female Radde's Rock Viper* Montivipera raddei *showing less contrasting colouration than that of the male. Near Kars (northeastern Turkey).* P. Geniez and A. Teynié.

The ground colour is beige-grey when young, gradually darkening with age to become lead-grey, almost black, in large individuals. A series of large orange marks, sometimes with black edges, along the back sometimes partially join together to form an irregular zigzag with rounded edges, sometimes regular with triangular edges. This lead-grey and orange colouration proves to be highly mimetic with the basalt rocks spotted with orange lichens that make up its habitat in most localities where it occurs. Also it is the only member of the genus *Montivipera* to have dorsal markings paler than the rest of the body. It has a series of vertical black bars along the flanks. The back of the head has a dark mask-shaped pattern, often preceded by 2 small black spots. There is a black mark under the eye, and a black temporal band from the eye to the corner of the mouth. The underside is dark grey to black, finely speckled with white, and there is a black spot on the external edge of 1 in every 2 ventral plates.

**Venom** Very effective venom, essentially cytotoxic. A bite from a Radde's Rock Viper can be fatal to a human if not treated and followed by hospitalisation. In most cases, the symptoms include a quick fall in blood pressure and some necrosis of the bitten area, which may require amputation. This is a shy snake that relies on its camouflage as its best defence. Worried, it will slide quite quickly under a rock or into scree. Cornered or captured, it will bite, like all vipers. Solenoglyphous fangs.

**Habitat** Sunny, rocky slopes with bushes, especially on basalt, rock crevices over rivers, scree, rocky plateaus, open oak woodland. Typically occurs in mountains between 1100 and 2400 m (e.g., on Mount Ararat, in extreme northeastern Turkey).

**Habits** An essentially diurnal, terrestrial snake that basks in the sun at length when the weather is not too hot.

Close-up of *Montivipera raddei. Note the preeminent supraocular scales, paler than rest of head, and the scales inserted between the supraoculars and the eye, all criteria special to the* M. raddei *group. Kreiner.*

Hibernates for up to 7 months. Locally abundant.

**Diet** Small rodents, ground-nesting birds, lizards and also crickets eaten by adults as well as the young (personal communication with Roozbeth Behrooz, observations in Iran). Immobilises large prey with its venomous bite.

**Reproduction** Ovoviviparous. Gives birth to 3–6 live young. They are born in early September after 3–3½ months gestation.

**Range** Northeastern and extreme eastern Turkey, Armenia, Nakhichevan territory (Azerbaijan) and northwestern Iran.

**Geographic variation** *Montivipera raddei* belongs to a group of genetically closely related species whose status is controversial; certain authors consider that there is just one species, with several subspecies; others say there are four distinct species, of which one, *M. raddei*, is divided into two subspecies. The *M. raddei* group contains five taxa:

▶ *Montivipera raddei raddei* (Boettger, 1890): throughout most of the species' range.

▶ *Montivipera raddei kurdistanica* (Nilson and Andrén, 1986): farther south, in extreme eastern Turkey and western Iran as far as the region of Lake Urmia. Genetically this subspecies is hardly distinct from *M. r. raddei*.

▶ *Montivipera albicornuta* (Nilson and Andrén, 1985): in northwestern Iran but farther east than *M. r. raddei*.

▶ *Montivipera kuhrangica* Rajabizadeh, Nilson and Kami, 2011: for the present, known from the central part of the Zagros chain, western Iran. It is the most southerly known population of any viper of the genus *Montivipera*.

▶ *Montivipera latifii* (Mertens, Darevsky and Klemmer, 1967): isolated in the upper Lar Valley (Elbourz chain, northern Iran).

Only *M. r. raddei* and *M. r. kurdistanica* are present in the geographic zone covered by this book. *M. r. kurdistanica* is not recognised by all authors. Very similar in general appearance to *M. r. raddei*, but more variable in its colouration; with experience it can be differentiated by the dorsal markings being closer to each other and by its more well-defined black border.

### Genus *Macrovipera*

As present knowledge exists, it has only one known member, the Blunt-nosed Viper. It is a sister-group to the genus *Montivipera*. The ranges of the two genera partially overlap in Anatolia, Armenia, the Levant countries and in Iran. When members of these genera are present in the same region, the Blunt-nosed Viper occupies dryer, more steppe-like or more Mediterranean-type habitat;, depending on the region, the genus

*Montivipera* does not occur at the highest elevations. Until recently, the Moorish Viper of North Africa was placed in the genus *Macrovipera* due to strong morphological and ecological similarities; recent genetic work has shown that *M. mauritanica* is really part of another lineage of large vipers related to the Palestine and Russell's Vipers and should be renamed *Daboia mauritanica*.

## Blunt-nosed Viper

*Macrovipera lebetina* (Linnaeus, 1758)
**Family:** Viperidae (vipers)
**Sub-family:** Viperinae (true vipers)
**F.:** Vipère lébétine or Vipère du Levant
**G.:** Levanteotter

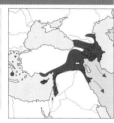

**Identification** A large, robust viper, often with a colouration very mimetic of the rocky substrate where it occurs. Length in most populations, 90–130 cm but can reach 150 cm, rarely longer than 2 m (records: 214 cm for *Macrovipera lebetina obtusa,* 230 cm for *M. l. euphratica* in southeastern Turkey). Thick, wide, triangular head very distinct from the body. Eyes have pupils with vertical slits, iris pale grey or slightly orange, marbled with dark in its lower part. Slightly prominent rounded snout;

there is a string of slightly larger and preeminent scales on the sides of snout and head as far as the eyes. Nasal plate complex and divided with a hollow pit as large as or larger than the eye, which contains the nostril. The top of the head is covered in small keeled scales. The supraoculars are divided into 2 or 3 parts and are thus not always distinct. 3 (sometimes 2¹/₂) rows of subocular scales between the lower edge of the eye and the supralabials. 11–18 scales around the eye. 10 or 11 (rarely 9 or

*Blunt-nosed Viper*
Macrovipera lebetina
lebetina, *juvenile in
Cyprus*. Cheylan.

12) supralabials on each side. Dorsal scales matte or only slightly shiny, usually arranged in 25 (rarely 27) rows at mid-body (but generally 23 on the isle of Milos). 128–181 ventral plates, anal scute undivided, subcordals divided and arranged in pairs. Generally quite variable colouration but dull and often with a remarkably "stony" aspect: pale grey, beige, sandy, yellowish, olive, greenish or bluish-grey, sometimes uniform red-brown or brick-coloured. On the back are 2 offset series of transverse markings that are sometimes attached to form an undulating bar, or on the contrary are joined in pairs to form well-spaced rectangular, transverse bars. These markings can be very faded, even totally absent (especially in *M. l. obtusa* and the red form of *M. l. schweizeri*). These markings are often browner, even redder, than the ground colour; they can have dark or pale borders. There is a series of dark vertical bars on the flanks that alternate with the back markings. The body is sometimes dotted with minuscule ochre or yellowish spots. There is no pattern on the top of the

head; head side often darker than the top; there is an often faded vertical dark mark under the eye and an only slightly contrasting dark temporal band from the eye to corner of the mouth. The belly is pale grey, whitish or grey-brown with much dark grey or blackish mottling.

**Venom** Very potent venom (mainly cytotoxic). Solenoglyphous fangs. A venomous bite is dangerous for humans and certainly necessitates rapid medical attention (an antivenin treatment). Being poisoned by a Blunt-nosed Viper is often serious and accompanied by significant necrosis that might require amputation of the bitten limb. Without treatment, bites can be fatal for humans. This snake is generally quick to flee, with a considerable rapidity despite its thickset stature; cornered, it can enlarge and flatten its neck in the manner of a cobra, but not as markedly. It can also defend itself by biting violently, projecting half of its body length towards the aggressor.

**Habitat** Sunny mountain and steppe slopes, scattered with scree; drystone walls, scrub, scree at the foot of cliffs. Readily occurs on slopes above watercourses. Also uneven garrigue next to the sea, as well as along the sides of wetlands bordered with rocks or walls. In mountains to 1500 m, locally as high as 2000 m.

**Habits** Robust terrestrial snake that can, however, climb bushes with skill. Basks in the sun in the morning as well as during the day in spring and autumn. During the hottest part of the year the Blunt-nosed Viper is mainly nocturnal. More strictly nocturnal in the southern part of its range, especially at low elevations. Winter inactivity lasts 2–5 months, depending on the region and elevation, but in the south it is more or less active throughout the year.

**Diet** Small mammals (rats, mice, gerbils, rabbits, young hares), birds, lizards, occasionally snakes and large insects. On islands, particularly Milos, it specialises in catching migrant birds that it awaits, hidden in a low bush, capturing them without letting go when they land. Kills victims with its venomous bite.

**Reproduction** There are ovoviviparous populations, while others are oviparous. *Macrovipera lebetina lebetina* is ovoviviparous on Cyprus; *M. l. obtusa*, in central Asia, lays 9–24 eggs, their incubation lasting between 2 and 2¹/₂ months. Laying sites can be communal, with up to 47 eggs. On the Milos archipelago, *M. l. schweizeri* is

oviparous; the development of the 4–11 eggs is already well advanced when they are laid.

**Range** Vast range, from southern and southeastern Turkey as far as northeastern Pakistan and extreme northwestern India, passing through a large part of the steppes and mountains of central Asia. Also present in the western parts of the Levant countries (western Syria, Lebanon and northern Israel, where it is thought to be extinct). Island populations on Cyprus and in the Milos archipelago (Greece).

**Geographic variation** The systematics of *Macrovipera lebetina* are still quite controversial, particularly due to a lack of genetic data on whole populations. If morphological variation and genetic structure are combined, it is possible to separate two groups, a southwestern group (*M. l. lebetina* and *M. l. schweizeri*) and a more eastern group (*M. l. obtusa, M. l. cernovi* and *M. l. turanica*), thus five subspecies:

▶ *Macrovipera lebetina lebetina* (Linnaeus, 1758): Cyprus and southern

Macrovipera lebetina lebetina, *adult of the "euphratica" population. Birecik (southeastern Turkey). Note the extremely cryptic colouration compared to the surrounding rocks.* P. Geniez.

Turkey. The Blunt-nosed Vipers of the western Levant (Lebanon and western Syria) probably belong to this subspecies. A very large viper (maximum 230 cm, but less on Cyprus: 153 cm); 25 rows of dorsal scales at mid-body on Cyprus, 25 (rarely 27) in Turkey; general aspect is matte and "stony"; resembles the colour of stones due to the presence of small, dark speckling on the scales; the dorsal bars are about the same width as the spaces between them and often a red-brown that can contrast with the grey or blue-grey ground colour (but on Cyprus the dorsal bars are often more blackish-grey); the side of the head is often darker than the top, sometimes a beautiful dark blue-grey; on Cyprus some individuals are entirely uniform brick-red. Several authors consider *M. l. euphratica* (Martin, 1838) to be the continental population found in southeastern Turkey and *M. l. lebetina* to be endemic to Cyprus.

▶ *Macrovipera lebetina schweizeri* (Werner, 1935): an island population completely isolated to the western part of the rest of the species' range, endemic to the Milos archipelago, in the Greek Cyclades: islands of Milos, Siphnos, Kimolos and Polyaïgos. Until quite recently, still considered to be a separate species, the Cyclades Blunt-nosed Viper *M. schweizeri* (Werner, 1935) but genetically very similar to *M. l. lebetina* (e.g., *M. l. obtusa* is more distinct from *M. l. lebetina* than it is from *M. l. schweizer*). Oviparous, it lays already well-incubated eggs and thus the incubation period is short. The colouration is very similar to that of *M. l. lebetina* and quite variable, which is rather surprising for a small isolated population: a general "stony" appearance; the dorsal bars are often fawn or orange, on an almost white, pale grey ground colour; there exist uniform brick-red specimens; mainly characterised by its small size (average 65–75 cm, more rarely almost 100 cm, rarely 107 cm) with a small number of dorsal scales: generally 23 rows at mid-body, rarely 19, 21 or 25.

▶ *Macrovipera lebetina obtusa* (Dwigubsky, 1832): more eastern distribution, northeastern Turkey,

Macrovipera lebetina obtusa, *subspecies of the Blunt-nosed Viper from northeastern Turkey and farther east. Note the dull colouration without reddish markings. There are, however, individuals with more contrasting colours, with brown or grey-brown transverse bars.* P. Geniez and A. Teynié.

*Cyclades Blunt-nosed Viper* Macrovipera schweizeri, *subspecies of the Milos archipelago, until recently considered to be a separate species, and genetically and morphologically very similar to* M. lebetina lebetina, *of Cyprus.* Trapp.

Transcaucasia, northwestern and western Iran. The isolated population in Jordan appears to belong to this subspecies. A large viper, often longer than 150 cm (rarely 214 cm): 25 rows of scales at mid-body, colouration less "stony", often with less contrast, from dull olive-brown to earth-beige, with the dorsal bars contrasting less with the ground colour; the sides of the head are not dark blue-grey. In central Asia, outside the area covered by this book, there are at least two other subspecies: *M. l. cernovi* and *M. l. turanica*.

**Remarks** The Blunt-nosed Viper has quite recently been identified in northern Algeria and Tunisia, and described as a distinct subspecies: *Macrovipera lebetina transmediterranea* (Nilson and Andrén, 1988). Known only from rare and old specimens kept in museums, with approximate data. These North African specimens have 25 rows of dorsal scales around the body (compared to 27 in the Moorish Viper) and a dorsal colouration with transverse bars. Much research in the field has so far proved unproductive. Some authors suggest it was introduced in North Africa by Romans from Asia Minor.

# Moorish Viper

*Daboia mauritanica* (Duméril and Bibron, 1848)
(formerly *Vipera mauritanica*)
**Family:** Viperidae (vipers)
**Sub-family:** Viperinae (true vipers)
**F.:** Vipère de Maurétanie
**G.:** Atlasotter

**Identification** A large, thickset viper similar to the Blunt-nosed Viper (which is genetically placed in a distinct genus). Length 100–150 cm, sometimes more, to a maximum of at least 181 cm. A wide, triangular head, quite distinct from the body. Eyes have pupils with a vertical slit; the iris is orangey or golden, darker in the bottom part. 27 rows of dorsal scales at mid-body. There are numerous small scales on the head, keeled on the rear part of the head; the supraoculars are divided into several small scales. 2 or 3 rows of subocular scales between the eye and the supralabials; generally 11, sometimes 12, supralabials on each side. 157–174 ventral plates, undivided anal scute, divided subcordals arranged in pairs.

Back quite variable in colour, from sandy-beige to pale or reddish-grey, with a very wide, dark wavy band or zigzag, with rounded or slightly angular sides. This band may be broken into slightly triangular, round or oval bands. These dorsal markings are so enlarged that they normally greatly reduce the spaces of the ground colour. There are large, more or less lozenge-shaped markings on the flanks. This colouration is very different from that of the Blunt-nosed Viper, in which the markings are more in the form of transverse bands on the back, vertical on the flanks and widely spaced from each other. It has a wide, dark, indistinct, sometimes almost invisible, temporal band from the eye to the corner of the mouth; a more or less triangular mark under the nostril, another wider one under the eye. The top of the head is uniform or nearly so, often greyer than the body and sides of the head. Some individuals have a very contrasting colouration, almost black and whitish-grey, or dark brown on pink, especially near the sea. Pale greyish belly with much dark mottling.

**Venom** Very efficient and potent venom (mainly cytotoxic), which can be inoculated in large quantities. Solenoglyphous fangs. Dangerous for humans, it is the species responsible for most snakebite victims in North Africa. If bitten, medical assistance and treatment with an antivenin are absolutely necessary. Even so, much necrosis caused by the haemolytic venom often results in the necessity to amputate, as well as irreversible consequences that can lead to death.

Daboia mauritanica mauritanica, *contrasting form. Note the back markings that are not really oval but rather angular. Between Tan-Tan and Guelmim (southwestern Morocco).* Crochet.

**Habitat** Occurs in well-exposed sites that are relatively arid with scattered rocks or dense buses. Mountain slopes with scattered stones, scree, foothills with cliffs, scrub, piles of stones, open forest. In mountains to 2000 m.

**Habits** A terrestrial snake that can, however, climb to 1½ m into bushes. The Moorish Viper is mainly active at dusk and at night, but basks in the sun in the early morning in cooler periods. Rests in cavities, crevices, under stones or deep in a rodent burrow during the day. Shy and always ready to flee but aggressive if cornered.

**Diet** Small mammals, birds and lizards.

**Reproduction** Oviparous; lays eggs with well-developed embryos. The young hatch from the eggs after 6–8 weeks.

**Range** Endemic to the Maghreb, present in a large part of Morocco except Saharan areas, northern Algeria, much of Tunisia and extreme northwestern Lybia.

**Geographic variation** *Daboia mauritanica* is represented by two subspecies until recently considered to be two separate species.

▶ *Daboia mauritanica mauritanica* (Duméril and Bibron, 1848): all of Morocco, northern Algeria and extreme northern Tunisia. Quite variable colouration; sometimes very contrasting, sometimes very pale in the most arid areas. The dorsal markings are often connected together, rounded or triangular on the sides, allowing the ground colour of the back to appear over a width of several scales, at least in places.

► *Daboia mauritanica deserti* (Anderson, 1892): range poorly defined and debated. At least the southern three-quarters of Tunisia and arid parts of northeastern Algeria. Until recently considered to be a separate species, Desert Viper *D. deserti* (Anderson, 1892), it has been reclassified as a subspecies on the basis of genetic work showing that there is very little difference between the two taxa. Even more recent research shows that *D. mauritanica* is structured in four main lineages, the "*deserti*" lineage extending from the west as far as the high Moroccan plateaus, thus supporting an earlier theory, the authors of which considered it to be present in this part of Morocco. In appearance, *D. m. deserti* is characterised by very wide, oval, quadrangular dorsal markings that are close together, reducing the ground colour of the back to thin, pale transverse half bars with slightly darker edges, which may alternate regularly or, on the contrary, be joined together to form slim rings. Other criteria given in the literature (that the scales on the top of the head between the eyes are smooth in *deserti*, keeled in *mauritanica*; that the rostral plate is higher than it is wide in *deserti*, wider than large in *mauritanica*) prove not to be pertinent if many *mauritanica* are examined. Individuals from the high plateaus of northwestern Algeria and eastern Morocco have non-angular, oval markings that in places allow the ground colour of the back to appear on several adjacent scales. In fact, they appear to be intermediate between *D. m. mauritanica* and the "real" *D. m. deserti*, from farther east.

Daboia mauritanica mauritanica, *a particularly contrasting individual. Near Aoulouz (southern Morocco)*. García-Cardenete.

*An enormous female* Daboia mauritanica mauritanica, *showing little contrast, the pattern somewhat reminiscent of that of* D. m. deserti. *Near Tan-Tan (southwestern Morocco)*. Cayuela.

*A Desert Viper* Daboia mauritanica deserti. *Note the very large back and flank markings that reduce the background colour to lines. Tunisia*. Fuchs.

# Palestinian Viper

*Daboia palaestinae* (Werner, 1938)
**Family:** Viperidae (vipers)
**Sub-family:** Viperinae (true vipers)
**F.:** Vipère de Palestine
**G.:** Palästinaviper

**Identification** A large, robust viper, quite thickset, reaching or commonly passing 100 cm, up to 140 cm. Large triangular head but round snout, very distinct from body. Eyes have pupils with vertical slits, a silvered golden iris, the rear half more or less dark. The top of the head is covered with small, keeled scales; the supraocular plates are well defined (the contrary to *Daboia mauritanica* and *Macrovipera lebetina*). Remarkably large nostrils are situated in a hollow in the nasal scale. 2 rows of scales between the eye and the supralabials, of which there are 9–11 on each side. The body scales are well keeled, arranged in 25 (rarely 24) rows at mid-body. 160–166 ventral plates, an undivided anal scute, subcordals divided and in pairs. Ground colour pale grey, ochre, yellowish, pale brown or orange-brown. There is a well marked, very

wide zigzag on the back composed of ovals or lozenges that are generally but not always interconnected; these markings are caramel or reddish-brown, sometimes nearly black; they have a thin black border with a thin white or whitish outer edge. The flanks have large rhomboidal markings, or elongated vertical bars that alternate with the back markings. The top of the head is marked with a large V, open at the rear and the same colour as the dorsal markings, bordered on the interior and exterior by 2 large pale Vs to quite beautiful effect. There is a large dark band on the side of the head that continues onto the snout, a second under the eye and a third, elongated, running from the eye to above the corner of the mouth; these markings contrast well with the ground colour, which ranges from white to pinkish-brown passing though golden yellow. Grey belly often with darker speckling. Very young Palestinian Vipers have a somewhat different colouration with, especially, blackish-brown dorsal markings without any reddish tint.

*Palestinian Viper* Daboia palaestinae, *adult found on the roadside at night. South of Kiriat Shmona (Israel).* Crochet.

335

**Venom** Large solenoglyphous fangs. Very potent venom, both cytotoxic and neurotoxic. As this species often occurs near urban or densely populated areas, it represents a real danger to humans. The venomous bite results in pain in the bitten limb and then in the abdomen, swelling that can be massive, nausea, vomiting and a quick reduction in blood pressure. Characteristic of *Daboia palaestinae* bites, significant haemorrhaging occurs that may affect the internal organs. For humans this can be fatal. In Israel and Jordan over two periods of 10 years, 167 cases of venomous snakebites were recorded, all attributed to the Palestinian Viper; 6.2% resulted in death. Thus, being bitten by this species is a serious event and necessitates immediate treatment in a hospital. Treatment with an antivenin is necessary but must be accompanied by additional care that might include a blood transfusion. The Palestinian Viper is rather indolent but capable of fleeing at speed. Cornered, it confronts the danger, lifting the front of its body, the neck folded in an S, ready to strike.

**Habitat** Originally in open oak forest, which is now a residual habitat within the species' range. Rocky hillsides with scrub, forest edges, scrub near arable land; fond of irrigated areas; sometimes penetrates into urban areas, gardens and orchards. Restricted to Mediterranean bioclimatic regions.

**Habits** Essentially nocturnal, terrestrial snake. May climb into bushes in the morning to bask in the sun, retiring to shade when the temperature rises.

**Diet** Small mammals, lizards and birds. Kills its prey with venomous bite.

**Reproduction** Oviparous. Lays 7–22 eggs at a time. Incubation lasts 6–8 weeks. At hatching, the young measure 19–23 cm in total length.

**Range** Restricted to the Mediterranean part of the Levant countries: Israel, Lebanon, northwestern Jordan and western Syria.

**Geographic variation** *Daboia palaestinae* is monotypic (no subspecies).

Daboia palaestinae, *juvenile photographed in captivity.* P. Geniez and J. Garzoni.

**Genus *Pseudocerastes***
A genus of only three species with a range stretching from the Sinai in the west to Pakistan, including Iran and the southeastern part of the Arabian Peninsula.

Their general appearance is somewhat similar to that of the horned vipers of the genus *Cerastes*, in particular due to the small horns above their eyes and their "desert" appearance, hence the generic name *Pseudocerastes*. However, from a genetic point of view, they are not at all closely related to the *Cerastes*. They are placed alongside the genus *Eristicophis* within a small clade that represents the sister-group of all ex-*Vipera* genera in the wide sense (*Vipera*, *Montivipera*, *Macrovipera* and *Daboia*).

## Field's Horned Viper

*Pseudocerastes fieldi* Schmidt, 1930
**Family:** Viperidae (vipers)
**Sub-family:** Viperinae (true vipers)
**F.:** Pseudocéraste de Field or Vipère à cornes de Field
**G.:** Westliche Trughornviper

**Identification** A viper of very arid areas with very pointed horns on the head above the eyes, each one composed of many small scales (contrary to *Cerastes* horned vipers, the horns of which are composed of just 1 pointed scale). Length 60–70 cm, occasionally to 89 cm. Wide, triangular head, very distinct from body. Body quite thickset, especially in adult females, and a very short tail. Quite small eyes with a vertical pupil, iris slightly orangey, darker in its lower part. Nostrils directed slightly upwards and outwards. The top of the head is covered with small granular scales, keeled only at the rear. 2 rows of scales between the rostral and nasal plates; 3 rows of scales between the eye and the supralabials; 12–14 supralabials on each side. Keeled dorsal scales, arranged horizontally, even on the flanks (contrary to *Cerastes* species) and arranged in 21 or 22 (rarely 23) rows at mid-body. 127–142 (usually 131–135) ventral plates, undivided ventral scute, 33–38 (rarely as many as 46) pairs of subcordals.

General aspect rough and "mineral", quite variable in colouration: general colour pale grey, yellowish-grey, sandy or pink-beige. There are 2 series of square markings on the back that can be alternate at the front of the body then attached in the centre to form transverse, rectangular bars, well-separated from each other. These markings are any colour between reddish-brown and blackish-grey. There are 2 series of blurred, alternating markings on the flanks, often indistinct, sometimes absent. The top of the head is generally quite pale, almost without patterning. There is a dark band, often indistinct, from the eye to the side of the head. In the lava deserts in northern Jordan and southern Syria, the species can be very dark, with an almost black back and reddish flanks covered in dark markings. The belly is greyish-white or pinkish-beige, sometimes with small dark spots or speckles.

**Venom** Quite a virulent venom. Any bite on a human requires treatment with an antivenim. Cases of death, if they occur, must be rare. Depending on the author, the species is slightly or quite aggressive; it hisses loudly before deciding to bite.

**Habitat** Semi-desert and quite arid areas, prefers more sandy soils with sparse scrub cover. Also on lava beds with numerous rocks. Finds refuge in rock crevices and small mammal burrows. A species of lower elevations, occurs to 1000 m in Jordan, probably higher in the Sinai.

**Habits** Mainly active during dusk and at night. Often advances using lateral movements, as do the *Cerastes* species. A very timid snake very quick to flee; avoids close contact with human habitations.

**Diet** Rodents, small birds and lizards. Kills its prey by venomous bite.

**Reproduction** Oviparous. Mating occurs in May and early June; 11–21 eggs, laid in July or early August, already contain well-developed embryos; they hatch in September.

**Range** Centred on the northwestern part of the Arabian Peninsula: Sinai (Egypt), Israel, Jordan, southern Syria, northwestern Saudi Arabia, western Iraq. Recently reported from two sites in southwestern Iran, north of the banks of the Persian Gulf.

**Geographic variation** Until quite recently, *Pseudocerastes fieldi* was considered to be a subspecies of the Persian Horned Viper *P. persicus* (Duméril, Bibron and Duméril, 1854), a species with a more eastern distribution, from the southeastern corner of the Arabian Peninsula and Iran as far as Pakistan. Among other

*Field's Horned Viper* Pseudocerastes fieldi, *juvenile. Near Shawbak (Jordan).* Cluchier.

*Close-up of* Pseudocerastes fieldi. *Note the horns, made up of several scales and not a single scale as in the* Cerastes *species (Horned and Arabian Horned Vipers). Cluchier.*

distinguishing characters, *P. persicus* can be separated from *P. fieldi* by having 23–25 rows of dorsal scales (compared to 21), a single row of scales (instead of 2) between the rostral and nasal plates, 148–158 ventral plates (against 127–142), quite distinct colouration, often more yellowish sandy-beige with narrower dorsal bars.

**Remarks** It is commonly stated in the literature that the Persian Horned Viper is present in northeastern Iraq, without precise confirmation. In Iran (thus outside the area covered by this book) the species is present in the Zagros chain but appears to be unknown farther west, in the depression that separates Iraq from Iran. There are records of *Pseudocerastes persicus* from southern Azerbaijan and even from extreme southeastern Turkey. These records, geographically isolated from the species' principle range, are considered as erroneous by most authors.

**Genus *Cerastes***

Small genus of three species of vipers totally adapted to desert life: the Horned Viper *Cerastes cerastes*, the Arabian Horned Viper *C. gasperettii* and the Sahara Sand Viper *C. vipera*. Their ranges cover the whole of the Sahara and the Arabian Peninsula. They are thickset, with a wide, triangular head, with matte, keeled dorsal scales; several rows of scales on the flanks are arranged obliquely and are desert-coloured.

The genera *Cerastes* and *Proatheris* form a separate clade within the family Viperidae.

# Horned Viper

*Cerastes cerastes* (Linnaeus, 1758)
**Family:** Viperidae (vipers)
**Sub-family:** Viperinae (true vipers)
**F.:** Vipère à cornes
**G.:** Hornviper

**Identification** A real desert snake characterised by the common presence of 2 horns on the top of the head. A thickset viper with a large head and of medium size (normally 50–60 cm, sometimes as long as 85 cm). Wide, triangular head, quite thick but well flattened on the top, very distinct from the body. There is a remarkable horn above each eye consisting of a single scale. These horns stick up vertically, slightly orientated forwards in most cases, not outwards. Some individuals have no horns, named the "mutila" form. The body scales are very matte, highly keeled, numerous

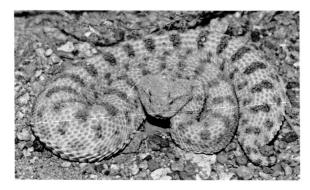

*Horned Viper* Cerastes cerastes cerastes, *an enormous 80-cm-long female. Between Ouarzazate and Skoura (Morocco).* P. Geniez.

or very numerous, arranged in 26–39 rows at mid-body. The flank scales are arranged obliquely and are slightly serrated like saw-teeth. The top of the head is covered with numerous keeled scales. 4 or 5 rows of subocular scales between the bottom of the eye and the supralabials. The supralabials are numerous, 11–15 on each side. 137–156 ventral plates form an angled ridge on each side; undivided anal scute, divided subcordals.

Their colouration is normally mimetic with that of their desert habitat: grey-beige, sandy yellow, pinkish-brown or reddish. There are 2 series of dark, often orange-brown, rectangular markings on the back, either alternating or joined to form transverse bands. There are smaller markings on the flanks that alternate with the back markings. Some individuals, particular younger ones, have a blue-grey (photo on page 342) or whitish band on the sides of the top of the head; this band continues on the side of the back in the form of bluish or white spots in alternation with the brown markings of the back. The

top of the head is often without any pattern, sometimes edged or mottled blue-grey; on its side there is an often ill-defined dark band from the eye to the corner of the mouth. The belly is porcelain-white, without markings. The tail extremity is black in females, yellow in males.

**Venom** A very effective venom, mainly cytotoxic. Large venomous fangs located at the front of the maxillae (as in all vipers). Discreet, often hidden. If removed from its hiding place, it can be very aggressive, especially if it is hot. Its bite is formidable; it strikes very rapidly and ejects a large quantity of venom. In most cases a bite from a Horned Viper results in humans having significant swelling of the bitten limb accompanied by dizziness and vomiting. There is necrosis of the affected area that may become gangrenous, resulting in amputation of the affected limb. In the worst cases, the victim may lose consciousness, but deaths are rare and usually concern children or the elderly. Treatment with an antivenin is highly recommended.

**Habitat** Totally dependent on desert habitat: partly sandy areas, extensive rocky areas, the bottom of cliffs. Also in dry savanna with spiny acacias. In extensive areas of large windblown dunes, the Horned Viper gives way to the Sahara Sand Viper *Cerastes vipera*. In mountain areas hardly occurs above 1500 m.

**Habits** During the day, readily hides in rodent burrows but may also occur under large stones or at the foot of a grass tussock, where it is often partly buried in the soft sediment. Becomes active at dusk and at night, sometimes travelling extensive distances looking for prey or a partner. Rapidly burrows into the desert sand, using the angular sided ventral scales, leaving just the eyes and horns in view. It advances quite quickly by lateral movements of the body that it repeatedly lifts up, then puts down a little further on, leaving a characteristic trail in the sand. When irritated it rubs together its keeled flank scales by moving the rings of its body, thus producing a rattling sound, as do the egg-eating snakes (genus *Dasypeltis*) and the *Echis* vipers. It is less active in winter, but it does not really hibernate: during bright, sunny days in January, the Horned Viper may be seen basking in the sun close to its hiding place.

**Diet** Small mammals (rodents and insectivores), lizards and similar, and small birds, particular migrant birds that rest or even sleep on the ground. Juveniles also feed on large insects. Kills its prey by venomous bite.

**Reproduction** Oviparous. Mating occur in April or May. Lays 10–20 eggs, sometimes stuck together in a mass, in July or August. Hatching occurs 6–8 weeks later.

**Range** Almost the whole of the Sahara and the northern part of the Sahelian belt, from Mauritania as far as the side of the Red Sea in Egypt and the Sudan. In the north, reaches the foothills of the mountains of Morocco, Algeria and Tunisia and the Mediterranean coast from eastern Tunisia to Egypt. There is an isolated population in the extreme southwestern part of the Arabian Peninsula. To the northeast, present in the Sinai (Egypt) and reaches southwestern Israel (Negev desert). Farther east it is replaced by another species, the Arabian Horned Viper.

**Geographic variation** Two subspecies are recognised.

▶ *Cerastes cerastes cerastes* (Linnaeus, 1758): in the majority of its range.
▶ *Cerastes cerastes hoofieni* Werner and Sivan, 1999; southwestern part of the Arabian Peninsula.

*Close-up of a Horned Viper, clearly showing the horns of a single, large, pointed scale. Aoussert (Atlantic Sahara, southwestern Morocco). Cayuela.*

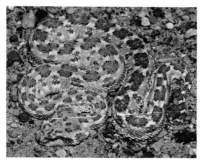

*A Horned Viper without horns ("mutila" form). Near Aghmakoum (northern Mauritania). P. Geniez.*

*A Young hornless Horned Viper ("mutila" form). Between Atar and Chinguetti (northern Mauritania). M. Geniez.*

The subspecies *C. c. karlhartli* Sochurek, 1979 (Sinai Peninsula) and *C. c. mutila* Doumergue, 1901 (over much of the species' range) are considered to be variants of *C. c. cerastes*.

# Arabian Horned Viper

*Cerastes gasperettii* Leviton and Anderson, 1967
**Family:** Viperidae (vipers)
**Sub-family:** Viperinae (true vipers)
**F.:** Vipère à cornes d'Arabie
**G.:** Arabische Hornviper

**Identification** Closely related to *Cerastes cerastes*, only recently considered as a separate species. General appearance similar, except more obviously flattened head. The horns are often absent, more often than in *C. cerastes*; when present, they slightly but distinctly point outwards and are slightly arced. 12–14 (rarely 15) scales around the eye (15–19, more rarely 14, in *C. cerastes*). On average a few more ventral plates than *C. cerastes* (146–164 compared to 137–156), anal scute undivided or divided. Maximum length: 72.5 cm in males, 78.5 cm in females.

Colouration somewhat different than that of *C. cerastes*: the band on the cheek is narrower and normally has a whitish or blue-grey band above that. In some populations the dorsal markings are quite small and more or less round. In others they form a zigzag on the back. The dorsal markings may also form transverse bands with a small pale spot on their outer edge. There may be a pale mark between each dark dorsal mark.

**Venom** Venom and defensive behaviour similar to that of *Cerastes cerastes*.

**Habitat** Extensive flat desert with or without much sand; sandy steppe with the occasional small bush.

**Habits** Similar to that of *Cerastes cerastes*. Spends much of its time in rodent burrows, is active at dusk and at night.

**Diet** Rodents, desert lizards and similar (e.g., Sandfish *Scincus conirostris* and Common Chameleon *Chamaeleo chamaeleon*) and small birds.

**Reproduction** Oviparous. Lays 8–23 eggs that hatch 8 or 9 weeks later.

**Range** More oriental than *Cerastes cerastes*: desert areas of most of the Arabian Peninsula. In the northwest as far as Israel and Jordan in Wadi Araba, Iraq to the northeast, Kuwait and extreme southwestern Iran. Except for the southwestern part of the Arabian Peninsula, where *C. c. hoofieni* occurs, it is the only *Cerastes* within its range.

**Geographic variation** Two subspecies have been described:

► *Cerastes gasperettii gasperettii* Leviton and Anderson, 1967: in most of the species' range.

► *Cerastes gasperettii mendelssohni* Werner and Sivan, 1999: endemic to the Arava Valley (Wadi Araba), southeastern Israel and southwestern Jordan. This subspecies, isolated from the rest of the species' range, has a more slender appearance in general and has a smaller maximum length (50 cm); the horns are always absent; the colouration is a little different: in particular, the dorsal markings are often smaller and, above all, the anal scute is undivided (divided in *C. g. gasperettii*).

*Arabian Horned Viper* Cerastes gasperettii gasperettii, *subadult. Bir Abraq (Kuwait). Note the head, which is more flattened than in* C. ceraste*s, the blue-grey spots on the sides of the dorsal markings and the absence of horns. There are individuals of* C. gasperettii *with horns, but they are less common than in* C. ceraste*s. Crochet.*

343

*Above:* Cerastes gasperettii mendelssohni. *Eilat, (Israel).* García-Cardenete.
*Right:* Cerastes gasperettii gasperettii, *adult without horns. Harib (Yemen).* Teynié.

Cerastes gasperettii gasperettii, *adult with horns. Khuzestan province (Iran).* Heidari.

# Sahara Sand Viper, Avicenna's Viper

*Cerastes vipera* (Linnaeus, 1758)
**Family:** Viperidae (vipers)
**Sub-family:** Viperinae (true vipers)
**F.:** Vipère de l'erg or Vipère des sables
**G.:** Avicennaviper

**Identification** The viper the best adapted to life in windblown Saharan sand dunes. Smaller than *Cerastes cerastes*, with which it locally cohabitates. Average length 30 cm, (maximum: 50 cm). Small, thickset desert snake. Wide, flat head obviously distinct from body covered in small keeled scales. It does not have enlarged supraoculars. The eyes are angled upwards (characteristic compared to the Horned Viper, which has laterally orientated eyes). Never has horns above the eyes. 12–14 scales around the eye: 3 or 4 rows of subocular scales between the bottom of the eye and the supralabials; 10–13 supralabials on each side. The body scales are well keeled, arranged in 23–29 rows at mid-body, compared to 26–39 in *C. cerastes* (a useful characteristic when a moulted skin is found in the desert). The scales at the top and centre of the flanks are oblique with a sawtooth keel. The ventral plates are highly angled on the sides, and less numerous than in other *Cerastes* species, 102–120; undivided anal scute; the subcordals are divided and arranged in pairs.

Colouration identical to that of the sand where it lives, often with little contrast; beige-grey, almost white,

yellowish-brown, pinkish-sandy colour, sometimes reddish-brown. There are generally no markings on the head. There are large indistinct brown, pinkish or reddish-brown markings on the back and flanks, often showing little contrast, sometimes absent. In some individuals, the top of the head and the back are covered with very small, dark speckles. The iris is usually sand-coloured, sometimes bright orange in certain individuals or populations. Belly uniform white or yellowish-white. The end of the tail is black in females, uniform yellow or with a few greenish-grey rings in males.

**Venom** Very effective venom, mainly cytotoxic, but its small size means that it is less dangerous for humans than the Horned Viper. Well-developed solenoglyphous fangs. After any venomous bite on a human medical assistance should be sought and the victim treated with an antivenin.

**Habitat** Prefers sandy deserts, sometimes with sparse low vegetation, in which case it can cohabit with the Horned Viper. Areas of large windblown desert dunes constitute its preferred habitat, where it is often the only snake. Rests buried in the sand, at the bottom of a dense

345

*Sahara Sand Viper* Cerastes vipera, *female (recognisable by its spindly black tail) burrowing into the sand. Tassili n'Ajjer (southern Algeria).* P. Geniez.

shrub or in the galleries of a rodent burrow.

**Habits** Essentially nocturnal, active also at dusk. Not very aggressive. Burrows into the sand with lateral body movements, like the Horned Viper, until the only visible part is the upwards-orientated eyes. Moves across the sand with lateral body movements. If worried, it gives a warning by producing a rattling sound by rubbing the keeled flank scales one against the other.

**Diet** Lizards and similar and small rodents. Kills its prey with its venomous bite.

**Reproduction** Ovoviviparous, giving birth to 3–5 young. The young are born within a transparent membrane, from which they immediately free themselves.

**Range** Very similar to that of *Cerastes cerastes* but absent from areas without any sand or that are too steep: the whole of the Sahara from Mauritania as far as Egypt and northern Sudan. However, extends less farther north in the Magreb (North Africa) due to an absence of extensive sandy areas. To the northeast, reaches the Sinai and southwestern Israel.

*Cerastes vipera,* male. *Negev desert (Israel). The easternmost population of the Sahara Sand Viper.* García-Cardenete.

*Extraordinary behaviour of a Sahara Sand Viper keeping its mouth wide open as a threat. Near Aoussert (Atlantic Sahara, southwestern Morocco). Aymerich.*

**Geographic variation** *Cerastes vipera* is considered to be monotypic. However, Sand Vipers from southwestern Morocco, on the sandy plateaus on the edge of the Atlantic Ocean are somewhat different: their eyes are less orientated upwards, there are generally 3 (sometimes 3½ or 4) rows of scales between the eye and the supralabials (generally 4 complete rows in other populations), often a pale bluish-grey band from the snout to the front of the eye that continues on the sides of the head to just above the temples. This blue-grey colour reappears on the sides of the back, between the dark markings (colouration comparable to that of certain Horned Vipers). The back and flank markings are generally well-defined, very wide, a dark brown colour, almost black, sometimes reddish-brown. The iris is bright orange, contrasting highly with the pinkish-grey or yellowish beige tint of the head scales.

**Remarks** Another species of *Cerastes* has recently been described, *Cerastes boehmei* Wagner and Wilms, 2010. The description of this new species is based on just one specimen, from eastern Tunisia. It has the same appearance as a *C. vipera*, but each eye has a strange crown of scales above it, cut flat at the top. The other characteristics given by the authors do not appear valid: the eyes supposedly orientated to the sides are hardly more so than in *C. vipera* of the Atlantic coast of Morocco, the general colouration is exactly the same as individuals found in Tunisia; and lastly the comparison of scale formations given for other *Cerastes* species are partly implausible. Until further information is available, we consider *C. boehmei* as a newer synonym of *C. vipera* and that the only known specimen is an example of *C. vipera* with a rare scale abnormality.

Cerastes vipera, *female of the southwestern Moroccan coastal population.* Near Tarfaya (southwestern Morocco). Aymerich.

*Comparaison between the* Cerastes vipera *of the large, windblown dunes of the Sahara (left: the Drâa Valley above Mhamid, Morocco), and the southwestern Moroccan coastal population (Khnifiss lagoon, southwestern Morocco). In this second photo, the eyes are orientated more to the sides, the iris is bright orange and the top of the head is edged with blue-grey. P. Geniez.*

## Genus *Echis*

The genus *Echis* contains small to medium-sized vipers occurring in the northern half of Africa, the Arabian Peninsula and central Asia, eastwards as far as northeastern India and Sri Lanka. The *Echis* vipers are characterised by having a more or less protruding short head, very distinct from the neck, with having obliquely arranged scales on the flanks that have toothed keels and by their undivided, undivided subcordal plates. All are highly venomous, and certain species (*E. carinatus* of Asia and *E. ocellatus* of Saharan Africa) are considered as among the most dangerous in the world. From a phylogenetic point of view the genus *Echis* constitutes a separate clade within the family Viperidae that may be a sister-group of the genus *Causus*. It is split into three groups: the *E. carinatus*, the *E. pyramidum* and the *E. coloratus* groups. In the geographic area covered by this book there are three species, one from each group.

# Carpet Viper, Egyptian Saw-scaled Viper

*Echis pyramidum* (Geoffroy Saint-Hilaire, 1827)
**Family:** Viperidae (vipers)
**Sub-family:** Viperinae (true vipers)
**F.:** Échide des pyramides or Vipère des pyramides
**G.:** Sahara-Sandrasselotter

**Identification** Smallish to medium-sized viper, with quite a short head, slightly triangular from above and quite distinct from the body. Average length of 50 cm, with a maximum of 65 cm in Egypt and 83 cm in West Africa. Short, rounded, protruding snout. The eyes are quite large, pupils with a vertical slit, the iris from orange to beige. Small nostrils orientated upwards. Quite a short tail. Back scales strongly keeled, arranged in 25–33 (most often 29 or 30) rows at mid-body. There are 3 or 4 rows of obliquely orientated scales on the flanks with serrated keels. The 2 rows at the bottom of the flanks, arranged horizontally, also have serrated keels (contrary to the Sahel Egg-eater *Dasypeltis sahelensis*, see page 204). The top of the head is covered with small keeled scales. No enlarged

*Egyptian Saw-scaled Viper* Echis pyramidum pyramidum. *The Siwa Oasis (Egypt). This population is considered to be special by some authors; it has been described as a separate subspecies,* Echis pyramidum lucidus. *Mazuch.*

Echis pyramidum pyramidum. *Siwa Oasis (Egypt)*. Mazuch.

supraocular plates. The snout plate (rostral) is wider than it is high; 13–19 scales around the eye; 2 or 3 rows of subocular scales between the eye and the supralabials. Normally 10 or 11 (more rarely 9 or 13) supralabials on each side, the 4th a little longer than the others. 158–189 ventral plates, anal scute undivided, 23–38 undivided subcordal plates, undivided (within the Western Palaearctic this character is diagnostic of vipers of the genus *Echis*, all other members of the other genera have divided subcordals arranged in pairs).

It is a snake with very striking colours that are difficult to describe. There is a series of 35–40 pale, sometimes white, lozenge-shaped transverse markings on the back, widely spaced and separated by wide dark brown bands. The pale markings on the back are exactly above dark brown more or less round markings on the flanks, each one with a white V-shaped border, pointed at the top, at its upper edge (see photo). Another series of smaller, dark markings, also with a pale upper edge, alternate with the larger flank markings. In individuals from the west, *Echis pyramidum leucogaster*,

the top of the flanks is often orange-brown (exactly like the Sahel Egg-eater *Dasypeltis sahelensis*, page 204), the bottom sandy beige. There is an elongated, slightly geometric dark mark on the top of the head. The snout and side of the head are beige or sand-coloured with a dark band from the eye to the cheek. Just below the eye is a more or less distinct dark mark. The belly is pure or yellowish-white, sometimes with indistinct dark spotting.

**Venom** The very effective, mainly cytotoxic venom is very dangerous for humans. Large solenoglyphous fangs. There are few recorded cases of death, but at least one case is known from Morocco. After a bite from a Carpet Viper, treatment with an antivenin and medical advice is absolutely necessary. It is, however, quite a timid snake. When feeling threatened, it coils itself up with its head in the centre and rubs together its flanks, producing a rattling sound with its serrated scales (as in the *Cerastes* species and the egg-eaters). It is only if provoked or captured that it will try to bite.

**Habitat** Dry savanna, semi-desert areas with spiny acacia. Sometimes found near isolated buildings, in crops or oases, where it searches for mice to prey on and a little humidity. Also in stony wadi beds and the base of rocky scree. It occurs mainly at low elevations in the northern part of its range.

**Habits** Hardly known. Active mainly at dusk and at night. Sometimes basks in the sun in the morning. Can advance with lateral body movements, sidewinding, something like the *Cerastes* species. Sometimes climbs up to 2 m in bushes looking for prey.

**Diet** Small mammals (mice and other rodents), birds, lizards and similar, small snakes, frogs and toads, also large arthropods (e.g., large scorpions).

**Reproduction** Little known. Oviparous. Lays 5–10 eggs at one time.

**Range** Vast range in sub-Saharan Africa, common within the Sahel belt, south of the Sahara. Otherwise isolated populations exist in southern and southwestern Morocco, in northeastern Algeria (the Aurès Massif), in the foothills of the mountain ranges of southeastern Algeria (Hoggar and Tassili n'Ajjer), in central Tunisia and extreme northwestern Libya. Farther east, it has been reported from Cyrenaica (northeastern Libya) and has a patchy presence in the northern half of Egypt. Unknown farther east, on the Asian continent.

**Geographic variation** Until recently, two species were recognised in this group: the Carpet Viper *Echis pyramidum* and the White-bellied Carpet Viper *E. leucogaster* Roman, 1972. Genetic studies have shown that these species are very closely related, with the result that most authors now consider them as two subspecies of *E. pyramidum*, one eastern, the other western. In the eastern part of the Horn of Africa, not within the Western Palaearctic, there are other subspecies,

Echis pyramidum leucogaster, *adult, near Aouïnet Torkoz (southwestern Morocco). The White-bellied Carpet Viper was until recently considered to be a separate species from the Carpet Viper* E. p. pyramidum. *Note the more orange colour of the eyes and dorsal markings.* Aymerich.

their exact status unclear. The same situation exists for *E. pyramidum lucidus* Cherlin, 1990 in Egypt.

► *Echis pyramidum pyramidum* (Geoffroy Saint-Hilaire, 1827): Egypt, Sudan and probably northeastern Lybia and eastern Chad. Appears to be devoid of an orange-red tint to the upper flanks. In some individuals, the white Vs on the flanks are joined and thus form a thin white zigzag along the length of the flanks. The belly is often spotted with grey.

► *Echis pyramidim leucogaster* Roman, 1972: all the rest of the species' range, situated more to the west, eastwards as far as Tunisia (and probably northeastern Libya) and Niger. Reddish tint to the top of the flanks; this tint sometimes continues to include the

*Young* Echis pyramidum leucogaster *swallowing a rodent. Southwestern Morocco. Aymerich.*

pale back markings that can thus be pale orange. Belly generally white without any markings.

## Arabian Saw-scaled Viper

*Echis coloratus* Günther, 1878
**Family:** Viperidae (vipers)
**Sub-family:** Viperinae (true vipers)
**F.:** Échide d'Arabie or Échide colorée
**G.:** Arabische Sandrasselotter

**Identification** Quite slim, beautiful viper, normally 40–50 cm long, to 83 cm (even 90 cm). Quite distinct from *Echis pyramidum* and *E. carinatus*. Longer head, more widened at the rear. The body is sometimes very thin in young individuals, more thickset in large females. The back scales are heavily keeled, arranged in 33–35 (rarely 30, 31 or 37) rows at mid-body. Flank scales arranged obliquely (except the lowest row) and with sawtooth keels. The top of the head is covered with numerous small scales. 15–21 scales around the eye, 3 or 4 rows of suboculars between the bottom of the eye and the supralabials (2 in the Carpet Viper); many supralabials, 12–17 on each side, the 4th more or less the same size as the others; 152–205 ventral plates, undivided anal and subcordal scales (not divided). The general colouration is often beautiful, very distinct from all other *Echis* species. On the back there are large pale, white, pale grey or bluish-grey markings with a very dark or black rim. These markings are often arranged alternately at the front of the back, then joined to form transverse bars. The ground colour of the back, between the markings, is also very bright: sandy beige, pinkish-beige, a real pale pink, orangey or brick-red. The flanks are grey or beige, patterned with dark makings topped with a vague pale V that's attached to the pale dorsal markings. There is often

a symmetric reddish pattern with dark edges on the top of the head that fades in adults. Finally, the sides of the head and often the underside are a beautiful dark blue-grey that highlights a pale band that goes from in front of the eye to the back of the top of the cheek. The iris is sandy beige or pale yellowish-grey, sometimes the lower part darker (in *E. pyramidum* it is generally uniform orange). The belly is white or greyish-white, often flecked with small dark spots with marbling on the sides.

**Venom** Quite strong haemolytic venom, but reputed to be less dangerous for humans than the other *Echis* species. Large solenoglyphous fangs. Most bites on humans are not very serious. However, swelling of the bitten area occurs that may become necrotic and even gangrenous, which necessitates amputation of the concerned limb. Some cases also result in bleeding of the gums, dizziness and vomiting. Cases of death exist but are rarer than with the other *Echis* species. When felt threatened it emits a rattling sound, using its keeled flank scales, as in other *Echis* species. Moves rapidly for a viper, but when cornered

*Arabian Saw-scaled Viper Echis coloratus coloratus, adult 70 cm long. Beer Abraq road (southeastern Egypt). P. Geniez.*

it can be unpredictable and irritable, striking with its head extremely rapidly, even biting.

**Habitat** Rocky desert, stony mountain sides with sparse vegetation of thorn bushes. Extremely arid sloping mountain valley sides, often with a few spiny acacias. Locally occurs to 2000 m in the Sinai, to 2600 m in the southwestern Arabian Peninsula.

**Habits** Essentially active at dusk and at night. On slopes where the Arabian Saw-scaled Viper can find support on rocky surfaces, it advances in an almost linear way. However, on flat ground and especially when trying to escape, it advances with lateral movements of the body.

**Diet** Small mammals (rodents), lizards, large arthropods, sometimes other snakes and anuran amphibians. Kills its prey with its fast-acting venom.

**Reproduction** Oviparous. Mating occurs in May or June. Lays mainly in July, sometimes August; 6–10 (rarely as many as 21) very adhesive eggs, more so than those of the majority of

*Close-up of* Echis coloratus coloratus, *southeastern Egypt.* P. Geniez.

other snake species, are stuck to a solid support (e.g., in rock crevices or under stones). Incubation is 43–61 days.

**Range** Egypt east of the Nile, northeastern Sudan, Sinai Peninsula, Israel, western edge of Jordan, Arabian Peninsula, except for its eastern edge.

**Geographic variation** *Echis coloratus* is represented by two subspecies.

▶ *Echis coloratus coloratus* Günther, 1878: the major part of the species' range.

▶ *Echis coloratus terraesanctae* Babocsay, 2003: a recently described subspecies, endemic to the northern Arava Valley, from the coast of the Dead Sea in the south as far as the latitude of Nazareth in the north (northeastern Israel and northwestern Jordan). This subspecies can be distinguished to some extent by having on average a few more ventral plates (with much overlap with *E. c. coloratus* but less than 185 ventral scales in males and less than 192 in females appears to be diagnostic

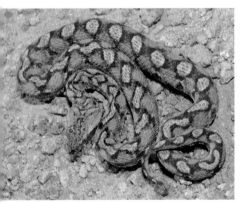

Echis coloratus coloratus, *juvenile. Jordan.* Cluchier.

of *E. c. terraesanctae*), on average a few less dorsal scales, and the pale markings on the back are surrounded by fewer dark scales. These differences are more obvious when comparing

Arabian Saw-scaled Vipers from northern Israel rather than those from the south; if the comparison is extended to the whole of the species' range, the differences are less clear.

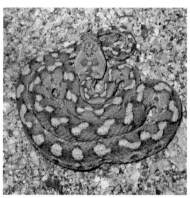

Echis coloratus terraesanctae, *subadult. Aqaba (southwestern Jordan)*. P. Geniez.

Echis coloratus terraesanctae, *adult. South of Hazeva (Israel)*. Crochet.

## Saw-scaled Viper

*Echis carinatus* (Schneider, 1801)
**Family:** Viperidae (vipers)
**Sub-family:** Viperinae (true vipers)
**F.:** Échide carénée
**G.:** Gemeine Sandrasselotter

**Identification** Description based on *Echis carinatus sochureki*, the only subspecies occurring in the Western Palaearctic. Small to medium viper (50–60 cm, can reach 73 cm), more thickset than *E. pyramidum* and *E. coloratus*. The head is particularly short and spherical, covered with

small, keeled scales. There are 1 or 2 slightly larger scales above the eyes, serving as supraoculars. 13–15 scales around the eye; generally 2 rows (sometimes only 1) of scales between the eye and the supralabials; 10–12 supralabials on each side, the 4th much longer than the others.

The dorsal scales are heavily keeled, arranged in 29–33 rows at mid body, many more (34–40) in the "*multisquamatus*" population (see "Geographic variation"). The 3rd–7th rows of scales on the flanks are heavily keeled and arranged obliquely. 153–188 ventral plates, undivided anal scute and subcordal plates.

Colouration usually not as bright as in other *Echis* species, normally of greyish, earth-brown and blackish and whitish tones. On the top of the head there is a pale mark with dark edges more or less in the form of a cross or an arrow. A wide, dark, elongated mark that goes from the eye to the rear of the temple often has a diffuse pale band above it. The transverse, often lozenge-shaped markings on the back are whitish, pale grey or pale brown with dark edges. These markings are well spaced; the ground colour between them is greyish or brownish. In some populations – e.g., in Iran and the southeastern part of the Arabian Peninsula – the ground colour of the

back can be more red-brown and thus reminiscent of the colouration of *E. pyramidum leucogaster*. The flanks have dark markings, each with an upside-down white V. In this species the white Vs are often joined, thus forming a white zigzag. The belly is whitish often spotted with dark on the sides.

**Venom** The large fangs at the front of the maxillae are capable of inoculating an extremely poisonous cytotoxic venom. When the Saw-scaled Viper is alarmed, it produces a rattling sound with its keeled scales. When cornered, it can strike violently, especially as it is a very rapid snake, muscular and capable of "leaping" in a single bound to bite its aggressor (or to capture a bird). Specialists who keep the species emphasise that even in captivity it can be dangerous to feed, and even more so to handle, as it is unpredictable and capable of suddenly throwing itself at its keeper. *Echis carinatus* is responsible for many deaths in India, as well as in Pakistan where *E. c. sochureki* occurs. A bite from a Saw-scaled Viper often provokes haemorrhaging (bleeding from the nose, ears, eyes, lips and nails) and can bring about significant renal complications, internal haemorrhaging and considerable necrosis of the bitten limb. Part of the treatment when bitten is the use of

*Saw-scaled Viper* Echis carinatus sochureki, *adult in captivity, probably originating from Pakistan. Note the short, spherical head and characteristic pattern on the head.* P. Geniez and J. Garzoni.

*Very young* Echis carinatus sochureki. P. Geniez and J. Garzoni.

an antivenin, but even so bites can be fatal for humans.

**Habitat** Quite arid areas of steppe: sandy and rocky wadi beds with sparse vegetation; also at the foot of scree, in hedgerows and bushy vegetation.

**Habits** Especially active at dusk and at night. During the day, often hidden under shrub roots. The Saw-scales Viper can travel long distances in search of prey.

**Diet** Toads and frogs, lizards, large arthropods. Can climb high into bushes in search of nestling birds.

**Reproduction** The Saw-scaled Viper is ovoviviparous (e.g., in India) or oviparous (the case of *E. c. sochureki*). The breeding season begins early in the year, from mid-February to late April. The female *E. c. sochureki* lays about 10 eggs and incubation lasts 45–50 days.

*Close-up of* Echis carinatus sochureki *adult. Note that the fourth supralabial is larger than the others*. P. Geniez and J. Garzoni.

**Range** Very large range in Asia from Iraq and the southeastern part of the Arabian Peninsula as far as eastern India. *Echis carinatus sochureki* is widespread in Iran and central Asia, from the eastern side of the Caspian Sea as far as northwestern India. This viper is also present in the southeastern corner of the Arabian Peninsula. In the area covered by this book, it is known from southeastern and central Iraq.

**Geographic variation** The systematics of *Echis carinatus* have almost been resolved, even if there remain questions on the status of certain populations. *E. carinatus sochureki* is the only subspecies that occurs in the Western Palaearctic. Research has shown that *E. multisquamatus*, a species of this group characterised by having a large number of dorsal and ventral scales and a more northern distribution, is hardly different form *E. c. sochureki* from a genetic point of view; it is now considered by most authors to be a synonym of the latter.

**Genus *Bitis***

This very distinct genus contains 17 tropical African species, of which one, the Puff Adder, is also present as a relict population in Morocco and the southern part of the Arabian Peninsula. The genus *Bitis* is particularly diversified morphologically, especially in South Africa: they have adapted to all sorts of habitats, from equatorial forest to the driest of deserts. Certain species resemble horned vipers, others sand vipers; they are also of small size with the same general shape and even upwards-pointing eyes, burrowing in the sand until only the eyes are visible.

It is one of the most remarkable examples of convergent evolution among snakes. It is quite thickset, with a large, flattened head that is quite distinct from the body. Some species are hardly ever longer than 30 cm, whereas others are enormous, reaching 190 cm (record: 205 cm for a Gaboon Viper *B. gabonica*, with a weight of 10 kg, circumference of 37 cm and body diameter of about 15 cm). Within the sub-family Viperinae, the genus *Bitis* represents a very particular clade that may be a sister-group of the genus *Atheris*.

# Puff Adder

*Bitis arietans* (Merrem, 1820)
**Family:** Viperidae (vipers)
**Sub-family:** Viperinae (true vipers)
**F.:** Vipère heurtante
**G.:** Gemeine Puffotter

**Identification** The most spectacular and distinctive of Western Palaearctic snakes. An enormous, very thickset viper, with a very fat body and a thick, short, rounded tail. In tropical Africa there are records of individuals over 180 cm long (record 191 cm). In Morocco it has a reputation for being smaller, around 100 cm; this is false as specimens at least 140 cm long have been recorded, with a diameter of 15 cm. The head is also very large, 15 cm by 10 cm for the largest specimens, triangular and flattened, well-separated from the neck, which is nonetheless quite thick; it is covered in small, heavily keeled scales without any differentiation of the supraoculars. The eyes are relatively small, slightly orientated upwards, with vertical pupils; the iris is golden brown, darker in the lower part. Wide, rounded snout; very large nostril openings, twice the size of the eyes and orientated upwards; 4 rows of scales between the eye and the supralabials; these are very numerous (13–15 on each side). Dorsal scales

*Puff Adder* Bitis arietans, *adult in intimidation posture, ready to strike. Atlantic Sahara (southwestern Morocco).* Cluchier.

*Puff Adder, adult. Wadi Assaka (southwestern Morocco). As seen here, the nostrils are as wide as the animal's eyes.* P. Geniez.

heavily keeled, arranged in 28–37 rows at mid-body; all the flank scales are arranged horizontally. 132–150 ventral plates, an undivided anal scute, divided subcordals that are arranged in pairs.

In Morocco, the general colour is sandy beige or pale grey with, on the back, very characteristic, narrow dark semi-circular bands that open at the front; these bands are pale-edged towards the rear; at the limit between the back and flanks, the pale edges form horizontal dashes. The flanks are mottled with dark brown or black and beige to reddish with pale triangles with dark tops at the base of the flanks. Large females show little contrast and are pale beige or greyish, whereas males and juveniles are more brightly coloured, with tints from dark green to brick-red passing through all tones of grey or brown. The pale markings can be white, cream, yellowish or pinkish. The top of the head is quite dark but sometimes shows little contrast, there is a transverse pale bar between the eyes, a few pale spots in the middle of the head and on each side a pale band from the transverse bar that ends just before the neck. On each side there is a dark mark below the nostril, an oblique bar under the eye and a wide, oblique band behind the eye, on the cheek. The belly is uniform cream-coloured or pale yellow but with a large, dark mark on the lateral extremities of every 3rd or 4th ventral plate.

**Venom** Disproportionally long, solenoglyphous fangs, reaching 4–5 cm, and connected to hypertrophied venom glands. The venom is highly cytotoxic, bringing about massive hemorrhaging and, if the victim does not die rapidly, massive swelling. This is followed by significant necrosis that often necessitates amputation of the bitten limb (the entire leg if the snake has

*Puff Adder darting its forked tongue. Atlantic Sahara (southwestern Morocco). Cluchier.*

bitten the foot or calf). Even with hospital attention and use of an antivenin, permanent damage will occur and death is still possible. Despite its bright, gaudy colouration, the Puff Adder trusts in its remarkable camouflage when lying in vegetation. Generally of a calm disposition and slow-moving, it can react very violently if stepped on and especially if cornered or provoked. It then holds up more than a third of the front of its body, the neck bent in an S, and hisses loudly. When it strikes, it throws itself abruptly at its aggressor, deploying its body over more than 1 metre, the mouth open at 180°, the fangs deployed forwards at right angles to the palate.

**Habitat** The Puff Viper typical inhabits savanna. However, in Morocco it is especially present in coastal desert near the Atlantic with a sea spray influence. Muddy desert steppe with numerous Sand Rat *Psammomys obesus* burrows in which it spends much of the day; steppe with succulent or bushy euphorbias, open Argan forest (tree species endemic to Morocco).

**Habits** The Puff Adder is principally nocturnal. It can sometimes be found in the sun in the morning or hidden under a bush in the shade. At night, it leaves its deep burrow in order to hunt. It is then that it is most likely to be seen – crossing a road, for example. In Morocco, the Puff Viper is severely threatened by road and housing development; it is also very much sought after by snake charmers, as is the Moroccan Cobra *Naja haje legionis*.

**Diet** Small and medium-sized mammals; birds sleeping on the ground; lizards, including quite large *Agama* species (e.g., Bibron's Agama *Agama impalearis*) and young of the Bell's Dabb Lizard *Uromastyx acanthinura*; sometimes other snakes.

**Reproduction** Ovoviviparous. Large females can give birth to tens of young.

Puff Adder, juvenile. Near Plage Blanche (southwestern Morocco). García-Cardenete.

**Range** An immense range over Africa, from the Sahal belt, the south of the Sahara as far as the Cape of Good Hope in South Africa. Also present in the southwestern part of the Arabian Peninsula. In Morocco, the only country in the Western Palaearctic where this species occurs with certainty, its presence is a true relict of a more tropical era, evidence that the African savanna once reached North Africa or the Sahara was not as dry as today. It is present from the Souss Valley (between Agadir and Taroudant, where it is now presumed extinct) as far as Cape Bojador; along the Atlantic Saharan coast, including Sidi Ifni, Guelmin, Tan-Tan and Laâyoune. The existence of the Puff Adder in the Hoggar massif in southern Algeria has been reported, or is assumed, by several authors. The photo of a specimen that was killed in the area in 1935 lends credit to its

presence there, with the Aïr massif in Niger and Morocco, it is the third relict zone where *Bitis arietans* survives to the north of the Sahel or the Sahara.
**Geographic variation** *Bitis arietans* is considered to be polytypic. Although it is considered that the nominative subspecies occurs in Morocco, this appears doubtful. The type locality (that is, the place of origin of the specimen first providing a species description) is the Cape of Good Hope, in South Africa, a region where the Puff Adder has very bright colours, usually lemon-yellow and black with many white spots. The subspecies found in Morocco is almost certainly the same subspecies (not formally described) that is found in the Sahel Belt of West Africa; it is probably quite similar to *B. a. somalica* Parker, 1949, an East Africa subspecies (Somalia, Kenya, Ethiopia).

# Halys Pit Viper

*Gloydius halys* (Pallas, 1776) (formerly *Agkistrodon halys*)
**Family:** Viperidae (vipers)
**Sub-family:** Crotalinae (pit vipers)
**F.:** Mocassin d'Halys, Mocassin d'Europe or Crotale d'Europe
**G.:** Halys-Grubenotter

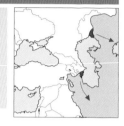

**Identification** The only pit viper to occur in the Western Palaearctic. This snake has a distinct appearance, quite slim, with a triangular head appearing flatter, more angular and more pointed than in other European vipers. Length 50–65 cm, sometimes to 75 cm. Relatively small eyes, pupil with vertical slit; iris golden orange-brown with well-defined dark bottom half. Very high, narrow, triangular rostral plate; no apical scales; 9 large bright plates on the top of the head, like colubrid snakes. On the side of the head, just in front of the eye, there is an opening that corresponds to a sensory pit that is sensitive to heat, an organ typical of pit vipers. 7–9 supralabials of very distinct shape and size, the 3rd very high in contact with the eye (the eye never touches the supralabials in the Viperinae); 2 elongated preoculars that overhang the nasal pit; a very elongated and concave, vertical postocular that touches the lower edge of the eye, a result of the fusion of a part of the postoculars and the suboculars.

*Halys Pit Viper*
Gloydius halys halys,
*in grassy steppe near Bozashchy (western Kazakhstan, eastern coast of the Caspian Sea).* Le Nevé.

*Close-up of* Gloydius halys halys. *Near Bozashchy (Kazakhstan). The sensory pit, typical of the pit vipers, just in front of the eye is visible here, it is wider than the nostril.* Le Nevé.

The dorsal scales are highly keeled, arranged in 23 rows at mid-body. 149–174 ventral plates, simple anal scute, divided subcordals arranged in pairs. Tail quite short, finishing with a pointed, horny scale.

The back is pale grey, beige, pale brown, pinkish or orangey, covered with 2 series of squarish, partly offset, half bars, most joined in pairs to form wide transverse bars; these marks are any colour between pale brown to orange or bright brick-red; they may or may not have dark edges. The background colour in the middle of the body is often very pale, accentuating the contrast of the markings and sometimes resembling the colouration of *Montivipera bulgardaghica*. There is a series of small dark markings along the flanks and, lower down, contrasting dark marks connected to the dark markings of the ventral plates. The top of the head is quite dark with more or less visible triangular markings, and edged on the sides by a pale line that accentuates the head's angular appearance; there is often a dark horseshoe-shaped mark, open at the rear, on the nape that partially or completely isolates a pale spot. A dark temporal band with pale edges goes from the eye towards the side of the neck. The lower labials, and sometimes the upper ones, are heavily edged with chocolate-brown and black on white.

*Gloydius halys caucasicus, a particularly red specimen. Gilan province (northern Iran). In contrast with G. h. halys, which occurs in low-elevation steppe, G. h. caucasius occurs in high mountains.* Fariborz Heidari.

Gloydius halys caucasicus, *Lar national park in the Elbourz range (northern Iran)*. Dab.

The belly is cream or pale grey finely speckled with brown or red, with quite large dark marks on the outer edges of the ventral plates.

**Venom** A snake armed with a venom apparatus, as in all pit vipers: large solenoglyphous fangs. The bite is reputed to be not very dangerous for humans but does provoke much pain and swelling of tissues. After being bitten, medical assistance is highly advisable.

**Habitat** Very varied habitats, from low plains to high mountains. Especially areas of steppe but also mountain slopes with scrub and scree, alpine meadows and riverbanks with thick vegetation. It hides in rocks crevices or rodent burrows. In mountains to 2500 m, as high as 3500 m in central Asia.

**Habits** Terrestrial snake active mainly at dusk. Readily basks in the sun in the early morning. Quite slow but aggressive if provoked; for protection it appears to rely mainly on its mimetic colouration. When threatened, the Halys Pit Viper can give a clicking sound as a warning; it makes this by vibrating the horny tip of the tail against the ground. In Iran it has been reported on several occasions that at night the pit viper, unlike the true vipers (*Montivipera latifi* and *Vipera eriwanensis ebneri*) will hide in camper's tents, under sleeping bags or luggage, probably attracted by the warmth of the human body, easily detected by the snake's sensory pits (personal communication with Roozbeh Behrooz). Inactive winter period of 4–6 months.

**Diet** Small mammals, birds and their eggs (especially ground-nesting species), other snakes. Juveniles eat lizards and orthoptera. Kills large prey with its venomous bite.

**Reproduction** Ovoviviparous. Gives birth to 3–12 young at a time. Mating occurs in April or May, birth during the summer or early autumn. Females can, after mating, store the spermatozoids for several years and then develop young without subsequent mating.

**Range** *Gloydius halys* is the pit viper with the largest range in the world: central and eastern Asia as far as northwestern China. To the west, reaches the eastern shore of the Caspian Sea, occurring northwards as far as the Ural River. In the European part of Kazakhstan, it has been cited by many authors as far as the banks of the Volga. However, its presence in Europe to the west of the Ural River (which forms the limit between Europe and Asia) is based on unconfirmed ancient records. Also present in northern Iran, along the whole of the Elburz range and farther east, the Kopet Dag range (Turkmenistan) and northwestern Afghanistan. Its distribution in Iran reaches as far west as the Talysh Mountains and extreme southeastern Azerbaijan and thus the Western Palaearctic, where it is apparently localised.

**Geographic variation** The systematics of *Gloydius halys* remain controversial and poorly established, due more to nomenclature rather than phylogenetic problems. Simply, opinion is divided as to the type locality of this species, which could be, according to a recent author, not in east-central Asia but in

the west, near the eastern banks of the Caspian Sea. Of the four recognised subspecies, two reach the Western Palaearctic:

▶ *Gloydius halys halys* (Pallas, 1776) = *Gloydius halys caraganus* (Eichwald, 1831): this is the subspecies of western Kazakhstan. Inhabits flat plains, hardly occurs above 100 m elevation. Large (at least 74 cm), 7–8 supralabials on each side, 23 (rarely 21) rows of scales at mid-body, 141–183 ventral plates.
▶ *Gloydius halys caucasicus* (Nikol'skij, 1916): southwestern Azerbaijan, northern Iran, Turkmenistan and northwestern Afghanistan. The mountain form can occur locally to 3000 m elevation. Smaller size than *G. h. halys* (maximum 66 cm), on average fewer (142–169) ventral plates.

Gloydius halys caucasicus. *Talysh Mountains, in southeastern Azerbaijan (thus in the Western Palaearctic).* Orlov.

# List of Snakes of the Western Palaearctic

| tick list | | Number of countries | Norway | Sweden | Finland | Denmark | United Kingdom | Netherlands | Belgium | Luxembourg | Germany | Estonia | Latvia | Lithuania | Poland | Belarus | Czech Republic | Slovakia | Switzerland | Austria | Hungary | France | Spain |
|---|---|---|---|---|---|---|---|---|---|---|---|---|---|---|---|---|---|---|---|---|---|---|---|
| | Xerotyphlops vermicularis (= Typhlops vermicularis) | 18 | | | | | | | | | | | | | | | | | | | | | |
| | Letheobia simonii (= Rhinotyphlops simoni) | 3 | | | | | | | | | | | | | | | | | | | | | |
| | Letheobia episcopa (= Rhinotyphlops episcopus) | 1 | | | | | | | | | | | | | | | | | | | | | |
| | Indotyphlops braminus (= Ramphotyphlops braminus) | 6 | | | | | | | | | | | | | | | | | | | | | |
| | Myriopholis macrorhyncha (= Leptotyphlops macrorhynchus) | 5 | | | | | | | | | | | | | | | | | | | | | |
| | Myriopholis algeriensis (= Leptotyphlops algeriensis) | 4 | | | | | | | | | | | | | | | | | | | | | |
| | Myriopholis cairi (= Leptotyphlops cairi) | 1 | | | | | | | | | | | | | | | | | | | | | |
| | Eryx jaculus | 21 | | | | | | | | | | | | | | | | | | | | | |
| | Eryx miliaris | 2 | | | | | | | | | | | | | | | | | | | | | |
| | Eryx jayakari (= Pseudogongylophis jayakari) | 3 | | | | | | | | | | | | | | | | | | | | | |
| | Eryx colubrinus (= Gongylophis colubrinus) | 1 | | | | | | | | | | | | | | | | | | | | | |
| | Hierophis viridiflavus (incl. gyarosensis) | 8 | | | | | | | | | | | | | | | | | 1 | | | 1 | |
| | Hierophis gemonensis (= Coluber laurenti) | 9 | | | | | | | | | | | | | | | | | | | | | |
| | Hierophis cypriensis | 1 | | | | | | | | | | | | | | | | | | | | | |
| | Dolichophis caspius | 13 | | | | | | | | | | | | | | | | | | | 1 | | |
| | Dolichophis schmidti | 6 | | | | | | | | | | | | | | | | | | | | | |
| | Dolichophis jugularis | 8 | | | | | | | | | | | | | | | | | | | | | |
| | Platyceps najadum | 19 | | | | | | | | | | | | | | | | | | | | | |
| | Platyceps collaris (= Coluber rubriceps) | 6 | | | | | | | | | | | | | | | | | | | | | |
| | Platyceps rogersi | 5 | | | | | | | | | | | | | | | | | | | | | |
| | Platyceps chesneii | 4 | | | | | | | | | | | | | | | | | | | | | |
| | Platyceps rhodorachis | 1 | | | | | | | | | | | | | | | | | | | | | |
| | Platyceps tessellatus (= Platyceps saharicus) | 7 | | | | | | | | | | | | | | | | | | | | | |
| | Platyceps sp. Israël | 2 | | | | | | | | | | | | | | | | | | | | | |
| | Platyceps florulentus | 1 | | | | | | | | | | | | | | | | | | | | | |
| | Platyceps elegantissimus | 3 | | | | | | | | | | | | | | | | | | | | | |
| | Platyceps sinai | 3 | | | | | | | | | | | | | | | | | | | | | |
| | Haemorrhois hippocrepis | 6 | | | | | | | | | | | | | | | | | | | | | 1 |
| | Haemorrhois algirus | 7 | | | | | | | | | | | | | | | | | | | | | |
| | Haemorrhois ravergieri | 9 | | | | | | | | | | | | | | | | | | | | | |
| | Haemorrhois nummifer | 9 | | | | | | | | | | | | | | | | | | | | | |
| | Spalerosophis diadema (incl. cliffordi) | 14 | | | | | | | | | | | | | | | | | | | | | |
| | Spalerosophis dolichospilus | 3 | | | | | | | | | | | | | | | | | | | | | |
| | Eirenis modestus | 5 | | | | | | | | | | | | | | | | | | | | | |
| | Eirenis aurolineatus | 1 | | | | | | | | | | | | | | | | | | | | | |
| | Eirenis levantinus | 5 | | | | | | | | | | | | | | | | | | | | | |
| | Eirenis barani | 2 | | | | | | | | | | | | | | | | | | | | | |
| | Eirenis coronella | 6 | | | | | | | | | | | | | | | | | | | | | |
| | Eirenis coronelloides | 4 | | | | | | | | | | | | | | | | | | | | | |
| | Eirenis lineomaculatus | 5 | | | | | | | | | | | | | | | | | | | | | |
| | Eirenis decemlineatus | 6 | | | | | | | | | | | | | | | | | | | | | |
| | Eirenis punctatolineatus | 3 | | | | | | | | | | | | | | | | | | | | | |
| | Eirenis thospitis (incl. hakkariensis) | 1 | | | | | | | | | | | | | | | | | | | | | |

| Malta | Slovenia | Croatia | Bosnia and Herz. | Serbia | Montenegro | Macedonia | Albania | Greece | Bulgaria | Romania | Moldova | Ukraine | Russia (Europe) | Kazakhstan (Europe) | Georgia | Armenia | Azerbaijan | Turkey | Cyprus | Morocco | Algeria | Tunisia | Libya | Egypt | Mauritania (north) | Mali (north) | Niger (north) | Chad (north) | Israel | Lebanon | Jordan | Syria | Iraq | Kuwait | Saudi Arabia (north) |
|---|---|---|---|---|---|---|---|---|---|---|---|---|---|---|---|---|---|---|---|---|---|---|---|---|---|---|---|---|---|---|---|---|---|---|---|
|  |  | ? |  | 1 | 1 | 1 | 1 | 1 | 1 |  |  |  | 1 |  | 1 | 1 | 1 | 1 | 1 |  |  |  |  | 1 |  |  |  |  | 1 | 1 | 1 | 1 | 1 |  |  |
|  |  |  |  |  |  |  |  |  |  |  |  |  |  |  |  |  |  |  |  |  |  |  |  |  |  |  |  |  | 1 |  | 1 | 1 |  |  |  |
|  |  |  |  |  |  |  |  |  |  |  |  |  |  |  |  |  |  | 1 |  |  |  |  |  |  |  |  |  |  |  |  |  |  |  |  |  |
|  |  |  |  |  |  |  |  |  |  |  |  |  |  |  |  |  |  |  |  |  |  |  | 1 | 1 |  |  |  |  |  |  |  |  | 1 | 1 | 1 |
|  |  |  |  |  |  |  |  |  |  |  |  |  |  |  |  |  |  | 1 |  |  |  |  | ? | 1 |  |  |  |  | 1 |  | 1 |  | 1 |  | ? |
|  |  |  |  |  |  |  |  |  |  |  |  |  |  |  |  |  |  |  |  | 1 | 1 | 1 |  | 1 |  |  |  |  |  |  |  |  |  |  |  |
|  |  |  |  |  |  | 1 | 1 | 1 | 1 | 1 |  |  | 1 |  | 1 | 1 | 1 | 1 |  | 1 | 1 | 1 | 1 | 1 |  |  |  |  | 1 | 1 | 1 | 1 | 1 |  | 1 |
|  |  |  |  |  |  |  |  |  |  |  |  |  | 1 | 1 |  |  |  |  |  |  |  |  |  |  |  |  |  |  |  |  |  |  |  |  |  |
|  |  |  |  |  |  |  |  |  |  |  |  |  |  |  |  |  |  |  |  |  |  |  |  | 1 |  |  |  |  |  |  |  |  | 1 | 1 | 1 |
| 1 | 1 |  |  |  |  |  | 1 |  |  |  |  |  |  |  |  |  |  |  |  |  |  |  |  | 1 |  |  |  |  |  |  |  |  |  |  |  |
| 1 | 1 | 1 | 1 | 1 | 1 | 1 | 1 |  |  |  |  |  |  |  |  |  |  |  | 1 |  |  |  |  |  |  |  |  |  |  |  |  |  |  |  |  |
|  |  | 1 |  | 1 | 1 | 1 | 1 | 1 | 1 | 1 | 1 | 1 | 1 |  |  | 1 | 1 |  |  |  |  |  |  |  |  |  |  |  |  |  |  |  |  |  |  |
|  |  |  |  |  |  |  |  |  |  |  |  | 1 | 1 | 1 | 1 |  |  |  |  |  |  |  |  |  |  |  |  |  |  |  | 1 | 1 |  |  |  |
| 1 | 1 | 1 | 1 | 1 | 1 | 1 | 1 | 1 |  |  |  | 1 |  | 1 | 1 | 1 | 1 | 1 |  |  |  | 1 |  |  |  |  |  | 1 | 1 |  | 1 | 1 |  |  |
|  |  |  |  |  |  |  | 1 |  |  |  |  |  |  |  |  |  |  | 1 |  |  |  |  |  |  |  |  |  | 1 | 1 | 1 | 1 |  |  |  |  |
|  |  |  |  |  |  |  |  |  |  |  |  |  | 1 |  |  |  | 1 | 1 |  | 1 | 1 |  |  |  |  | 1 |  | 1 | 1 |  | 1 | 1 | 1 |  |
|  |  |  |  |  |  |  |  |  |  |  |  | 1 |  | 1 | 1 | 1 | 1 |  |  |  |  |  |  |  |  |  |  |  | 1 |  |  | ? | 1 | 1 | 1 |
|  |  |  |  |  |  |  |  |  |  |  |  |  |  |  |  | 1 |  |  |  |  | 1 |  | 1 | 1 |  |  |  | 1 |  | 1 |  |  |  | 1 |
|  |  |  |  |  |  |  |  |  |  |  |  |  |  |  |  |  | ? |  |  |  |  | 1 |  |  |  |  | 1 |  | 1 |  |  |  |  |
|  |  |  |  |  |  |  |  |  |  |  |  |  |  |  |  |  |  |  |  |  |  | 1 |  |  |  |  |  | 1 |  | 1 |  |  | 1 |
|  |  |  |  |  |  |  |  |  |  |  |  |  |  |  |  |  |  |  | 1 | 1 | 1 |  |  |  |  |  |  |  |  |  |  |  |  |
|  |  |  |  |  |  |  |  |  |  |  |  |  |  |  |  |  | 1 | 1 | 1 | 1 | 1 | 1 |  |  |  |  |  |  |  |  |  |
|  |  |  |  |  |  |  |  |  | 1 |  |  |  | 1 | 1 | 1 | 1 |  |  |  |  |  |  |  | 1 | 1 | 1 | 1 |  |
|  |  |  |  |  |  |  |  |  |  |  |  | 1 |  |  |  | 1 | 1 |  |  |  |  |  | 1 | 1 | 1 | 1 | 1 |
|  |  |  |  |  |  |  |  |  |  |  |  | 1 | 1 | 1 | 1 | 1 | 1 | 1 | ? | 1 |  | 1 | 1 | 1 | 1 | 1 | 1 | 1 |
|  |  |  |  |  |  |  |  |  |  |  |  | 1 | 1 | 1 |  |  |
|  |  |  |  |  |  |  | 1 |  |  | 1 | 1 | 1 | 1 |  |
|  |  |  |  |  |  |  |  | 1 |  |  |  |  | 1 |  | 1 | 1 |  | 1 |
|  |  |  |  |  |  |  |  | 1 | 1 |  |  | 1 |  |  |  |  | 1 |
|  |  |  |  |  |  |  |  |  |  |  | 1 |  |  | 1 | 1 | 1 | 1 | 1 |
|  |  |  |  |  |  |  | 1 |  | 1 | 1 | 1 | 1 |  |  |
|  |  |  |  |  |  |  |  | 1 |  |  |  |  |  |  |  |  |  |  |
|  |  |  |  |  |  |  | 1 | 1 |  |  | 1 |  |  | 1 | 1 |  | 1 |
|  |  |  |  |  |  |  | 1 |  |  |  |  | 1 |
|  |  |  |  |  |  |  |  |  |  |  | 1 |  | 1 | 1 | 1 |  | 1 |
|  |  |  |  |  |  |  | 1 |  |  | 1 | 1 | 1 |
|  |  |  |  |  |  |  |  |  | 1 |  |  |  |  |  | 1 | 1 | 1 |
|  |  |  |  |  |  |  |  |  | 1 |  |  |  |  | 1 | 1 | 1 | 1 | 1 |
|  |  |  |  |  |  |  | 1 | 1 | 1 |
|  |  |  |  |  |  |  | 1 |

| tick list | Number of countries | Norway | Sweden | Finland | Denmark | United Kingdom | Netherlands | Belgium | Luxembourg | Germany | Estonia | Latvia | Lithuania | Poland | Belarus | Czech Republic | Slovakia | Switzerland | Austria | Hungary | France |
|---|---|---|---|---|---|---|---|---|---|---|---|---|---|---|---|---|---|---|---|---|---|
| *Eirenis collaris* | 6 | | | | | | | | | | | | | | | | | | | | |
| *Eirenis eiselti* | 2 | | | | | | | | | | | | | | | | | | | | |
| *Eirenis rothi* | 5 | | | | | | | | | | | | | | | | | | | | |
| *Eirenis persicus* (= *Pseudocyclophis persicus*) | 3 | | | | | | | | | | | | | | | | | | | | |
| *Rhynchocalamus melanocephalus* | 6 | | | | | | | | | | | | | | | | | | | | |
| *Rhynchocalamus satunini* | 4 | | | | | | | | | | | | | | | | | | | | |
| *Rhynchocalamus barani* | 1 | | | | | | | | | | | | | | | | | | | | |
| *Elaphe quatuorlineata* | 10 | | | | | | | | | | | | | | | | | | | | |
| *Elaphe sauromates* | 14 | | | | | | | | | | | | | | | | | | | | |
| *Elaphe dione* | 5 | | | | | | | | | | | | | | | | | | | | |
| *Zamenis longissimus* | 24 | | | | | | | | | 1 | | | | | | | 1 | 1 | 1 | 1 | 1 | 1 |
| *Zamenis lineatus* (= *Elaphe longissima romana*) | 1 | | | | | | | | | | | | | | | | | | | | |
| *Zamenis persicus* | 1 | | | | | | | | | | | | | | | | | | | | |
| *Zamenis situla* | 13 | | | | | | | | | | | | | | | | | | | | |
| *Zamenis hohenackeri* | 9 | | | | | | | | | | | | | | | | | | | | |
| *Rhinechis scalaris* | 3 | | | | | | | | | | | | | | | | | | | | 1 |
| *Coronella austriaca* | 39 | 1 | 1 | | | 1 | 1 | 1 | 1 | 1 | 1 | 1 | 1 | 1 | 1 | 1 | 1 | 1 | 1 | 1 | 1 |
| *Coronella girondica* | 7 | | | | | | | | | | | | | | | | | | | | 1 |
| *Macroprotodon cucullatus* | 3 | | | | | | | | | | | | | | | | | | | | |
| *Macroprotodon brevis* | 4 | | | | | | | | | | | | | | | | | | | | |
| *Macroprotodon abubakeri* | 2 | | | | | | | | | | | | | | | | | | | | |
| *Macroprotodon mauritanicus* | 3 | | | | | | | | | | | | | | | | | | | | |
| *Telescopus fallax* | 21 | | | | | | | | | | | | | | | | | | | | |
| *Telescopus nigriceps* | 5 | | | | | | | | | | | | | | | | | | | | |
| *Telescopus hoogstrallii* | 3 | | | | | | | | | | | | | | | | | | | | |
| *Telescopus tessellatus* (ssp. of *Telescopus fallax*?) | 1 | | | | | | | | | | | | | | | | | | | | |
| *Telescopus dhara* | 4 | | | | | | | | | | | | | | | | | | | | |
| *Telescopus obtusus* | 2 | | | | | | | | | | | | | | | | | | | | |
| *Telescopus tripolitanus* | 4 | | | | | | | | | | | | | | | | | | | | |
| *Lytorhynchus diadema* (incl. *kennedyi*) | 12 | | | | | | | | | | | | | | | | | | | | |
| *Dasypeltis sahelensis* | 1 | | | | | | | | | | | | | | | | | | | | |
| *Dasypeltis* cf. *scabra* (extinct?) | 1 | | | | | | | | | | | | | | | | | | | | |
| *Natrix natrix* (incl. *Natrix megalocephala*) | 49 | 1 | 1 | 1 | 1 | 1 | 1 | 1 | 1 | 1 | 1 | 1 | 1 | 1 | 1 | 1 | 1 | 1 | 1 | 1 | 1 |
| *Natrix tessellata* | 31 | | | | | | | | | 1 | | | | | | | 1 | 1 | 1 | 1 | 1 | |
| *Natrix maura* | 9 | | | | | | | | | | | | | | | | | | 1 | | | 1 |
| *Boaedon fuliginosus* (= *Lamprophis fuliginosus*) | 1 | | | | | | | | | | | | | | | | | | | | |
| *Lycophidion capense* (extinct?) | 0 | | | | | | | | | | | | | | | | | | | | |
| *Malpolon monspessulanus* | 6 | | | | | | | | | | | | | | | | | | | | 1 |
| *Malpolon insignitus* | 25 | | | | | | | | | | | | | | | | | | | | |
| *Rhagerhis moilensis* (= *Malpolon moilensis*) | 13 | | | | | | | | | | | | | | | | | | | | |
| *Psammophis schokari* | 13 | | | | | | | | | | | | | | | | | | | | |
| *Psammophis aegyptius* | 3 | | | | | | | | | | | | | | | | | | | | |
| *Psammophis lineolatum* | 0 | | | | | | | | | | | | | | | | | | | | |
| *Psammophis sibilans* | 2 | | | | | | | | | | | | | | | | | | | | |
| *Atractaspis engaddensis* | 3 | | | | | | | | | | | | | | | | | | | | |
| *Micrelaps muelleri* (incl. *tchernovi*) | 4 | | | | | | | | | | | | | | | | | | | | |
| *Naja haje* | 5 | | | | | | | | | | | | | | | | | | | | |
| *Naja nubiae* | 1 | | | | | | | | | | | | | | | | | | | | |
| *Walterinnesia aegyptia* | 4 | | | | | | | | | | | | | | | | | | | | |

| Malta | Slovenia | Croatia | Bosnia and Herz. | Serbia | Montenegro | Macedonia | Albania | Greece | Bulgaria | Romania | Moldova | Ukraine | Russia (Europe) | Kazakhstan (Europe) | Georgia | Armenia | Azerbaijan | Turkey | Cyprus | Morocco | Algeria | Tunisia | Libya | Egypt | Mauritania (north) | Mali (north) | Niger (north) | Chad (north) | Israel | Lebanon | Jordan | Syria | Iraq | Kuwait | Saudi Arabia (north) |
|---|---|---|---|---|---|---|---|---|---|---|---|---|---|---|---|---|---|---|---|---|---|---|---|---|---|---|---|---|---|---|---|---|---|---|---|
|  |  |  |  |  |  |  |  |  |  |  |  |  | 1 |  | 1 | 1 | 1 | 1 |  |  |  |  |  |  |  |  |  |  |  |  |  |  |  | 1 |  |
|  |  |  |  |  |  |  |  |  |  |  |  |  |  |  |  |  |  | 1 |  |  |  |  |  |  |  |  |  |  |  |  |  | 1 |  |  |  |
|  |  |  |  |  |  |  |  |  |  |  |  |  |  |  |  |  |  | 1 |  |  |  |  |  |  |  |  |  |  | 1 | 1 | 1 | 1 |  |  |  |
|  |  |  |  |  |  |  |  |  |  |  |  |  |  |  |  | 1 |  | 1 |  |  |  |  |  |  |  |  |  |  |  |  |  |  | 1 |  |  |
|  |  |  |  |  |  |  |  |  |  |  |  |  |  |  |  |  |  | 1 |  |  |  | 1 |  |  |  |  |  |  | 1 | 1 | 1 | 1 |  |  |  |
|  |  |  |  |  |  |  |  |  |  |  |  |  |  |  |  | 1 | 1 | 1 |  |  |  |  |  |  |  |  |  |  |  |  |  |  | 1 |  |  |
|  |  |  |  |  |  |  |  |  |  |  |  |  |  |  |  |  |  | 1 |  |  |  |  |  |  |  |  |  |  |  |  |  |  |  |  |  |
|  | 1 | 1 | 1 | 1 | 1 | 1 | 1 | 1 | 1 | 1 |  |  |  |  |  |  |  |  |  |  |  |  |  |  |  |  |  |  |  |  |  |  |  |  |  |
|  |  |  |  |  |  |  |  |  | 1 | 1 | 1 | 1 | 1 | 1 | 1 | 1 | 1 | 1 |  |  |  |  |  |  |  |  |  |  | 1 | 1 |  | 1 | ? |  |  |
|  |  |  |  |  |  |  |  |  |  |  |  | 1 | 1 | 1 |  |  | 1 | 1 |  |  |  |  |  |  |  |  |  |  |  |  |  |  |  |  |  |
|  | 1 | 1 | 1 | 1 | 1 | 1 | 1 | 1 | 1 | 1 | 1 | 1 | 1 |  | 1 | ? |  | 1 |  |  |  |  |  |  |  |  |  |  |  |  |  |  |  |  |  |
|  |  |  |  |  |  |  |  |  |  |  |  |  |  |  |  |  | 1 |  |  |  |  |  |  |  |  |  |  |  |  |  |  |  |  |  |  |
| 1 | 1 | 1 | 1 | 1 | 1 | 1 | 1 | 1 | 1 |  |  | 1 |  |  |  |  |  | 1 |  |  |  |  |  |  |  |  |  |  |  |  |  |  |  |  |  |
|  |  |  |  |  |  |  |  |  |  |  |  |  | 1 |  | 1 | 1 | 1 | 1 |  |  |  |  |  |  |  |  |  |  | 1 | 1 |  | 1 | 1 |  |  |
|  | 1 | 1 | 1 | 1 | 1 | 1 | 1 | 1 | 1 | 1 | 1 | 1 | 1 |  | 1 | 1 | 1 | 1 |  |  |  |  |  |  |  |  |  |  |  |  |  |  |  |  |  |
|  |  |  |  |  |  |  |  |  |  |  |  |  |  |  |  |  |  |  |  | 1 | 1 | 1 |  |  |  |  |  |  |  |  |  |  |  |  |  |
|  |  |  |  |  |  |  |  |  |  |  |  |  |  |  |  |  |  |  |  |  |  | 1 | 1 |  |  |  |  |  | 1 |  |  |  |  |  |  |
|  |  |  |  |  |  |  |  |  |  |  |  |  |  |  |  |  |  |  |  | 1 | 1 |  |  |  |  |  |  |  |  |  |  |  |  |  |  |
|  |  |  |  |  |  |  |  |  |  |  |  |  |  |  |  |  |  |  |  | 1 | 1 |  |  |  |  |  |  |  |  |  |  |  |  |  |  |
|  |  |  |  |  |  |  |  |  |  |  |  |  |  |  |  |  |  |  |  |  |  | 1 | 1 |  |  |  |  |  |  |  |  |  |  |  |  |
| 1 | 1 | 1 | 1 | 1 | 1 | 1 | 1 | 1 | 1 |  |  | 1 |  |  | 1 | 1 | 1 | 1 | 1 | 1 | 1 | 1 |  |  |  |  |  |  | 1 | 1 |  | 1 | 1 |  |  |
|  |  |  |  |  |  |  |  |  |  |  |  |  |  |  |  |  |  | 1 |  |  |  |  |  | 1 |  |  |  |  |  | 1 | 1 | 1 | 1 |  |  |
|  | 1 | 1 | 1 | 1 | 1 | 1 | 1 | 1 | 1 | 1 | 1 | 1 |  |  | 1 | 1 | 1 | 1 | 1 |  |  |  |  | 1 |  |  |  |  | 1 | 1 |  | 1 | 1 |  |  |
|  | 1 | 1 | 1 | 1 | 1 | 1 | 1 | 1 | 1 | 1 | 1 | 1 | 1 | 1 |  | 1 | 1 | 1 | 1 |  |  |  |  | 1 |  |  |  |  | 1 | 1 | 1 | 1 | 1 |  |  |
|  |  |  |  |  |  |  |  |  |  |  |  |  |  |  |  |  |  |  |  | 1 | 1 | 1 | 1 |  |  |  |  |  |  |  |  |  |  |  |  |
|  |  |  |  |  |  |  |  |  |  |  |  |  |  |  |  |  |  |  |  | 1 |  |  |  |  |  |  |  |  |  |  |  |  |  |  |  |
|  |  |  |  |  |  |  |  |  |  |  |  |  |  |  |  |  |  |  |  |  |  |  |  | ? |  |  |  |  |  |  |  |  |  |  |  |
|  |  |  |  |  |  |  |  |  |  |  |  |  |  |  |  |  |  |  |  | 1 | 1 |  |  |  |  |  |  |  |  |  |  |  |  |  |  |
|  | 1 | 1 | 1 | 1 | 1 | 1 | 1 | 1 | 1 |  |  | 1 |  |  | 1 | 1 | 1 | 1 | 1 | 1 | 1 | 1 | 1 |  |  |  |  |  | 1 | 1 | 1 | 1 | 1 |  |  |
|  |  |  |  |  |  |  |  |  |  |  |  |  |  |  |  |  |  |  |  | 1 | 1 | 1 | 1 | 1 | 1 |  | 1 |  | 1 |  | 1 | 1 | 1 | 1 | 1 |
|  |  |  |  |  |  |  |  |  |  |  |  |  |  |  |  |  |  |  |  | 1 | 1 | 1 | 1 | 1 | 1 |  |  |  | 1 |  | 1 | 1 | 1 | 1 | 1 |
|  |  |  |  |  |  |  |  |  |  |  |  |  |  |  |  |  |  |  |  |  |  |  |  | ? |  |  | 1 | 1 | 1 |  |  | ? |  |  |  |
|  |  |  |  |  |  |  |  |  |  |  |  | ? |  |  |  |  |  |  |  |  |  |  |  |  |  |  |  |  |  |  |  |  |  |  |  |
|  |  |  |  |  |  |  |  |  |  |  |  |  |  |  |  |  |  |  |  |  |  | 1 | 1 |  |  |  |  |  |  |  |  |  |  |  | ? |
|  |  |  |  |  |  |  |  |  |  |  |  |  |  |  |  |  |  |  |  |  |  |  |  | 1 |  |  |  |  | 1 |  | 1 |  |  |  |  |
|  |  |  |  |  |  |  |  |  |  |  |  |  |  |  |  |  |  |  |  | 1 | 1 | 1 | 1 | 1 |  |  | ? |  |  |  |  |  |  |  |  |
|  |  |  |  |  |  |  |  |  |  |  |  |  |  |  |  |  |  |  |  |  |  |  |  | 1 |  |  |  |  |  |  |  |  |  |  |  |
|  |  |  |  |  |  |  |  |  |  |  |  |  |  |  |  |  |  |  |  |  |  |  |  | 1 |  |  |  |  | 1 |  | 1 |  |  |  | 1 |

| tick list | | Number of countries | Norway | Sweden | Finland | Denmark | United Kingdom | Netherlands | Belgium | Luxembourg | Germany | Estonia | Latvia | Lithuania | Poland | Belarus | Czech Republic | Slovakia | Switzerland | Austria | Hungary | France | Spain |
|---|---|---|---|---|---|---|---|---|---|---|---|---|---|---|---|---|---|---|---|---|---|---|---|
| | *Walterinnesia morgani* | 5 | | | | | | | | | | | | | | | | | | | | | |
| | *Hydrophis cyanocinctus* | 3 | | | | | | | | | | | | | | | | | | | | | |
| | *Hydrophis gracilis* | 2 | | | | | | | | | | | | | | | | | | | | | |
| | *Hydrophis ornatus* | 2 | | | | | | | | | | | | | | | | | | | | | |
| | *Enhydrina schistosa* | 1 | | | | | | | | | | | | | | | | | | | | | |
| | *Vipera ursinii* (incl. *macrops, rakosiensis, moldavica*) | 13 | | | | | | | | | | | | | | | | | 1 | 1 | 1 | | |
| | *Vipera renardi* (incl. *lotievi, shemakhensis*) | 3 | | | | | | | | | | | | | | | | | | | | | |
| | *Vipera eriwanensis* (incl. *ebneri*) | 4 | | | | | | | | | | | | | | | | | | | | | |
| | *Vipera graeca* | 2 | | | | | | | | | | | | | | | | | | | | | |
| | *Vipera anatolica* | 1 | | | | | | | | | | | | | | | | | | | | | |
| | *Vipera darevskii* (incl. *olguni*) | 2 | | | | | | | | | | | | | | | | | | | | | |
| | *Vipera kaznakovi* (incl. *magnifica, orlovi*) | 3 | | | | | | | | | | | | | | | | | | | | | |
| | *Vipera dinniki* | 3 | | | | | | | | | | | | | | | | | | | | | |
| | *Vipera berus* (incl. *barani, nikolskii*) | 34 | 1 | 1 | 1 | 1 | 1 | 1 | 1 | | 1 | 1 | 1 | 1 | 1 | 1 | 1 | 1 | 1 | 1 | 1 | 1 | |
| | *Vipera seoanei* | 3 | | | | | | | | | | | | | | | | | | | | 1 | 1 |
| | *Vipera aspis* | 5 | | | | | | | | | 1 | | | | | | | | 1 | | | 1 | 1 |
| | *Vipera latastei* | 4 | | | | | | | | | | | | | | | | | | | | | 1 |
| | *Vipera monticola* | 1 | | | | | | | | | | | | | | | | | | | | | |
| | *Vipera ammodytes* (incl. *transcaucasiana*) | 15 | | | | | | | | | | | | | | | | | 1 | 1 | | | |
| | *Montivipera xanthina* | 2 | | | | | | | | | | | | | | | | | | | | | |
| | *Montivipera bulgardaghica* (incl. *albizona*) | 1 | | | | | | | | | | | | | | | | | | | | | |
| | *Montivipera bornmuelleri* | 3 | | | | | | | | | | | | | | | | | | | | | |
| | *Montivipera wagneri* | 1 | | | | | | | | | | | | | | | | | | | | | |
| | *Montivipera raddei* | 3 | | | | | | | | | | | | | | | | | | | | | |
| | *Macrovipera lebetina* (incl. *schweizeri*) | 12 | | | | | | | | | | | | | | | | | | | | | |
| | *Daboia palestinae* | 4 | | | | | | | | | | | | | | | | | | | | | |
| | *Daboia mauritanica* (incl. *deserti*) | 4 | | | | | | | | | | | | | | | | | | | | | |
| | *Pseudocerastes fieldii* | 6 | | | | | | | | | | | | | | | | | | | | | |
| | *Cerastes cerastes* | 10 | | | | | | | | | | | | | | | | | | | | | |
| | *Cerastes gasperettii* | 5 | | | | | | | | | | | | | | | | | | | | | |
| | *Cerastes vipera* | 9 | | | | | | | | | | | | | | | | | | | | | |
| | *Bitis arietans* | 1 | | | | | | | | | | | | | | | | | | | | | |
| | *Echis pyramidum* (incl. *leucogaster*) | 8 | | | | | | | | | | | | | | | | | | | | | |
| | *Echis coloratus* | 4 | | | | | | | | | | | | | | | | | | | | | |
| | *Echis carinatus* (incl. *sochureki*) | 1 | | | | | | | | | | | | | | | | | | | | | |
| | *Gloydius halys* | 1 | | | | | | | | | | | | | | | | | | | | | |
| | | | | | | | | | | | | | | | | | | | | | | | |
| | Number of species per country | | 3 | 3 | 2 | 3 | 3 | 3 | 3 | 2 | 6 | 3 | 3 | 3 | 3 | 3 | 3 | 5 | 5 | 9 | 7 | 12 | 15 |

A record of those species that have been seen in the wild in the Western Palaearctic can be kept by making a mark in the left-hand column.
(=): a widely and recently used synonym
(incl.): indicating *included*: mention of a subspecies considered as a separate species by some authors.

Countries are more or less grouped by regional affinities. Neither Iceland nor Ireland have snakes and therefore do not appear in the table.

Those species present on the Greek Dodecanese islands off the coast of Turkey are not included in the total for Greece.

Those species present on the Italian island of Lampedusa are not included in the total for Italy.

| Malta | Slovenia | Croatia | Bosnia and Herz. | Serbia | Montenegro | Macedonia | Albania | Greece | Bulgaria | Romania | Moldova | Ukraine | Russia (Europe) | Kazakhstan (Europe) | Georgia | Armenia | Azerbaijan | Turkey | Cyprus | Morocco | Algeria | Tunisia | Libya | Egypt | Mauritania (north) | Mali (north) | Niger (north) | Chad (north) | Israel | Lebanon | Jordan | Syria | Iraq | Kuwait | Saudi Arabia (north) |
|---|---|---|---|---|---|---|---|---|---|---|---|---|---|---|---|---|---|---|---|---|---|---|---|---|---|---|---|---|---|---|---|---|---|---|---|
|  |  |  |  |  |  |  |  |  |  |  |  |  |  |  |  |  |  | 1 |  |  |  |  |  |  |  |  |  |  |  |  |  | 1 | 1 | 1 | 1 |
|  |  |  |  |  |  |  |  |  |  |  |  |  |  |  |  |  |  |  |  |  |  |  |  |  |  |  |  |  |  |  |  |  | 1 | 1 | 1 |
|  |  |  |  |  |  |  |  |  |  |  |  |  |  |  |  |  |  |  |  |  |  |  |  |  |  |  |  |  |  |  |  |  | 1 | 1 | ? |
|  |  |  |  |  |  |  |  |  |  |  |  |  |  |  |  |  |  |  |  |  |  |  |  |  |  |  |  |  |  |  |  |  | 1 | 1 | ? |
|  |  |  |  |  |  |  |  |  |  |  |  |  |  |  |  |  |  |  |  |  |  |  |  |  |  |  |  |  |  |  |  |  | ? | 1 | ? |
|  |  | 1 | 1 | 1 | 1 | 1 | 1 |  | 1 | 1 | 1 |  |  |  |  |  |  |  |  |  |  |  |  |  |  |  |  |  |  |  |  |  |  |  |  |
|  |  |  |  |  |  |  |  |  |  |  |  | 1 | 1 | 1 |  |  |  |  |  |  |  |  |  |  |  |  |  |  |  |  |  |  |  |  |  |
|  |  |  |  |  |  |  |  |  |  |  |  |  |  |  | 1 | 1 | 1 | 1 |  |  |  |  |  |  |  |  |  |  |  |  |  |  |  |  |  |
|  |  |  |  |  |  |  |  |  | 1 | 1 |  |  |  |  |  |  |  |  |  |  |  |  |  |  |  |  |  |  |  |  |  |  |  |  |  |
|  |  |  |  |  |  |  |  |  |  |  |  |  |  |  |  |  |  | 1 |  |  |  |  |  |  |  |  |  |  |  |  |  |  |  |  |  |
|  |  |  |  |  |  |  |  |  |  |  |  |  |  |  | ? | 1 |  | 1 |  |  |  |  |  |  |  |  |  |  |  |  |  |  |  |  |  |
|  |  |  |  |  |  |  |  |  |  |  |  | 1 |  |  | 1 |  |  | 1 |  |  |  |  |  |  |  |  |  |  |  |  |  |  |  |  |  |
|  |  |  |  |  |  |  |  |  |  |  |  | 1 |  |  | 1 |  | 1 | 1 |  |  |  |  |  |  |  |  |  |  |  |  |  |  |  |  |  |
| 1 | 1 | 1 | 1 | 1 | 1 | 1 | 1 | 1 | 1 | 1 | 1 | 1 | 1 |  |  |  |  | 1 |  |  |  |  |  |  |  |  |  |  |  |  |  |  |  |  |  |
|  | ? |  |  |  |  |  |  |  |  |  |  |  |  |  |  |  |  |  |  |  |  |  |  |  |  |  |  |  |  |  |  |  |  |  |  |
| 1 | 1 | 1 | 1 | 1 | 1 | 1 | 1 | 1 | 1 |  |  |  |  |  |  |  | 1 |  |  |  |  |  |  |  |  |  |  |  |  |  |  |  |  |  |  |
|  |  |  |  |  |  |  |  |  | 1 |  |  |  |  |  |  |  | 1 |  |  |  |  |  |  |  |  |  |  |  |  |  |  |  |  |  |  |
|  |  |  |  |  |  |  |  |  |  |  |  |  |  |  |  |  | 1 |  |  |  |  |  |  |  |  |  |  |  |  |  |  |  |  |  |  |
|  |  |  |  |  |  |  |  |  |  |  |  |  |  |  |  |  |  |  |  |  |  |  |  |  |  |  |  |  | 1 | 1 |  | 1 |  |  |  |
|  |  |  |  |  |  |  |  |  |  |  |  |  |  |  |  |  | 1 |  |  |  |  |  |  |  |  |  |  |  |  |  |  |  |  |  |  |
|  |  |  |  |  |  |  |  |  |  |  |  |  |  |  | 1 | 1 | 1 |  |  |  |  |  |  |  |  |  |  |  |  |  |  |  |  |  |  |
|  |  |  |  |  |  |  |  | 1 |  |  |  | 1 |  |  | 1 | 1 | 1 | 1 | 1 | ? | ? |  |  |  |  |  |  |  | 1 | 1 | 1 | 1 | 1 |  |  |
|  |  |  |  |  |  |  |  |  |  |  |  |  |  |  |  |  |  |  |  |  |  |  |  |  |  |  |  |  |  | 1 | 1 | 1 | 1 |  |  |
|  |  |  |  |  |  |  |  |  |  |  |  |  |  |  |  |  |  |  |  | 1 | 1 | 1 | 1 |  |  |  |  |  |  |  |  |  |  |  |  |
|  |  |  |  |  |  |  |  |  |  |  |  |  |  |  |  |  |  |  |  |  |  |  |  |  | 1 |  |  |  |  | 1 |  | 1 | 1 | 1 |  | 1 |
|  |  |  |  |  |  |  |  |  |  |  |  |  |  |  |  |  |  |  |  | 1 | 1 | 1 | 1 | 1 | 1 | 1 | 1 | 1 | 1 |  |  |  |  |  |  |
|  |  |  |  |  |  |  |  |  |  |  |  |  |  |  |  |  |  |  |  |  |  |  |  |  |  |  |  |  | 1 |  | 1 |  | 1 | 1 | 1 |
|  |  |  |  |  |  |  |  |  |  |  |  |  |  |  |  |  |  |  |  | 1 | 1 | 1 | 1 | 1 | 1 | 1 |  |  | 1 |  |  |  |  |  |  |
|  |  |  |  |  |  |  |  |  |  |  |  |  |  |  |  |  |  |  |  | 1 | ? |  |  |  |  |  |  |  |  |  |  |  |  |  |  |
|  |  |  |  |  |  |  |  |  |  |  |  |  |  |  |  |  |  |  |  | 1 | 1 | 1 | 1 | 1 | 1 | 1 |  |  | 1 |  |  |  |  |  |  |
|  |  |  |  |  |  |  |  |  |  |  |  |  |  |  |  |  |  |  |  |  |  |  |  | 1 |  |  |  |  |  | 1 |  | 1 |  |  | 1 |
|  |  |  |  |  |  |  |  |  |  |  |  |  |  |  |  |  |  |  |  |  |  |  |  |  |  |  |  |  |  |  |  | 1 |  |  |  |
|  |  |  |  |  |  |  |  |  |  |  |  | ? |  |  |  |  | 1 |  |  |  |  |  |  |  |  |  |  |  |  |  |  |  |  |  |  |
| 4 | 13 | 14 | 13 | 15 | 14 | 16 | 17 | 20 | 17 | 10 | 8 | 10 | 22 | 7 | 19 | 22 | 24 | 52 | 11 | 27 | 26 | 20 | 19 | 37 | 10 | 3 | 4 | 3 | 43 | 24 | 37 | 36 | 36 | 13 | 18 |

**1**   presence natural or presumed natural

**1**   species endemic to or presumed endemic to a single country

**1**   extinct or presumed extinct species

**1**   species introduced or presumed introduced by man

**1**   sea snake

**?**   presence uncertain or probable

# Recommended References

The choice of references is personal. Here are mainly those the author consulted when writing the present book. References are listed based on geographic region or country. Those that were written years ago use now-redundant systematics but do include precise and significant information, and the author considers them worth consulting or acquiring. Many are in English.

## Europe

**Arnold, N., and D. Ovenden** (2010). *Le Guide Herpéto: 228 Amphibiens et Reptiles d'Europe.* Delachaux et Niestlé: Paris (France), 290 pp.

**Engelmann, W. E., J. Fritzsche, R. Günther, and F. J. Obst** (1993). *Lurche and Kriechtiere Europas.* Neumann Verlag: Radebeul (Germany), 440 pp.

**Gasc, J.-P.,** ed. (1997). *Atlas of the Amphibians and Reptiles in Europe.* Societas Europaea Herpetologica and Muséum National d'Histoire Naturelle: Paris (France), 494 pp.

**Glandt, D.** (2010). *Taschenlexikon der Amphibien und Reptilien Europas: Alle Arten von den Kanarischen Inseln bis zum Ural.* Quelle and Meyer: Wiebelsheim (Germany), 633 pp.

**Kwet, A.** (2009). *Guide Photographique des Reptiles et Amphibiens d'Europe: 130 Espèces et 60 Sous-espèces.* Delachaux et Niestlé: Paris (France), 252 pp.

**Speybroek, J., W. Beukema, B. Bok, and J. Van Der Voort** (2016) *Field Guide to the Amphibians and Reptiles of Britain and Europe.* Bloomsbury Publishing PLC: London, 432 pp.

**Stumpel, T., and H. Strijbosch** (2006). *Veldgids Amfibieën en Reptielen.* KNNV: Utrech (Netherlands), 318 pp.

## Former USSR

**Ananjeva, N. B., N. L. Orlov, R. G. Khalikov, I. S. Darevsky, S. A. Ryabov, and A. V. Barabanov** (2006). *The Reptiles of Northern Eurasia.* Pensoft: Sofia (Bulgaria), 245 pp.

## Mediterranean Basin

**Cox, N., J. Chanson, and S. Stuart** (2006). *The Status and Distribution of Reptiles and Amphibians of the Mediterranean Basin.* World Conservation Union (IUCN): Gland (Switzerland) and Cambridge (U.K.), 42 pp. plus one CD.

## Western Palaearctic and Eurasia

**Dobiey, M., and G. Vogel** (2007). *Venomous Snakes of Europe, Northern, Central and Western Asia. Giftschlangen Europas, Nord-, Zentral- und Westasiens.* Chimaira: Frankfurt am Main (Germany).

**Sindaco, R., A. Venchi, and C. Grieco** (2013). *The Reptiles of the Western Palaearctic. Vol. 2. Annotated Checklist and Distributional Atlas of the Snakes of Europe, North Africa, Middle East and Central Asia, with an Update to Vol. 1.* Edizioni Belvedere: Latina (Italy), 543 pp.

## World

**Mattison, C.** (2008). *The New Encyclopedia of Snakes.* Octopus Publishing Group: London (UK), 272 pp.

## Armenia

**Arakelyan, M. S., F. D. Danielyan, C. Corti, R. Sindaco, and A. E. Leviton** (2011). *Herpetofauna of Armenia and Nagorno-Karabakh.* Society for the Study of Amphibians and Reptiles, 149 pp.

## Austria

**Cabela, A., H. Grillitsch, and F. Tiedemann** (2001). *Atlas zur Verbreitung und Ökologie der Amphibien und Reptilien in Österreich.* Umweltbundesamt: Vienna (Austria), 880 pp.

## Cyprus

**Baier, F., D. J. Sparrow, and H. J. Wiedl** (2009). *The Amphibians and Reptiles of Cyprus.* Chimaira: Frankfurt am Main (Germany), 364 pp.

## Czech Republic

**Necas, P., D. Modrý, and V. Zavadil** (1997). *Czech Recent and Fossil Amphibians and Reptiles: An Atlas and Field Guide*. Chimaira: Frankfurt am Main (Germany), 94 pp.

## Egypt

**Baha El Din, S.** (2006). *A Guide to the Reptiles and Amphibians of Egypt*. American University in Ciaro Press: Cairo (Egypt) and New York (USA), 359 pp.

## France

**Delaugerre, M., and M. Cheyla**, eds. (1992). *Batraciens et Reptiles de Corse*. Parc Naturel Régional de Corse et École Pratique des Hautes Études: Ajaccio et Montpellier (France), 128 pp.

**Geniez, P., and M. Cheylan** (2012). *Les Amphibiens et les Reptiles du Languedoc-Roussillon et Régions Limitrophes. Atlas Biogéographique*. Biotope and Muséum National d'Histoire Naturelle, Mèze and Paris (France), 448 pp. This book, although local in its focus, is recommended because it addresses the latest advances in molecular systematics and situates the snakes of the south of France within a global context. But there are other magnificent regional atlases for most regions of France.

**Lescure, J., and J.-C. de Massary**, eds. (2012). *Atlas des Amphibiens et Reptiles de France*. Biotope and Muséum National d'Histoire Naturelle: Mèze and Paris (France), 272 pp.

**Vacher, J.-P., and M. Geniez**, eds. (2010). *Les Reptiles de France, Belgique, Luxembourg et Suisse*. Biotope and Muséum National d'Histoire Naturelle: Mèze and Paris (France), 544 pp.

## Germany

**Günther, R.**, ed. (1996). *Die Amphibien und Reptilien Deutschlands*. Gustav Fischer: Jena (Germany), 825 pp.

## Hungary

**Puky, M., P. Schád, and G. Szövényi** (2005). *Magyarország Herpetológiaia Atlasza: Herptological Atlas of Hungary*. Varangy Akciósoport Egyesület: Budapest (Hungary), 207 pp.

## Italy

**Bernini, F., G. Doria, E. Razzetti, and R. Sindaco** (2009). *Atlante degli Anfibi e dei Rettili d'Italia*. Polistampa (Italy), 792 pp.

## Jordan

**Disi, A. M.** (2002). Jordan Country Study on Biological Diversity. *The Herpetofauna of Jordan*. Ahmad M. Disi and GCEP: (Jordan), 288 pp.

**Disi, A. M., D. Modrý, P. Necas, and L. Rifa** (2001). *Amphibians and Reptiles of the Hashemite Kingdom of Jordan*. Chimaira: Frankfurt am Main (Germany), 408 pp.

## Morocco

**Bons, J., and P. Geniez** (1996). *Amphibiens et Reptiles du Maroc (Sahara Occidental Compris): Atlas Biogéographique*. Asociación Herpetológica Española: Barcelona (Spain), 320 pp.

**Geniez, P., J. A. Mateo, M. Geniez, and J. Pether** (2004). *The Amphibians and Reptiles of the Western Sahara*. Chimaira: Frankfurt am Main (Germany), 229 pp.

## Near and Middle East

**Leviton, A. E., S. C. Anderson, K. Adler, and S. A. Minton** (1992). *Handbook to Middle East Amphibians and Reptiles*. Society for the Study of Amphibians and Reptiles: Oxford, Ohio (USA), 252 pp.

## North and Saharan Africa

**Dobiey, M., and G. Vogel** (2007). *Venomous Snakes of Africa*. Giftschlangen Afrikas. Chimaira: Frankfurt am Main (Allemagne), 149 pp.

**Schleich, H. H., W. Kästle, and K. Kabisch** (1996). *Amphibians and Reptiles of North Africa*. Koeltz Scientific Publishers: Koenigstein (Germany), 629 pp.

**Trape, J.-F., and Y. Mané** (2006). *Guide des Serpents d'Afrique Occidentale: Savane et Désert*. IRD Éditions: Paris (France), 226 pp.

## Portugal

**Loureiro, A., N. Ferrand de Almeida, M. A. Carretero, and O. S. Paulo** (2010). *Atlas dos Anfíbios e Répteis de Portugal*. Espera do Caos Editores: Lisboa (Portugal), 255 pp.

**Saudi Arabia**
**Gasperetti, J.** (1988). *Snakes of Arabia*. In *Fauna of Saudi Arabia*, vol. 9: pp. 169–450. NCWCD: Riyadh (Saudi Arabia).

**Spain**
**Pleguezuelos, J. M., R. Márquez, and M. Lizan**, eds. (2002). *Atlas y Libro Rojo de los Anfibios y Reptiles de España*. Ministerio de Medio Ambiante: Madrid (Spain), 585 pp. (also available on CD and online, updated).
**Salvador, A., and J. M. Pleguezuelos** (2013). *Guía de Reptiles de España: Identificación, Historia Natural y Distribución*. Canseco Editores: Talavera de la Reina (Spain), 462 pp.

**Switzerland**
**Hofer, U., J.-Cl. Monney, and G. Dušej** (2001). *Die Reptilien der Schweiz: Verbreitung–Lebensräume – Schutz. Les Reptiles de Suisse: Répartition–Habitats–Protection. I Rettili della Svizzera: Distribuzione–Habitat–Protezione*. Birkhäuser: Basel (Switzerland), 202 pp.

**Turkey**
**Baran, I., Ç. Ilgaz, A. Avcı, Y. Kumlutas, and K. Olgun** (2005). *Türkiye Amfibi ve Sürüngenleri*. Tübitak Kitaplar Müdürlügü: Ankara (Turkey), 204 pp.
**Franzen, M., M. Bussmann, T. Kordges, and B. Thiesmeier** (2008). *Die Amphibien und Reptilien der Südwest-Türkei*. Laurenti Verlag: Bielefeld (Germany), 328 pp.

**References Concerning Snake Groups**
**Broadman, P.** (1987). *Die Giftschlangen Europas und die Gattung Vipera in Afrika und Asien*. Kümmerly+Frey: Bern (Switzerland), 148 pp.
**Ineich, I.** (2004). *Les Serpents Marins*. Institut Océanographique: Paris (France) and Monaco, 320 pp.
**Schulz, K.-D.**, ed. (2013). *Old World Ratsnakes*. A collection of papers. Bushmaster Publications: Berg (Switzerland), 432 pp.

# Index of English Names

# Index of Scientific Names

# List of Photographers

The photographer's name is also provided in the relevant caption: Benjamin Adam (Biotope), Michel Aymerich (Groupe d'Étude et de Recherches des Écologistes Sahariens), Sherif Baha El Din, Felix Baier, Baumgart, Roozbeh Behrooz, Fabien Bettex, Maxime Briola (Regard du Vivant), Hugo Cayuela, Marc Cheylan (EPHE-BEV), Alexandre Cluchier (Eco-Med), Cramm, Pierre-André Crochet (CEFE-CNRS), Koosha Dab, Grégory Deso, Philippe Évrard, Michael Franzen, D. Fuchs, Luis García-Cardenete, Jean Garzoni, Michel Geniez (Biotope), Philippe Geniez (EPHE-BEV), Bayram Göçmen, Juan-Pablo González de la Vega, Ulrich Gruber, Cornelius de Haan, Guy Haimovitch, Heckes, Fariborz Heidari, Ç. Ilgaz, Ulrich Joger, König, Guido Kreiner, Layer, Alfred Limbrunner, Arnaud Lyet, Mägdefrau, Jérôme Maran, Gabriel Martínez del Mármol Marín, Tomáš Mazuch, Konrad Mebert, David Modrý, Arnaud Le Nevé, Jean Nicolas, Nikolai Orlov, Olivier Peyre (Naturalia), Gilles Pottier, Hans Reinhard, Christoph Riegler, Francisco Rodríguez Luque, W. Rohdich, Xavier Rufray, Mostafa Saleh, Josef Friedrich Schmidtler, Klaus Dieter Schulz, Mario Schweiger, Boaz Shacham, Roberto Sindaco, Synatzschke, Rania Talbi, Alexandre Teynié (Alcide d'Orbigny), Jean-François Trape, Benny Trapp, Yehudah L. Werner, Wolfgang Wüster.

# Venomous Snakes

Adder or Common Viper
*Vipera berus*

Asp Viper
*Vipera aspis*

Nose-horned Viper
*Vipera ammodytes*

Blunt-nosed Viper
*Macrovipera lebetina*

Horned Viper
*Cerastes cerastes*

Halys Pit Viper
*Gloydius halys*